Rocketeers and Gentlemen Engineers:

A History of the American Institute of Aeronautics and Astronautics ... and What Came Before

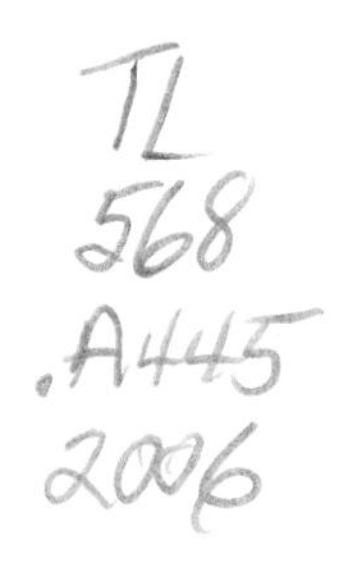

Rocketeers and Gentlemen Engineers:
A History of the American Institute of Aeronautics and Astronautics ... and What Came Before

By
Tom D. Crouch

American Institute of
Aeronautics and Astronautics
1801 Alexander Bell Drive, Suite 500
Reston, VA 20191–4344

American Institute of Aeronautics and Astronautics, Inc., Reston, Virginia

1 2 3 4 5

Library of Congress Cataloging-in-Publication Data

Crouch, Tom D.
Rocketeers and gentlemen engineers : a history of the American Institute of Aeronautics and Astronautics—and what came before \ by Tom D. Crouch.
p. cm.
Includes bibliographical references and index.
ISBN 1-56347-668-1
1. American Institute of Aeronautics and Astronautics—History. 2. Aeronautics—Research—United States—History. 3. Rocketry—United States—History. I. Title.

TL568.A445C76 2006
629.10973—dc22

2005032035

Cover design by Dorothy Banzon
Interior design by Gayle Machey

This book is dedicated to the Founders:
David Lasser, G. Edward Pendray,
Lester Durand Gardner, and Jerome Hunsaker

Foreword

It was with considerable personal pride that I accepted the invitation to introduce, *Rocketeers and Gentlemen Engineers*. Having joined the Institute of the Aeronautical Sciences in 1960, when I was just beginning my doctoral work at MIT, I have now been a member of AIAA or its predecessors for over forty-five years. I have presented papers in the *Journal of Spacecraft and Rockets*, and at AIAA meetings and conferences. As an outlet for technical work, a place to share ideas with colleagues, and a forum for offering informed commentary on aerospace issues, I can testify to the positive effect that the Institute has had on my career.

But my personal connection to the history of the Institute runs much deeper than that. As Dr. Tom Crouch explains in the pages of this book, my father, Edwin Eugene Aldrin, Sr., was one of the leading "movers and shakers" in American aviation during the years between the wars. As an Army pilot, he served as an aide to the legendary General Billy Mitchell. He was also an early graduate of the MIT program in aeronautical engineering, a contributor to some of the early and important textbooks in the field, the founder of what became the Air Force Institute of Technology, and a student and correspondent of rocket pioneer Robert H. Goddard. Having achieved those things, he pioneered a role for businessmen in aviation as a "flying executive" with Standard Oil.

It was scarcely surprising, then, that he was one of the first industry leaders whom Jerome Hunsaker and Lester Gardner recruited in 1932, when they set out to create the Institute of the Aeronautical Sciences. He was a member of the first governing council that shaped the IAS, and played a key role in arranging quarters for the new organization in Rockefeller Center, where his own offices were located. He maintained a lifelong commitment to AIAA and its predecessors as an effective means of encouraging progress in air and space.

Since 1930, the American Rocket Society, the Institute of the Aeronautical Sciences, and the American Institute of Aeronautics and Astronautics have defined the community of aerospace scientists and engineers in the Untied States. I am proud that my father helped to establish that tradition, and proud to have played my own part in carrying it forward into the space age. I commend this fascinating history of the organization to you. In addition to underscoring the work of the men and women who have built AIAA on the foundation established by my father and so many others, this book offers much encouragement to the belief that the Institute will remain at the forefront of aerospace progress in the decades to come.

Buzz Aldrin
Los Angeles, California

Table of Contents

Introduction and Acknowledgments

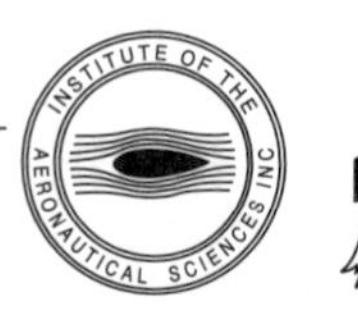

AIAA

It was a big year for the American Institute of Aeronautics and Astronautics (AIAA). The American Interplanetary Society, the oldest predecessor organization, had been founded 51 years before, in the early spring of 1930. The Institute of the Aeronautical Sciences (IAS), the other major founder society, was organized in the fall of 1932. Splitting the difference, AIAA celebrated its 50th anniversary in 1981.

The festivities included dinners honoring the founders of both organizations, old-timers' banquets celebrating long-time members, special awards, and a ceremony in which a record 50 members of the Institute were elevated to the status of AIAA Fellow. These ceremonial occasions provided the opportunity to read the letters of congratulation that had poured into the New York headquarters from across the nation and around the world.

There was a special publication, *AIAA at 50*, a short, glossy, and very well illustrated overview of Institute history published in the size and format of an issue of the organization's magazine, *Astronautics & Aeronautics*. It contained reprints of two articles originally published a decade before by Daniel Rich, who received his Ph.D. in political science from the Massachusetts Institute of Technology and specialized in science and public policy issues. He had become interested in the AIAA in 1968, while serving as an assistant to Robert C. Seamans, Jr., then President of the Institute. Recognizing how little the present-day staff knew of the history of their own organization, Rich offered profiles of both the American Rocket Society (ARS) and the IAS. A third article by AIAA Executive Director Jim Harford on the development of the organization since the merger in 1963 rounded out the 50th anniversary publication.[1,2]

Throughout 1981, *Astronautics & Aeronautics* reprinted classic articles from earlier ARS/IAS publications. The May issue of the magazine featured a commentary by David Lasser, founder of the ARS, and a short memoir by William Pickering describing the events surrounding the 1963 merger that created the modern Institute. There was a reprint of Hugh Dryden's 1963 appreciation of the life and work of Theodore von Kármán; Robert T. Jone's thoughts on the nature of progress; and three articles by staff members of the National Air and Space Museum (NASM), Richard Hallion, Frank Winter, and me.[3–5]

Not long out of graduate school and fascinated by the roots of professionalism in the aerospace science and engineering, my contribution to that issue described the foundation of the IAS. It marked the beginning of my long and happy relationship with the AIAA. During the years that followed, I served several terms on the History Technical Committee, presented papers at AIAA conferences, and even published a paper in the prestigious *AIAA Journal*, a rare thing for a professional historian. I served as an AIAA Distinguished Lecturer in the 1980s and again in the 21st century. The organization was kind enough to make me a Senior Member and to award me both the AIAA History Manuscript Prize in 1977 and the Gardner–Lasser Aerospace History Literature Prize in 2005 (Ref. 6).

Back in 1981, Chairman H. Norman Abramson and the members of the 50th Anniversary Committee suggested that AIAA commission a history of the organization. For a variety of reasons, that did not happen. With the approach of the 75th anniversary, staff members

Rodger Williams and Emily Springer, and Tony Springer, Chair of the History Technical Committee, were determined not to let a second opportunity pass them by. The three of them obtained approval to commission a history of the AIAA and challenged me to take up the task at the Reno Meeting in January 2003.

They did not have to ask twice. Having observed the AIAA for two and one-half decades and participated in its activities, it often struck me that a real history of the organization would be an important contribution to our understanding of the evolution of flight and its impact on society. The moment when engineers in an emerging field band together is a defining moment in the history of any technology, after all, and few professions can match the role that aerospace engineers have played in shaping the course of history.

The Scottish physician and journalist Samuel Smiles published the first volume of his classic, *Lives of the Engineers*, in 1861. He celebrated the achievements of the men who had built the canals, railroads, bridges, steamships, machine tools, and factories that were the foundation of the industrial revolution. These men were models of character, determination, and personal initiative. More than any politician or general, he argued, the engineers were the real heroes of the Victorian age.

A latter-day Samuel Smiles seeking to portray the most compelling achievements of technical professionals over the past century would do well to focus on aerospace engineers. Millennia from now, when the citizens of the future look back on the 20th century, they will surely remember it as the time when human beings first broke the chains of gravity and took to the sky. The members of the IAS, the ARS, and the AIAA have led the way in making that so. I can only hope that this book is worthy of the organization that they built.

My debts to those who helped with the project are many. First, my thanks to Rodger Williams and to Tony and Emily Springer, who conceived this project, sold it to the powers-that-be, and honored me with their confidence and support. The three of them suggested sources, provided materials, and introduced me to the records held at the AIAA headquarters in Reston, Virginia. They have never tired of providing encouragement, answering questions, or helping me to locate information or individuals. Special thanks also to Merrie Scott, my oldest friend at AIAA, and to the other members of the staff, most of whom must surely have wondered about the identity of the guy who was perpetually hogging their copy machine with his old documents.

As always, I owe a special debt to the archivists and librarians who have assisted with this project. First and foremost are my friends and colleagues of the NASM Archive, who have preserved the personal papers of an extraordinary number of individuals who appear in these pages: Tom Soapes, Marilyn Graskowiak, Dana Bell, Dan Hagedorn, Kate Igoe, Allan Janus, Mark Kahn, Kristine Kaske, Melissa Keiser, David Schwartz, Paul Silbermann, Mark Taylor, Barbara Weitbrecht, Patricia Williams, and Larry Wilson. Thanks also to the staff of the NASM Library, who never let me down: Bill Baxter, Phil Edwards, Carol Heard, and Leah Smith. Leonard Bruno and the staff of the Manuscript Division of the Library of Congress have grown accustomed to my research in their AIAA History Collection over the years. The staffs of Princeton University's Seely G. Mudd Manuscript Library and the Johns Hopkins University Archives offered genuine hospitality to a visiting scholar and extended themselves far beyond the normal call of duty.

James Harford; Cort Durocher; Dr. Jerry Grey; Frederic C. Durant, III; and Frederick I. Ordway, III, all gave generously of their time and insight during the course of long oral history interviews and subsequent questions. Published memoirs by each of these individuals were also of enormous value. Likewise, Dr. John Anderson and Hal Andrews were among those whose memories and insights were essential in shaping this volume. Long discussions with these individuals and others helped to flesh out the bare bones of the documentary record and brought the people whose lives made up the history of the ARS, IAS, and AIAA to life for me. I can only hope that I have communicated some of that to readers.

Peter Jakab and Ted Maxwell, the Chairman of the NASM Aeronautics Division and the NASM Associate Director for Collections and Research, respectively, encouraged the work and provided an environment conducive to research and writing. Finally, my thanks to a handful of scholars who have passed this way before me and whose insights have contributed to my own understanding of this story: Richard Hallion, Michael Neufeld, Daniel Rich, Tony Springer, and Frank Winter. I am especially grateful to Tony Springer, Rodger Williams, Fred Durant, and Jerry Grey, each of whom read all or part of the manuscript and offered valuable suggestions for its improvement.

I have no doubt where my greatest debt lies. Thanks to my wife Nancy, who renovated our new home while her husband, who has a tough time pounding nails straight, was struggling to finish this book on time.

Tom D. Crouch
National Air and Space Museum
Smithsonian Institution

August 2005

References

[1] Daniel Rich, "AIAA Yesterday and Today: Part 1: Profile of the American Rocket Society," *Astronautics & Aeronautics*, Jan. 1971.

[2] Daniel Rich, "Profile of the Institute of the Aerospace Sciences," *Astronautics & Aeronautics*, Feb. 1971.

[3] Robert T. Jones, "The Idea of Progress" *Astronautics & Aeronautics*, May 1981.

[4] Frank H. Winter, "ARS-Founders—Where Are They Now?" *Astronautics & Aeronautics*, May 1981.

[5] Richard Hallion, "The Rise of Air and Space," *Astronautics & Aeronautics*, May 1981.

[6] Tom D. Crouch, "IAS: The Origin and Early Years of the Institute of the Aeronautical Sciences," *Astronautics & Aeronautics*, May 1981, pp. 82–86.

Chapter 1

Creating a Profession: The Prehistory of the Institute of the Aeronautical Sciences, 1881–1928

The Birth of a Profession

In the early spring of 1952, Jerome Clarke Hunsaker (1886–1984) and Lester Durand Gardner (1876–1956) began circulating a draft of the first chapter of their proposed history of the Institute of the Aeronautical Sciences (IAS). The two men had known one another for 40 years and had much in common. Both were graduates of the Massachusetts Institute of Technology (MIT). Hunsaker had created the first program in aeronautical engineering at an American university just a year before World War I broke out in Europe. Three years later, Gardner founded a weekly magazine that became the "bible" of the American aviation industry. They had worked together in the fall of 1932 to establish the IAS, a national organization to serve the interests of technical and scientific professionals involved in aviation. Although their names were not as familiar to the public as those of the test pilots who were flying faster than sound, any aeronautical engineer worth his salt knew who Hunsaker and Gardner were and respected their achievements.[1]

So far as we know, the attempt to produce a joint history of the IAS ended with that single draft of one chapter. Perhaps it was just as well. The authors provided a useful and concise account of their own efforts leading to the founding of the Institute but were less successful in identifying the deeper historic roots of the organization. They devoted a dozen paragraphs to the early history of flying clubs in the United States from the Boston Aeronautical Society (1895), through the Aero Club of New England (1902), the Aero Club of America (1907), and the Aeronautical Society of America (1908), although the wealthy sportsmen and armchair aviators who dominated those organizations had little in common with the dedicated professionals who made up the membership of the IAS. They covered the history of the leading American engineering societies, on the other hand, in only two sentences.[2]

In fact, the IAS was deeply rooted in a tradition dating to the late 18th century, when English engineers began to define themselves as a distinct, self-governing profession like medicine or law through membership in organizations such as the Society of Civil Engineers (1771), the Institution of Civil Engineers (1819), and the Institution of Mechanical Engineers (1848). Their American counterparts followed suit a bit later. The American Society of Civil Engineers (ASCE) led the way in 1852, followed by the mining engineers in 1871, the mechanical engineers in 1880, the electrical engineers in 1884, and the chemical engineers in 1908.

During the 19th century, when most rational human beings regarded the notion of a winged flying machine as the very definition of the impossible, engineers took a much more positive view. The Aeronautical Society of Great Britain, founded in 1866, attracted leading technicians, including fellows of the Royal Society and past presidents of the British Association for the Advancement of Science, the Institution of Civil Engineers, and the Society of Engineers. Charles Bright, planner of the Atlantic cable; Charles William Siemens, who pioneered the dynamo and the telegraph; and James Nasmyth, the inventor of the steam

Jerome Clarke Hunsaker (1886–1984), IAS President for 1933. (NASM Archive, Smithsonian Institution, 89-12612.)

hammer, served on the Council of the Society and offered papers at early meetings. Although the total membership numbered only 65 at the end of 1867, some of Britain's finest engineers were on board.

The same thing was true in America, where Octave Chanute (1832–1910), who had served as president of both ASCE in 1891 and the Western Society of Engineers in 1901, was the most enthusiastic of the professional technologists. He arranged aeronautical lectures at meetings of professional societies, wrote an important book on aeronautics, conducted his own flying experiments, and befriended younger enthusiasts, including Wilbur and Orville Wright.[3]

"In spite ... of the continued failures," Chanute had no doubt that "... man may hope to navigate the air ... at no very distant day ... ." This achievement, he was sure, would "... bring nothing but good into the world; that it shall abridge distance, make all parts of the globe accessible, bring men into closer relation with each other, advance civilization, and hasten the promised era in which there shall be nothing but peace and good-will among all men."[4]

Nor was he alone. Mechanical flight was the subject of the very first presidential address offered to the newly organized American Society of Mechanical Engineers (ASME). Speaking in 1881, President Robert Henry Thurston, professor of engineering at the Stevens Institute of Technology, suggested that there was "... no department of engineering in which the art of the mechanic has opportunity for greater achievement." Although balloons and airships were all fine and well, he argued that the "real promise lies in the direction of flying machines lifted by their own power, not buoyed up by gas." Thurston concluded by suggesting that someone in his audience might "... in our own day win the fame that awaits the first successful builder of a flying machine."[5]

Thurston was only a bit optimistic. At the time of his speech, Wilbur Wright was 13 years old; his brother Orville was 10. Although neither of them would ever spend a day in a college class, that was also true of Octave Chanute and many other late 19th century engineers who had started at the bottom and worked their way up in the profession. The Wright brothers' solid high school training in mathematics and the scientific process enabled them to understand and evaluate the meager technical literature on the subject of flight. In addition, no amount of university training could have improved their extraordinary ability to move from the abstract to the concrete, the ease with which they were able to recognize a good idea in one technology and apply it to another, or their intuitive grasp of an effective process for solving difficult technical problems.

A young Lester Gardner. (NASM Archive, Smithsonian Institution, ZB09335.)

Indeed, the invention and early development of the airplane were the work of intuitive, often self-trained, engineers, brilliant innovators who achieved the dream of the ancients and changed the world in the process. For all of that, the business of aircraft construction remained an empirical process, which, although based on often-brilliant engineering research and data gathering, was pursued without reference to measured drawings, a theoretical foundation, or standard engineering procedures.

The eye of the craftsman and the ear of the mechanic were more important tools than the slide rule or the drawing board. Louis Blèriot, a professionally trained engineer, was the first man to fly an airplane across the English Channel. Nevertheless, Ferdinand Colin, his chief mechanic, recalled that "not once have I held in my hands the smallest drawing or sketch of Blèriot's, neither I nor any of my co-workers. We were driven hard to invent and construct parts with nothing more than the objective indicated and that only in general terms."[6]

It was the same in America. Grover C. Loening (who graduated from Columbia University in 1910 and received an M.A. in 1911), the first American to earn a degree in aeronautics, recalled that Orville Wright, his employer and friend, was accustomed to making rough drawings for a new part in the dust of a loading dock. Even when Loening prepared measured engineering drawings, Orville would sometimes make changes after consulting a favorite workman rather than his chief engineer.[7]

The aviator, not the engineer, was the hero of the day. As the editor of *Flight*, the English language journal of record in the field, explained to readers in January 1909, the men who designed and built airplanes "deserve a degree of credit for their work, which is far higher than the uninitiated are apt to accord them," but, "... their names must of course always stand second to those intrepid pioneers who have actually practiced the art of flight."[8]

As late as 1915, established practices in use in other branches of engineering remained rare in the aircraft industry on either side of the Atlantic. Donald Wills Douglas (1892–1981), a young engineer fresh from MIT, arrived in Los Angeles that year to take a job with the

The Harold Erskin sculpture recognizing Lester Gardner as "the first aerial hitchhiker." (NASM Archive, Smithsonian Institution, ZB09379.)

Glenn L. Martin Company. "At this time," he noted, "there was almost no engineering." At Martin, one of the oldest and most successful of American aircraft manufacturers, airplanes were built without the aid of drawings. Structural testing consisted of placing a machine on wooden supports, climbing aboard, and "jouncing around."

It was a dangerous approach to an unforgiving technology. So many deaths resulted from the collapse of monoplane wings before 1914 that both the French and British governments grounded such machines for a time. The rising death toll among American military airmen led Army officials in 1913 to ban the use of old-style, pusher machines with forward elevators in favor of the new tractor biplanes. "It was a constant gamble," English designer T.O.M. Sopwith remarked of this period when structural analysis was a matter of judgment, not mathematical calculation. "Some of us were lucky, and some of us were not."[9]

That would begin to change during a decade dominated by war and technical achievement. More powerful engines brought increased speeds and imposed much higher aerodynamic forces on an airplane in flight. The requirement to carry heavier loads and to maneuver in combat created further problems. Germany pointed the way to the future through the use of all-metal aircraft and underscored the need for a more rigorous approach to the analysis of flight structures. No factor would have a greater impact on progress in the air, however, than a revolution in aerodynamics that had been a long time coming.

The airplane was invented by practical engineers who could not fully explain why the thing flew. By the end of the 19th century, physicists had achieved genuine breakthroughs in fluid dynamics, including the development of equations that enabled them to calculate complex aspects of flow. Unfortunately, the mathematical tools, theoretical insights, and even some of the experimental data on fluid dynamics were of little value in solving the practical problems that barred the way to winged flight.

The Wright brothers stood at the end of a long line of experimenters stretching back to the 18th century who had used various engineering instruments, from the whirling arm to the wind tunnel, to measure the forces operating on solid bodies immersed in fluid streams. Their goal was to apply those data to the design of practical devices, whether windmill blades or wings that would carry them into the sky. It was an approach that produced the first practical flying machines but revealed very little about the basic scientific principles underlying flight.

The steam engine, a product of practical men pursuing empirical research strategies, inspired the birth of an engineering science, thermodynamics, which provided a theoretical framework that pointed the way to fundamental improvements in design. In the same way, the fact that airplanes were actually in the air attracted the attention of a generation of mathematicians and scientists who would give birth to aerodynamic theory, opening the door to the future of aviation. The twin streams of fluid dynamics and practical aeronautical engineering would not fully converge until the early 20th century, when German university professor Ludwig Prandtl (1875–1953) and others drew the disparate scientific threads into a coherent theory that explained the circulation of the air around a wing.

As early as 1894, English researcher Frederick Lanchester (1868–1946), drawing on the work of Isaac Newton (1642–1727), Daniel Bernoulli (1700–1782), and other pioneers of fluid dynamics, suggested how the complex circulation of air around a wing might generate lift. Between 1902 and 1906, German mathematician Wilhelm Kutta and Russian Nikolai Zhukovsky, working independently, began to establish the mathematical foundation for a theory of lift based on the motion of fluid around a solid body.

Ludwig Prandtl and his graduate students carried the work forward. Born in Freising, Germany, Prandtl earned his doctorate in engineering and became interested in fluid dynamics while studying the flow of air through industrial vacuum systems. In 1904, he presented an eight-page document that has been described as one of the most important papers in the history of fluid dynamics.[10]

Prandtl had discovered the boundary layer. He theorized that friction held a very thin layer of fluid motionless on the surface of a wing. Beyond the boundary layer, the flow was scarcely affected by surface friction. With this notion in place, the elements of a circulation theory of lift that could be presented in elegant mathematical form began to fall into place. A professor of applied mathematics at Germany's historic Göttingen University since 1904, Prandtl supervised the development of a wind tunnel and laboratory facilities second to none. His reputation as the world leader in aerodynamic research attracted a steady stream of talented graduate students, including a young Hungarian named Theodore von Kármán, who would extend Prandtl's work and spread his influence around the globe. The birth of theoretical aerodynamics would gradually change the way in which engineers were trained and aircraft designed.

By 1913, European engineers and scientists were blazing a technological path toward the future of aviation. The nation that had given flight to the world, however, had fallen far behind the cutting edge of aeronautical progress. The reason for that was clear. With the prospect of war looming on the horizon, European nations had invested large sums in all phases of aviation, from the establishment of flight research laboratories to lavish spending in support of domestic airframe and engine producers. Americans, isolated from the problems of Europe, had spent very little on the new technology. That year, the United States stood 14th among the world's nations in terms of government expenditures on aviation, trailing such military powerhouses as Japan, Bulgaria, Greece, Spain, and Brazil. In August 1913 the U.S. Army could field a total of six serviceable aircraft, two less than Japan and none of them up to European standards.[11]

Captain Washington Irving Chambers, the officer responsible for naval aviation in the office of the Secretary of the Navy, spoke for many when he explained that "the crude efforts of the pioneer inventors" would have to give way to "the methods of scientific engineers" if the United States were ever to gain strength in the air. No one could have agreed more than Richard Cockburn Maclaurin (1870–1920). A Scot, Maclaurin was a graduate of the University of New Zealand and Cambridge University, where he won prizes in both mathematics and law. In 1899 he accepted a post as professor of mathematics at Victoria University College in Wellington, New Zealand. Nine years later he was named professor of mathematical physics at New York's Columbia University, where he supervised Grover Loening's work on the first American graduate degree in aeronautics.[12]

Just a year later, in 1909, he accepted the presidency of the MIT. Located on Boston's Copley Square, MIT was chartered in 1861 and recognized as a land grant college under the provisions of the Federal Morrill Act of 1862, but it did not graduate its first class until 1868. In the words of founding President William Barton Rogers, it was to be a place where students were trained to discover and apply new knowledge for the benefit of society. By the early years of the 20th century, however, MIT had fallen on hard times.

Determined to transform moribund and nearly bankrupt "Boston Tech" into a great technological university, Richard Maclaurin began raising the money with which to build a new campus across the Charles River in Cambridge. Long before MIT moved into its new quarters in 1916, however, the president had begun to revise the curriculum. As part of that effort, he set out to establish a flight technology program that would meet what he recognized as a very real need for trained engineers in the emerging aircraft industry.

Interest in aeronautics preceded Maclaurin's arrival at MIT. Albert A. Wells had designed and built a simple wind tunnel powered by the building ventilation system in 1896 to study the aerodynamic qualities of flat plates as part of his research for his graduate thesis in mechanical engineering. In 1909, Henry Morse, an MIT engineering professor, surveyed the aeronautical courses offered by European universities. That year a group of students established the Aero Tech Club and set to work on a glider, followed by their own version of a pusher biplane based on the airplanes of Glenn Hammond Curtiss. One of the members, Frank Caldwell (class of 1912), would become the single most important international figure in the history of the aircraft propeller.[13]

Traveling to England in 1910, Maclaurin discussed ideas for an aeronautics course with his friend Richard Glazebrook, head of the National Physical Laboratory (NPL), a center of English aeronautical research. As a means of gauging interest in such a program, Albert A. Merrill, a founder of the Boston Aeronautical Society, was invited to give a series of lectures on flight at MIT. Some 90 students attended. In 1913, the Executive Committee of the MIT Corporation approved $3,500 for the construction of a wind tunnel and other laboratory equipment to support a graduate course in aeronautics. With funding in place and planning underway, one faculty member suggested that other MIT courses involving work in fluid dynamics, such as naval architecture and marine engineering, could provide a foundation and a source of instructors for the new program in aeronautical engineering. President Maclaurin agreed. He had just the man in mind to shape such a program.

Jerome C. Hunsaker, a native of Creston, Iowa, grew up in Detroit and Saginaw, Michigan, where his father was the owner/editor of local newspapers. He entered the U.S. Naval Academy in 1904, graduating four years later at the head of the class. Rejecting the accepted wisdom that the life of a line officer offered the clearest route to promotion, the young graduate was determined to train as a naval architect and engineer and serve with the Bureau of Construction and Repair. After completing his required one year of service at sea aboard the armed cruiser USS *California*, Hunsaker was ordered to the Boston Navy Yard and joined the dozen or so young naval officers assigned to attend a three-year graduate course in naval engineering and construction at MIT.

He prospered in Cambridge, distinguishing himself in courses ranging from mechanics and ship design to metallurgy and electrical engineering before beginning work on his master's thesis in 1912, which involved instrumenting the rudder of the destroyer USS *Sterett* to gather data that could be used to design improved steering mechanisms. Hunsaker forged ahead with work on his Ph.D. with scarcely a pause to catch his breath. In spite of a busy academic schedule, he helped to apply the scientific management principles of engineer Frederick Winslow Taylor to work at the Boston Navy Yard and still found time to meet and marry Alice Porter Avery, a student painter enrolled at the Boston Museum of Fine Arts.

Hunsaker turned his attention to aviation after attending the Harvard–Boston Aero Meet in 1910. "Considering the climate of the times, and that airplane flying was new," he commented half a century later, "I wondered what kept them up, what made them steerable, what made them go out of control."[14,15] The answers to those questions, he knew, lay in an emerging branch of fluid research—aerodynamics.

The young engineer scoured the MIT library for information on the subject. Most of what he found was badly dated—the papers and research reports of such late 19th century experimenters as Samuel Pierpont Langley, Octave Chanute, and Otto Lilienthal. One more recent article, E. Lapointe's "Aviation in the Navy," which appeared in a 1910 issue of the *United States Naval Institute Proceedings*, was of some value. Lapointe discussed mathematical procedures for calculating the stability of an aircraft and the design of efficient propellers. More important, he introduced Hunsaker to the work of Alexandre-Gustave Eiffel.

One of the best known civil engineers of his generation, Eiffel had designed a series of internationally known bridges, innovative buildings combining glass and cast iron, the armature for the Statue of Liberty, and the world-famous tower that bore his name. Fascinated by the problems of flight, he had begun his own aerodynamic experiments, dropping test items from the second deck of the Eiffel Tower. By 1909 he had constructed an aeronautical laboratory complete with a wind tunnel on the Champs de Mars near the tower.

Eiffel's importance to the field can be seen in the fact that he actually coined the term, "wind tunnel." Earlier tunnels had been identified as "channels," or "tubes," or simply as pieces of "apparatus." More important, his wind tunnel would establish the pattern for others around the globe. He published the first fruits of his research in 1909, as *La résistance de l'air et l'aviation: experiences effecturées au Labotatoire du Champs-de-Mars, (The Resistance of the Air and Aviation: Experiments Conducted at the Champs-de-Mars Laboratory)*.[16]

Eiffel's work was essential for a variety of reasons. Having begun with drop tests and continued with the wind tunnel, he could provide experimental confirmation of the assumption that it did not matter whether a test object was moving through the air or the air was moving over the test object. He was the first to conduct pressure measurements on the surface of an airfoil and to demonstrate that lift was principally the result of an area of lower pressure on top of the surface. Eiffel was also the first to test models of complete aircraft in his wind tunnel.[17]

Recognizing that Eiffel's book, complete with the data on individual airfoils, had established both a pattern and a foundation for future aerodynamic research, Hunsaker published an English translation, complete with an appendix from Eiffel that described his new laboratory and wind tunnel in the Paris suburb of Auteuil. The book was released in June 1913 to very favorable reviews in the United States and England.

In April 1913, confident that Hunsaker was his man, President Maclaurin offered him the opportunity to create a graduate program in aeronautical engineering and a laboratory to go with it. From the Chief of the Bureau of Construction and Repair to Secretary of the Navy, Hunsaker's superiors approved his request for an additional three-year special assignment to MIT. Recognizing the value of an officer who was a master of aeronautical science, Captain Chambers also agreed to launch the project by dispatching Hunsaker on a tour of aeronautical laboratories and teaching programs in Europe.

Unwilling to risk offending any of the factions jockeying for control of an official U.S. flight research organization, Washington Irving Chambers arranged for Albert Francis Zahm to accompany Hunsaker to Europe. A controversial pioneer who had worked with Octave Chanute in the 1890s and testified against the Wright brothers in their long-running patent infringement suit against Glenn Hammond Curtiss, Zahm had recently been named head of the Smithsonian aeronautical effort.

The pair boarded the North German Lloyd liner *Kronprinz Wilhelm* on July 29, 1913. They began in Paris, where Gustave Eiffel gave the Americans a warm welcome, spent two weeks showing them around the laboratory at Auteuil, and invited them to assist in tests of the latest aircraft in the world's most advanced wind tunnel. They moved on to the new aeronautics laboratory funded by philanthropist Henri Deutsch de la Meurthe, operated by the University of Paris in suburban St. Cyr, where researchers sent specially instrumented wings, propellers, and even full-size aircraft speeding three-quarters of a mile down a track on electrically powered cars.

The pair then traveled to Germany's Göttingen University, where Ludwig Prandtl was laying the mathematical foundation of theoretical aerodynamics. He had furnished his laboratory with a new closed-circuit wind tunnel, which continuously recirculated a volume of air rather than drawing outside air in at one end and exhausting it at the other. His approach produced a smoother flow and allowed the operators to control the pressure, temperature, and humidity.

Nothing that they saw on the trip impressed Hunsaker more than the next stop, the Deutsche Versuchsanstalt fur Flugwesen Luftfahrt (DVL), a lavishly funded, well-equipped official German flight research center established in 1911 in the Berlin suburb of Aldershof. In addition to a wind tunnel and special facilities for testing aircraft structures and power plants, the DVL pursued "... full-scale experimentation and testing—with a flying field, test pilots and airplanes that were better instrumented than at any other place in the world." If Prandtl was the international leader in theoretical aerodynamics, the engineers at Aldershof operated the world's best equipped facility for practical work in aeronautics.[18–20]

Hunsaker and Zahm were especially anxious to learn what they could about Count Ferdinand von Zeppelin's famous airships. They were treated to a flight aboard the *Sachsen*, an airship operated by DELAG, a Zeppelin-owned company that offered sightseeing rides over German cities. They visited the Johannisthal flying field, the Berlin base for the DELAG operation, and were on hand for the landing of L.2, a German naval airship that had made the long flight from corporate headquarters at Friedsrichshafen on Lake Constance. Although Zeppelin company officials were not eager to share information with the visiting Americans, August von Parseval wined and dined the visitors and openly discussed the design, manufacture, and operation of his kite balloons and nonrigid airships.

Their final stop was England, where the two Americans visited facilities that represented the intersection of theoretical aerodynamics and practical engineering. At the NPL, chief of aerodynamics Leonard Bairstow was studying aircraft stability. Professor G. L. Bryan, of Cambridge University, had published equations describing the forces operating on an aircraft in flight: pitch, roll, yaw, lift, drag, and side forces. The NPL team used their wind tunnel to discover the coefficients that could be plugged into those equations to predict the flying characteristics of an airplane that had yet to be flown. The trip concluded with a visit to the Royal Aircraft Factory at Farnborough, where the government conducted research on all aspects of flight and designed and built prototype aircraft, the best of which would be farmed out to private firms for production.

Those few short weeks spent in Europe marked the birth of professional aeronautical engineering in America. Hunsaker returned to MIT in November 1913 inspired with a vision of the future of American aviation. "We are at the point where the inventor can lead us but little farther," he commented in December 1913. In the future, "it is to the engineer and scientist that we must look for the perfection of our aircraft and the development of a new industry growing out of their manufacture and operation." He set out both to create a curriculum that would prepare students to employ the quantitative methods pioneered in Europe and to establish the finest aeronautical laboratory that President Maclaurin's money could buy.

The aeronautical laboratory was the first MIT building to be built on the site of the new campus in Cambridge. The centerpiece of the facility was a wind tunnel based on drawings provided by Richard Glazebrook, who also arranged for an English firm to produce the all-important balance, the delicate instrument on which the test surface was mounted that provided aerodynamic data.

Hunsaker developed a curriculum that mixed classes in aerodynamic and hydrodynamic theory with courses in design and practical experience in the laboratory. Initially, the program was a special graduate-level course. Most of the early students were serving officers following in Hunsaker's footsteps by preparing themselves to meet the technological needs of the military in a new era.

A significant number of these young engineers would help to shape American aviation. One, Major Virginius Clark, earned his M.S. in aeronautical engineering in 1914 and was immediately appointed chief engineer of the Aviation Section of the Army Signal Corps. He would remain an important technological contributor and business leader for decades to come. Alexander Klemin, an English immigrant who entered the program in 1915, would spend 20 years as Director of the Guggenheim Aeronautical Laboratory at New York University.[21]

The most distinguished early product of the MIT program, however, was not a student. Donald Wills Douglas (1892–1981), a Brooklyn native, had dropped out of the U.S. Naval Academy during his third year, noting that he "... just didn't like the idea of discipline over me."[22] Fascinated by aeronautics, he enrolled at MIT in 1912 and completed a four-year degree in mechanical engineering in only two years. Broke and with no prospect of a job, Douglas accepted a temporary position as Hunsaker's assistant, helping to install and operate the wind tunnel and assisting in developing the course work. When aircraft builder Glenn Luther Martin approached MIT in search of a chief engineer, Hunsaker suggested his assistant. Douglas took the job, launching one of the most important careers in the history of American aviation.

Hunsaker's grand tour had also convinced him of the critical need for state support of aeronautical research. It was apparent that profits in this new industry would be low while costs would be high. Should not the federal government be willing to conduct a research program that would support an industry producing machines that might be critically important to the defense of the nation?

The British government, for example, funded two mutually supporting flight research laboratories. In addition, aeronautical leaders served on a government-sponsored Advisory Committee on Aeronautics, an organization that coordinated the efforts of all British flight research programs. Hunsaker had seen the quality of the work undertaken at the German DVL. Other European governments had adopted similar, if less comprehensive, plans.

Whether the result of private enthusiasm and philanthropy, like the Eiffel laboratories, or state support, the well-equipped aviation research facilities established in Europe before 1914 were the early centers of technological progress. There was general agreement that the creation of such a facility in the United States was a critically important next step. Precisely how such a flight laboratory should be organized was quite another matter.

Captain Washington Irving Chambers supported the proposal of Smithsonian Secretary Charles Doolittle Walcott to establish a federal aeronautical research facility to be named the Langley Aeronautical Laboratory in honor of his predecessor, Samuel Pierpont Langley. Captain David W. Taylor, who was already operating a hydrodynamic research facility at the Washington Navy Yard, argued that it made better sense to place the military in charge of such a laboratory. Academics, including both Richard Maclaurin and William F. Durand of Stanford University, suggested that universities were in the best position to operate an aeronautical research institution.

In the end, Congress and the Executive branch would move in their own direction. On March 3, 1915, President Woodrow Wilson signed a naval appropriations bill into law, complete with an amendment chartering a new and independent government organization, the National Advisory Committee for Aeronautics (NACA). The NACA would "... supervise and direct the scientific study of the problems of flight with a view to their practical solution, and to determine the problems of flight which should be experimentally attacked and to discuss their solution and their application to practical questions."[23]

Exactly how that might be accomplished was far from clear. Initially, the NACA spent considerable time and effort conducting a survey of the state of aeronautics in the United States and advising Congress on political issues related to aviation. Not until June 11, 1920, did the new organization dedicate a three-building complex, complete with a wind tunnel, in Hampton, Virginia. Named the Langley Memorial Aeronautical Laboratory in honor of Smithsonian Secretary and flying machine experimenter Samuel Pierpont Langley, the research conducted at the new laboratory would support both the military services and the American aircraft industry.

Langley was the conduit through which new ideas, including the elements of circulation theory, arrived from Europe. One of Ludwig Prandtl's most promising students, Max Monk, emigrated to the United States following World War I and went to work at NACA Langley for a few short, stormy, and very important months. In the decades to come, a wealth of aerodynamic data flowed out of the Langley wind tunnels. NACA researchers produced revolutionary design ideas and critically important information on materials and aircraft structures. In 1939, with another European war looming on the horizon, Congress approved the creation of two additional NACA research centers. A new aerodynamic facility named in honor of Joseph Ames, the long-time chairman of the NACA, took shape at Moffett Field, a naval air station south of San Francisco, California. A second facility, located in Cleveland, Ohio, was dedicated to improving aircraft propulsion and was named in honor of George W. Lewis, NACA chief of research.[24]

As the business of designing and operating aircraft became more complex, the need for professionally trained engineers grew. Philanthropist Daniel Guggenheim would help to fill that need. A talented entrepreneur, Guggenheim had built his family's mining empire into one of the world's great fortunes and then turned his attention from making money to performing good works.

His son, Harry Frank Guggenheim, a graduate of Yale and Oxford, had served as a naval aviator during World War I. As a sportsman pilot on Long Island, Guggenheim met Alexander Klemin in 1925. An MIT graduate, Klemin was charged with establishing an aeronautical laboratory and a teaching program in aeronautical engineering at New York

University. Intrigued, the younger Guggenheim mentioned the project to his father, who liked the idea and announced that he would give $500,000 to establish a Guggenheim School of Aeronautics at New York University.

Recognizing that aviation was an area where focused philanthropy could have an enormous positive impact, Daniel Guggenheim began to consider expanding his involvement, spending several million dollars on the creation of a fund that would support civil aviation. Father and son, the Guggenheims discussed the idea with everyone from Orville Wright to Secretary of Commerce Hoover and President Coolidge. Founded in June 1926 with an initial $500,000, the Daniel Guggenheim Fund for the Promotion of Aeronautics would eventually distribute some $3 million.

The original focus was on engineering education. Over the next four years, the Fund created six more Guggenheim Schools of Aeronautics at the California Institute of Technology, Stanford University, the University of Michigan, the University of Washington, MIT, and the Georgia School (later Institute) of Technology. Later, the Guggenheims would help to establish major aeronautical research programs at Northwestern University, Syracuse University, Harvard University, and the University of Akron. The impact of the Guggenheim schools on the history of American aviation is simply incalculable, both in the work of the graduates of those programs and the research sponsored and conducted at those institutions. The day of the engineer who could work his way to the top of the profession on the basis of experience alone was vanishing.

The Fund would support a great many other projects. Guggenheim money fueled the development of improved radio navigation systems and cockpit instruments that would enable an aviator to fly "blind," in bad weather and at night; a "safe" airplane competition aimed at identifying and rewarding aircraft designs that were exceptionally stable and capable of low-speed operation and short-distance takeoffs and landings; and a model airline, complete with a weather forecasting operation and good communications between air and ground. By 1930, Daniel Guggenheim noted, "... with commercial aircraft companies assured of public support and aeronautical science equally assured of continued research, the further development of aviation in this country can best be fulfilled in the typically American manner of private business enterprise." Well, that and the vast sums of money that would be poured into the industry by the federal government.[25]

The Detroit Conspiracy

By 1930, the era of cut-and-try empiricism was vanishing from American aviation. The knowledge and skills of the professional engineer were the keys to the progress of aviation. For such individuals, membership in an engineering society was an important mark of professional status.

Because many of the technicians entering the new industry were trained as mechanical engineers, they naturally gravitated to the ASME, which had maintained a lively interest in flight since Professor Robert Thurston's first presidential address in 1881. Just 11 years later, William Frederick Durand presented his first paper on the theory of propeller design. The focus was on ship propulsion, but Durand would carry his study into the new realm of flight in the early years of the new century when he emerged as one of the leading American professors of aeronautical engineering. In 1908, Major George O. Squier, who was responsible for Army aviation, brought members of the ASME up to date on the aeronautical activities of governments around the world. In 1912 and 1913, MIT's Albert A. Merrill presented two aeronautical papers to members of the society. By the spring of 1914, the ASME was sponsoring special technical meetings devoted to aeronautics.[26]

As might be expected, the number and quality of aeronautical papers published by the ASME increased after World War I. Sanford Moss of General Electric; Frank Caldwell of Westinghouse; and S. D. Heron, a consultant to the U.S. Army Air Service, described their research on superchargers, propeller design, and engine cooling methods for ASME readers. Helicopters, aerial photography, airships, the use of metal in aircraft construction, seaplane design, air-cooled vs. liquid-cooled engines, and the impact of aircraft design on operating costs were among the topics addressed in ASME publications during the decade following World War I.

The ASME established an Aeronautic Division in 1920 and published an aeronautical dictionary in 1929, complete with terms in English, French, and German. Between 1929 and 1931, the division published a quarterly technical journal, *Aeronautical Engineering*. By 1945, 3,500 of the 18,500 members of ASME were registered with the division. Postwar developments in turbojet propulsion and manufacturing, areas of special interest to mechanical engineers, remain a focal point of what is now known as the Aerospace Division. The ASME continues to support its members working in the aerospace industry with conferences, special publications, and activities aimed at increasing support for the industry in Congress and the Executive branch.

By definition, however, the ASME focused its attention on supporting mechanical engineers employed in aviation. The Society of Automobile Engineers (SAE), organized in New York City in 1905, took a much broader approach, attempting to sign up all technical professionals working in a single industry. The leaders of the SAE, who held their first technical meeting and published their first journal in 1906, thought it only natural to stretch their organization to include engineers working in aviation, the other new enterprise requiring internal combustion engines. On April 30, 1910, the SAE Council proposed an amendment to the constitution, opening affiliate membership to "any firm engaged in the manufacture of self-propelled vehicles for navigating the land, water or air . . . ."[27] The first SAE logo, unveiled in 1909, was "mathematically developed" to symbolize the connection between land, sea, and air transport.

By 1916, the SAE had 1,800 members. With war raging in Europe and an obvious need to improve the state of things in America aviation, leaders of the automotive industry seized the opportunity to apply their experience in the large-scale production of motor vehicles to aviation, then, with the exception of the Curtiss Aeroplane and Motor Company, the domain of relatively small scale manufacturers. From their point of view, aircraft builders feared a "Detroit conspiracy" aimed at taking over their budding industry.[28]

An important first step for the automobile manufacturers was to attract the knowledge and experience of the men who were actually designing and building aircraft into the SAE. In 1915, Secretary of the Navy Josephus Daniels asked inventor Thomas Edison to head a Naval Consulting Board to advise on scientific and technical matters. Edison, in turn, asked for assistance in selecting two aeronautical experts to serve on the board. Howard E. Coffin, President of the Hudson Motor Car Company and a founder of the SAE, responded by calling a meeting in Washington to discuss the need of the Army and Navy for aircraft engines. Those attending Coffin's meeting established a new organization, the American Society of Aeronautic Engineers, the first task of which would be to select the aviation representatives to Edison's board.[28,29]

Although neither of them had attended the meeting, both Hunsaker and Grover Loening were initially enthusiastic about the potential of the group. Loening wrote an optimistic note informing Orville Wright of his hopes for the new organization. "For a long time," he remarked, "it has been one of the great ambitions of the technical men in this movement, who have had actual experience in aeroplane work, to form a society founded on technical merit which would distinctly strive to pursue a course of advancement in a technical way and as much as possible eliminate the 'Aero Club' type of activity . . . ."[30]

It quickly became apparent, however, that the Society of Aeronautic Engineers was not the organization that Loening had hoped for. The new organization would be governed by a total of 9 officers and 27 directors. Although a great many men with real aeronautical experience had been included, he believed that their names had been "appropriated without permission." That fact, together with the large number of directors, suggested to Loening that they were mere figureheads.

Without consulting Loening or any of those he regarded as genuine aeronautical engineers, the Society immediately named Henry Wise Wood and Elmer Sperry as the aeronautical representatives to the Naval Consulting Board. Although both men were certainly competent engineers, neither had any experience in aircraft design or manufacture. Loening immediately informed Orville Wright of the situation and of the fact that the Army and Navy officers directly responsible for aviation had decided to ". . . withhold their names or any connection of their subordinates, with this Society, until it is clearly in the control of men more qualified to be termed Aeronautical Engineers . . . ."[31]

The Society of Aeronautic Engineers had, Loening feared, fallen into the hands of "Aero Club propagandists." He was particularly offended by the presence of Henry Woodhouse (1884–1970) on the board of directors. Born Mario Casalegno in Turin, Italy, Woodhouse immigrated to the United States in 1905, found employment as a cook, and was almost immediately involved in an altercation that led to the death of a coworker. Released from prison in 1910, he began to use the Anglicization of his name, Woodhouse, and launched a career as a writer on aviation issues. In 1911, he was named managing editor of a new magazine, *Flying*, bankrolled by publisher Robert J. Collier and Henry Wise Wood, an inventor serving on the board of governors of the Aero Club of America.

Loening took his concerns to Secretary Daniels, who explained that there was little he could do. Edison, who had organized the Consulting Board at his request, accepted the candidates appointed by the Society as members in good standing. If Loening and his friends were so worried that the organization was moving in the wrong direction, Daniels suggested, they should wrest control away from the "propagandists of the Woodhouse order."

"The reason that the Woodhouse-Wood group have been able to acquire their notorious hold as representatives of the industry," Loening suggested to Orville Wright, "is because no one of us have troubled to work to get it in our hands." Orville Wright sympathized, noting that "I shall be very glad if you are successful in bringing it [the Society of Aeronautic Engineers] into the hands of real aeronautic engineers, but I am afraid that you are up against a hard proposition."[32]

In the end, Loening was unable to organize a coup. He watched in disgust as the members of the Aeronautic Society of New York, organized in 1908 by enthusiasts who regarded the Aero Club of America as insufficiently active, battled for a role in the Society of Aeronautic Engineers and for representation on the Naval Consulting Board. It seemed that the amateur enthusiasts were destined to win the day. To make matters worse, in December 1915, Alan Hawley, president of the Aero Club, launched an attack on what he regarded as the Wilson administration's half-hearted approach to building strength at sea and in the air. Henry Wise

Wood followed suit, resigning from the Naval Consulting Board so that he could join in the attack. Far from developing as a reputable professional organization, the Society of Aeronautic Engineers was soon floundering in rough political waters.

Then, in June 1916, Henry Wise Wood and Elmer Sperry petitioned SAE to take their aeronautical brethren into the fold. The leaders of the SAE agreed and changed their name to the Society of Automotive Engineers as a tip of the hat to the aeronautical newcomers.

Loening and his friends chose to support the plan as the lesser of two evils. They would make common cause with Howard Coffin and the SAE in order to prevent "scoundrels" like Woodhouse from dominating and politicizing a purely aeronautical organization. Still, the engineers who came from an aeronautical background continued to grumble about a "Detroit conspiracy." As Loening noted, "technical aeronautics by the end of 1916 was officially in the hands of the automobile engineers ...."[28]

In view of the small scale of the aeronautical industry in America and the embryonic state of aeronautical engineering, it was probably all for the best. Just 10 years old, the SAE was a rapidly maturing technical society in the mold of the older organizations representing mechanical and civil engineers. In addition to sponsoring regular technical meetings, publishing a refereed journal, and offering awards and other forms of professional recognition, the SAE organized projects that were essential to the growth of the industry.

None of those was more important than the effort to establish engineering standards for the new industry. At the time, the U.S. Army and Navy were still asking contractors and suppliers to comply with standards for nuts, bolts, and screws that had first been established in 1868. The program began in 1917 with the establishment of the first international aeronautical engineering standard for sparkplugs, developed in cooperation with engineering organizations in Britain, France, and Italy.

By the spring of 1920, Robert Burnett, manager of the SAE Standards Department, had enlisted a large team of the best and most experienced engineers in the field to produce an aeronautical handbook. In the tradition of such engineering reference works, the document would include some historical references, baseline data on elementary aerodynamics (formulae and tables covering the density of air with variations in altitude and temperature and information on the lift and resistance of various surfaces and shapes), conversion tables, and tabulated data on weights and volumes.

The core of the handbook would consist of very detailed standards of materials to be used in aeronautical construction. These were to be developed by teams of engineers organized by the SAE and approved by government officials charged with the purchase of aircraft and aviation-related items. In the case of wood, for example, the handbook would include tables of specific gravity and weights of various species along with acceptable standards for strength, hardness, and elasticity; information on the acceptability of defects (knots, cross and spiral grain, pitch pockets); appropriate levels of moisture; and the effects of different methods of drying and finishing treatments. The same sort of standards would be established for metals used in aircraft construction: fabrics, thread, and cord; paints, dopes, and varnishes; and the whole range of miscellaneous materials such as wire, glues, welding and soldering procedures and equipment, and shock absorbers. From information on the effect of sunlight on fabrics and dopes to methods of testing, the planners of the handbook attempted to provide specific answers to the important questions.

That was only the beginning. Other sections of the handbook would address aircraft engines, propeller design and construction, and aircraft structures in similar detail. The engineer could find everything from tables of inertia, moment equations, and formulae for calculating stresses to the weight of standard metal fittings. Standards were established for specific parts: screw threads, turnbuckles, wires, cables, and tubing. There were chapters on computing everything from range to stability. Acceptable factors of safety were established, as were standards for testing. Questions relating to flight instruments, armament, aerial photographic equipment, parachutes, landing fields, and hangars would be considered.

The preparation of the standards required a considerable managerial effort. The ubiquitous Alexander Klemin would serve as project editor. Robert Burnett recruited a broad range of engineering specialists to serve on a series of committees that would suggest all of the required standards. Representatives of the military services and other government agencies responsible for aeronautics or technological research served on a Subdivision for approving existing standards and other organizations that would pass judgment on the work of the technical committees.

Lieutenant Commander Jerome Hunsaker received a Ph.D. in aeronautical engineering from MIT in July 1916, the first ever granted by an American university, and immediately reported to Washington, DC, determined to demonstrate the value of science and technology to the national defense. He would head the Aircraft Division of the Bureau of Construction and Repair, under the command of the legendary Admiral David W. Taylor. In his new position, Hunsaker would be responsible for the design and acquisition of each one of the roughly 1,000 Naval aircraft that saw service in World War I. Following the war, he headed the effort to fly the Atlantic with Navy's NC flying boats and designed the supervised construction of the first rigid airship built in the United States, the ZR-1 *Shenandoah*.[33]

In addition to his normal workload, Hunsaker played an important role in shaping the SAE standards. He selected the Navy representative to the SAE Subdivision for Approving Existing Standards and personally served on the committees establishing standards for Iron and Steel; Performance and Testing; Wire and Fittings (Chair); Materials Testing; Dope, Cement, Varnish, and Glue; Structural Tubing; Tires and Wheels; and Lighter-Than-Air (Chair). He was also a member of a 22-man SAE Aeronautic Division, originally charged with overseeing the entire operation. That group included Virginius Clark, Alexander Klemin, and Donald Douglas—all Hunsaker protégés.

The project was eventually streamlined. The oversight group, the Aeronautic Division, was discarded as too cumbersome. The SAE staff and the editor would coordinate the work of the individual standards committees. There was simplification there, as well. Hunsaker himself suggested disbanding the Lighter-Than-Air Committee, arguing that every airship constructed was an experiment composed of items already covered by the various material standards.

Hunsaker's involvement suggests the amount of effort required to produce the first *SAE Aeronautical Handbook*, complete with standards, that was issued in 1924 and constantly updated thereafter. The project was a major contribution to the growth of the aviation industry in America, assuring that everyone, from government acquisitions officer to the engineer sitting at a drawing board, would share a common understanding of accepted standards for virtually all materials included in aircraft construction.

Like the ASME, SAE has remained heavily involved in the aerospace enterprise. It continues to sponsor a range of specialized meetings aimed at engineers employed in flight-related industries, especially aeronautical propulsion. A serial publication, *The Journal of Aerospace*, is an element of the SAE Transactions series. Other journals in the series (*Journal of Engines*, *Journal of Fuels and Lubricants*, and *Journal of Materials and Manufacturing*) also cover topics of interest to aerospace engineers.

Grover Loening's fear that the SAE was part of a "Detroit Conspiracy" aimed at taking over the aircraft industry proved groundless. Far from representing a danger, the organization provided an early professional home for aeronautical engineers at a time when their own industry was not large enough to support its own technical society. The SAE established an essential baseline through its handbook and set of standards. Still, many aeronautical professionals remained uncomfortable with the notion of sharing a professional home with earthbound automotive engineers. When queried recently as to how he felt about continued SAE involvement in flight, a long-time leader of the American Institute of Aeronautics and Astronautics (AIAA) simply remarked that he wished they would stick to automobiles.[34]

Major Gardner's Vision

Lester Durand Gardner (1876–1956) was a big man in every sense. He was a tall, formidable figure with a broad face, ruddy cheeks, and a strong cleft chin. He parted his thinning hair with great care as a younger man and combed it almost straight back as he grew older. Always impeccably dressed, a pince nez gave way to glasses with severely thin frames as the years went by. In those few photographs in which he is not smiling, he looks as though he is about to break into a grin. His personality expanded to fill whatever space he was in.

"He was an autocrat by nature," his friend J. Lawrence Prichard explained. "A superb, kindly, understanding host, a great and persistent organizer for the great cause, and friend to all who mattered in aviation."[35] The quality and quantity of his global circle of friends was legend. The friendships that he made, Gardner insisted, "were the best reward of all."[36,37] American aviation's unofficial ambassador to the world, he is reputed to have had more aeronautical acquaintances than anyone else in his generation. The series of elephant quarto-sized scrapbooks in which he preserved the flight-themed Christmas cards that he received each year for decades are evidence of that.

Lester Gardner was a native New Yorker but grew up in Detroit, where he attended public schools. He was nothing if not active. He sang in a school chorus and participated in athletics at the YMCA. His legendary wanderlust began at an early age. He attended the 1892 Republican National Convention in Minneapolis, visited Chicago's World's Columbian Exposition, and observed sessions of the House of Representatives and the Senate when his high school class traveled to Washington, DC. In 1893, he toured Europe with his family, visiting London, Antwerp, and Cologne.

Determined to document his life, young Mr. Gardner preserved tickets, photos, scorecards, programs, and news clippings in the first of what would become a long series of autobiographical scrapbooks. There is his membership card in the Detroit "Y," a program to the first ballet that he attended at Chicago's Auditorium Building, a ticket to a gymnastic exhibition in which he participated, a press clipping revealing that he finished fourth from last in a field of 19 cyclists entered in a road race, and news articles recounting his role in capturing the thieves who had robbed his parent's home.[37]

Always sure of himself, Gardner applied for admission to MIT in 1894, having completed only two years of high school. He decided to spend the summer riding his bicycle the roughly 800 miles from Detroit to Cambridge, Massachusetts, via New York City. True to form, he stopped by the newspaper offices in small towns and large cities along the way to announce himself. Most of the local newsmen seem to have been suitably impressed. Gardner's favorite notice came from an Albany reporter, who commented on the young athlete's "white duck trousers and otherwise fashionable dress." It was, Gardner explained in his scrapbook, "My first style note."[38]

During his first semester at "the Tech," Gardner took courses in algebra, geometry, mechanical drawing, and German as well as the instruction in military tactics and drill required of all members of the MIT Corps of Cadets. Although he did well enough in the classroom, it is fair to say that he focused considerable attention on extracurricular activities. He played baseball; sang first tenor in the glee club; ran for class office and occasionally won; took part in all of the activities of the Corps of Cadets; and earned good reviews for his performance as "a new woman," complete with a wig and bloomers, in a student operetta.

Scrapbook mementos of his college career include membership cards in both the Athletic and Football Associations, grade reports, dance cards, athletic scorecards, and article after article clipped from student publications and the Boston papers. Souvenirs of trips to New York during his student years range from tickets to both the Metropolitan Opera and Koster and Bial's vaudeville house to a theater seat number (914) requisitioned to commemorate a particularly memorable night at Huntington Hall.

The articles also record his involvement in more boisterous college activities. Gardner was arrested for shouting college yells in the street during his freshman year. As a sophomore he was implicated in a gag that involved "fixing" a freshman class election. He was named as a participant in a no-holds barred battle between the classes of 1897 and 1898 to see who could put a flag on top of a greased pole at the south end of the baseball field. Then there was the "riot" occasioned when Harvard supporters of presidential candidate William McKinley encountered the pro-William Jennings Bryan crowd from MIT on Boston's Park Square. The police arrived too late to make an arrest. In his senior year, the Boston papers reported that Gardner was one of five students who broke up a "medium show" by rushing onto the stage and pulling aside a curtain to reveal "the Reverend Concannon" as he was changing out of the "spiritual clothes" that he wore when impersonating a ghost. Small wonder that he recorded June 3, 1898, as the date of his first "great thrill." It was the day on which he was informed that the MIT faculty had recommended him for graduation.

Gardner never forgot his MIT years. In 1915, he headed a committee of Tech grads who suggested the beaver as a suitable MIT mascot. "Of all the animals in the world," Gardner explained, quoting the words of zoologist William Hornaday, "the beaver is noted for his engineering skill and habits of industry. His habits are nocturnal. He does his best work at night."[39] By 1920, he was serving as president of the Technology Club, a New York alumni organization boasting a thousand members. At the end of his life, Gardner led the campaign to raise $500,000 with which to establish an endowed chair at MIT named in honor of his old colleague, Jerome Hunsaker. In 1954, his alma mater memorialized his contributions with a Lester Gardner Scholarship Fund.

As Gardner was pedaling his way to Cambridge in 1894, he planned to study electrical engineering. During his time at MIT, however, he became far more interested in writing and publishing. He wrote articles for *The Tech*, a student newspaper, and served on the editorial board. By his junior year, he was supplying the *Boston Post* with a steady stream of articles on MIT athletics and student antics. In 1898, he published articles in the *Post*, the *Philadelphia Inquirer*, *Detroit News*, and *Minneapolis Journal*.

After graduation, he studied administrative law at Columbia University for a year and then spent seven years working for a string of New York newspapers: the *Times*, *Evening Mail*, *Herald-Tribune*, and *Sun*. Becoming interested in new discoveries in radiation, he took a leave of absence from the *Sun*, visited Marie and Pierre Curie in Paris, and began importing radium into both the United States and England for use in medicine and other applications, from the manufacture of luminous dial watches to glow-in-the-dark theatrical costumes. When he returned to publishing, he focused on magazine production, working on the staff of *Collier's* and *Everybody's*. In 1912, the American Lithographic Company introduced a gravure process for printing photographs on newsprint. Gardner took charge of the operation and made the acquaintance of W. D. Moffat, with whom he would establish the Gardner–Moffat Publishing Company. He married Margaret Kettle and seemed to be settling into the life of a successful businessman.

However, Lester Gardner had discovered his true passion in 1910, when, like Jerry Hunsaker, he attended the Harvard–Boston Air Meet. Fascinated, he began to read everything he could find on flying. A range of periodicals had popped up to report on aeronautical activity and to explain the latest developments in this rapidly changing field. None of them met Gardner's exacting standards. An experienced newsman and publisher, he was determined to produce a magazine that would match the best of the European aviation journals.

The first issue of *Aviation and Aeronautical Engineering* was dated August 1, 1916, and had a cover price of five cents. "The future of the aeroplane," Gardner noted in the first line of the first issue, "will depend largely on the use that is made of the technical information that is gathered in all parts of the world." He hoped that his new enterprise would stimulate "... the whole aeronautical profession, the members of which will find in the new publication a continuous source of reliable information as well as a medium worthy of receiving and transmitting to the aeronautical world the results of their valuable experiments, researches, constructional developments and matured view on the many controversial aspects of this great branch of engineering."[40]

In that he would succeed. The first issue offered the first installment of a 24-part course in aeronautical engineering prepared by Alexander Klemin, who had replaced Hunsaker as MIT's instructor in aeronautics. There was an editorial on worldwide government appropriations for aviation and articles on the latest developments around the globe. *Aviation and Aeronautical Engineering* was, C. G. Grey, editor of the English journal *Aeroplane* remarked, "a welcome American newcomer."[41] The biweekly publication, priced at five cents per issue, quickly became the American journal of record for aviation.

Always the patriot, Gardner was one of many businessmen who signed up for the famous reserve officer training camp at Plattsburgh, New York in 1915. Aware of the publisher's contacts in the aviation community, Major General George Squire, commander of Army aviation, offered him a commission in the event of war. Gardner accepted. Enlisting as a lieutenant in the Signal Officers Reserve Corps in August 1917, he was quickly promoted to captain in the regular army. He was initially ordered to Kelly Field, Texas, where he organized the local Liberty Loan campaign.

Captain Gardner then took several Air Service companies to South Carolina, where they went into quarantine before being shipped overseas. Rather than traveling to Europe with his men, however, Gardner was next ordered to Waco, Texas, to command a larger training camp for troops scheduled for service at the front. Finally, he was reassigned to Washington, DC, and a seat on the Air Service Control Board. While serving in the nation's capital, he learned to fly and was promoted to the rank of Major, a title that he would continue to use for the rest of his career.

The Major returned to life as an editor/publisher and president of the Gardner–Moffet Company following the Armistice. The company opened a new printing plant in Highland

Park, New York, in 1920. Gardner continued to preach the gospel of aviation in the pages of his magazine. He addressed what he saw as the critical issues, such as building and maintaining a positive public image for aviation. In the immediate post-war years, he saw the trend toward "barnstorming" as a potential problem for the industry.

"One of the most interesting phases of present aviation activities," he noted in October 1919, "is the great number of small companies engaged in exhibition flights and in passenger flights of short duration." While "such work" had not yet achieved the "dignity" of the commercial air transport ventures sprouting in Europe, Gardner argued that the aerial gypsies hopscotching across the nation did have an impact on public attitudes toward aviation. He warned that "stunting should be avoided."

> Even if the passenger should ask for stunting, he will not enjoy it. He will come down congratulating himself on being a brave man but with the feeling he has had a very serious experience. The passenger should come down feeling as though he has had a perfectly safe and normal experience which he would like to repeat.[42]

Gardner certainly did his part to demonstrate that flying was a safe and comfortable way to travel. During these years he earned a reputation as American aviation's flying ambassador to the world. During their 1926 vacation, Lester and Margaret covered 21,000 miles during a 53-day flying tour of 27 European, North African, and Middle Eastern nations. At the conclusion of the trip, the Society of British Aircraft constructors held a special lunch to honor Gardner and his achievement.

He made another European aerial tour in 1929. Seven years later, he flew the airship *Hindenburg* to Europe and then traveled on to Moscow by airplane. By 1938, he had earned a reputation as one of the world's premiere air travelers. He was also famed for his ability to talk air carriers into conveying him from place to place at no charge. In 1938, sculptor Harold Erskine presented Gardner with a bronze statue of a winged forearm, with its fist and thumb positioned in the time-honored symbol of the hitchhiker. "The First Aerial Hitch-hiker," the Associated Press noted, had been "Immortalized in Bronze."[43]

Gardner was expanding his small publishing empire. He was now the owner/publisher of another trade paper, *Rubber Age*, and aviation reference works such as *Who's Who in American Aviation*. His flagship publication, *Aviation*, continued to evolve. In its early years, when technical information on aeronautics had been hard to come by, the magazine sought recognition as an engineering journal. As the NACA, SAE, ASME, and other organizations and government agencies took up that task, Gardner transformed *Aviation* into a trade paper carrying the latest insider news of the aeronautical industry. By 1923 he had acquired full control of the publishing firm. He supervised the work of a series of editors until he hired Earl D. Osborn, treasurer of Aeromarine Airways, as vice-president of Gardner Publishing and editor of *Aviation*. Osborn became publisher in 1924, when Gardner assumed directorship of the company. Gardner sold the firm in 1926. By March 1929, McGraw–Hill had taken over the publication.

In 1927 President Calvin Coolidge appointed Gardner the U.S. delegate to the Fourth International Aviation Conference in Rome. The following year The Major represented the United States at the International Conference on Civil Aeronautics in Washington, DC. In 1928 he was also named president of both Aeronautical Industries, Inc., an investment trust, and the Aeronautical Chamber of Commerce of America (ACCA), the industry trade association. During his year in office as head of the ACCA, he was also giving considerable thought to another sort of aeronautical organization. He had in mind a society that would be a bit more than simply a professional association for aeronautical engineers and scientists. His model would be the oldest and most prestigious organization in the world of flight—the Royal Aeronautical Society (RAeS).

The Model Organization

Founded as the Aeronautical Society of Great Britain in 1866, the RAeS was intended to distinguish serous technical professionals interested in flight from the naïve enthusiasts and wild-eyed flying machine cranks that abounded in the mid-19th century. Inspired by the older engineering societies, the group would support research, recognize important contributions of its members, and work to convince the public that winged flight was not a foolish dream but a real possibility that might be achieved by professional technicians who banded together to lay a foundation for success.

Many of Britain's leading engineers became active members of the RAeS. The group arranged lectures and technical meetings to share information and in 1867 established the *Annual Report of the Aeronautical Society*, a journal by engineers for engineers seeking to extend professional standards into the new field. They also sought to educate the public. The members staged the world's first public exhibition of flying apparatus at the Crystal Palace, Sydenham, for 10 days beginning on June 25, 1868. The exhibition featured a hodgepodge of aircraft models, lightweight power plants, and other bits and pieces of aeronautical paraphernalia. As the promoters had hoped, the show drew attention to the serious prospect of winged flight.

The real business of the Aeronautical Society of Great Britain, however, was to encourage progress toward powered flight. No member of the group contributed more to the achievement of that goal than Francis Herbert Wenham. A native of Kensington, born in 1824, he was a talented engineer with professional interests ranging from photography to microscopy, scientific instrument design, and the development of high-pressure steam and internal combustion engines. "When I was a little chap," Wenham remarked many years later, "I was fond of making kites and flying them." His first aeronautical experiments involved propeller design in 1859. Over the next six years he progressed to flight tests with a full-scale manned glider featuring long, narrow, or high-aspect ratio, "Venetian blind" style wings.

Wenham recognized that aircraft designers were operating in the dark. "A series of experiments is much needed," he explained, "to provide data for construction." In his first paper to the Aeronautical Society, he announced that he would "... shortly ... try a series of experiments by the aid of an artificial current of air of known strength, and to place the Society in possession of the results."

The result was the wind tunnel, an invention that would play a critically important role in the history of flight. Designed by Wenham and constructed by John Browning with a grant from the Aeronautical Society, the world's first wind tunnel was operated at Greenwich and London, 1870–1872. It was a hollow box, 10–feet long and open at both ends. A fan moved a constant stream of air over a "balance," an instrument mounted in the tunnel that measured the forces generated by the model wing being tested.

Wenham's primitive balance was not even sensitive enough to record measurements on a test surface operating at the relatively small angles of attack where an airplane would actually fly. One of his few genuine discoveries was to note that the center of pressure on a flat plate moves toward the leading edge with a decrease in the angle of attack. Still, it was a beginning.

Horatio Phillips (1845–1912), a member of the Aeronautical Society who had attended Wenham's lectures, developed a much-improved tunnel and balance in the 1880s, gathering lift and drag data on a series of six cambered airfoils that he patented in 1884.

The Aeronautical Society had an enormous impact, but it can scarcely be said to have prospered. By 1896, the treasury held only five guineas. The new secretary, Captain B.F.S. Baden-Powell, brother of the founder of the Boy Scouts, was determined to reinvigorate the organization. He began to build a world-class library of books on flight and launched a quarterly serial, the *Aeronautical Journal*, which succeeded the *Annual Report* as the Society's principal technical publication. For the next three years, Baden-Powell would fund the new publication out of his own pocket.

Even with the invention of the airplane and the birth of a new industry, the RAeS (so honored by King George V in 1917) continued to struggle. Membership fell from 1,100 in 1919, to 800 in 1923, and little more than 600 in 1925. The Society could no longer pay its secretary, Lockwood Marsh, who resigned. Salvation came in the person of J. Laurence Pritchard, a Cambridge graduate who had coauthored the standard text on aircraft structures. Named editor of the *Aeronautical Journal* in 1919, Pritchard accepted only a token salary during the hard times of the early 1920s. He kept body and soul together working as a journalist specializing in aviation issues.

In addition to his service as a first-rate editor, Pritchard worked hard to get the RAeS back on its feet. He helped to organize the first International Air Congress in 1923, with H.R.H. the Duke of York as president. He played an important role in acquiring 250-pound sterling annual donations from both the Air Ministry and the Society of British Aircraft Constructors (S.B.A.C.), an industry trade group organized in 1916. Pritchard paid repeated visits to the offices of corporate executives in search of financial support. By 1930 his tireless efforts had increased membership to more than one thousand.

The RAeS was also subtly redefining its mission. During the postwar years, the organization created grades of membership and accepted the task of setting professional standards and developing the courses of study and examinations that would qualify an individual as an associate member of the organization and a recognized professional engineer. The Society took over the functions of the short-lived Institution of Aeronautical Engineers in 1927 and established its first branches. By 1933 the S.B.A.C. had agreed that aircraft manufacturers would give hiring preference to engineers who passed those examinations and qualified as associate members of the RAeS. For aeronautical professionals, membership in the organization was a mark of genuine distinction, visible proof of having entered the elite upper circles of British aviation.

Fascinated by the history of flight, Pritchard sought to forge a new kind of technical society that would meet the needs of engineers while at the same time preserving and celebrating the culture and traditions of aeronautics. Central to that goal was the creation of the finest aeronautical library and archive in the world, a source of information that would both meet the needs of the members and become an international center for research into the past and present of aviation. In 1923 the Carnegie International Trust donated 500 pounds sterling for the purchase of rare books and aeronautical materials. Two years later, the Guggenheim Fund for the Promotion of Aeronautics gave the RAeS 1,000–pounds sterling to be spent on the library and publications. J. E. Hodgson, the Honorary Librarian of the Society, was himself a distinguished collector of books, manuscripts, engravings, models, and other materials documenting the history of flight. In 1929, Pritchard and Hodgson created a large and impressive exhibition on the history of flight at the annual Olympia aviation show.

Across the Atlantic, Lester Gardner took note of the emergence of the RAeS as an elite professional organization that also sought to preserve and define aeronautical culture. An anglophile, an organization man, and a genuine lover of the lore and legend of flight, he liked what he saw.

References

[1] J. C. Hunsaker and Lester D. Gardner, "Background and Incorporation of the Institute of Aeronautical Societies," in Box 2.53, Folder 1952, the Papers of Hugh L. Dryden, Special Collections and Archives Division, Johns Hopkins Univ. Library.

[2] The leading U.S. professional engineering societies were known as the Founders because other specialist societies spun off them. They were the American Society of Civil Engineers (1852), the American Institute of Mining and Metallurgy (1871), the American Society of Mechanical Engineers (1880), the American Institute of Chemical Engineers (1876), and the Institute of Electrical Engineers (1884).

[3] For more on Octave Chanute, see Tom D. Crouch, *A Dream of Wings: Americans and the Airplane, 1875–1905*, W.W. Norton, Inc., New York, 1981.

[4] Octave Chanute, *Progress in Flying Machines*, The American Engineer and Railroad Journal, New York, 1894.

[5] Thurston quoted in Harrison F. Reeve, "AME Role in Powered Flight," *Mechanical Engineering*, Dec. 1953, pp. 987–999.

[6] Louis Blèriot and Edward Ramond, *La Gloire des Ailes*, Les Editions de France, Paris, 1927, p. 26.

[7] Grover Loening (1888–1976) was born in Bremen, Germany, the son of the U.S. Counsel–general. After receiving his professional engineering education at Columbia University, he worked as chief engineer for the Queen Aircraft Company (1911–1913), owned by Chicagoan Willis McCormick; chief engineer and special assistant to Orville Wright and manager of the Wright Company Factory (1913–1914); and chief aeronautical engineer with the U.S. Army Signal Corps (1914–1917). Loening founded his own company following World War I and remained a major figure in American aeronautics until his death.

[8] Anon, "First Paris Aeronautical Salon," *Flight*, Jan. 2, 1909, p. 2.

[9] T.O.M. Sopwith quoted in, Lee Kennett, *The First Air War, 1914–1918*, New York, Free Press, 1991, p. 26.

[10] My comments on the early history of aerodynamics are based on a reading of John Anderson, *The History of Aerodynamics and its Impact on Flying Machines*, Cambridge Univ. Press, Cambridge, England, U.K., 1997.

[11] *Aeronautics in the Army*, Hearing before the Committee on Military Affairs, House of Representatives, 63rd Congress, First Session, U.S. Government Printing Office, Washington, DC, 1913.

[12] Henry Greenleaf Pearson, *Richard Cockburn Maclaurin: President of the Massachusetts Institute of Technology*, Macmillan, New York, 1937.

[13] On the history of MIT, see Samuel C. Prescott, *When MIT Was "Boston Tech," 1861–1916*, MIT Press, Cambridge, MA, 1954.

[14] William Trimble, *Jerome Hunsaker and the Rise of American Aeronautics*, Smithsonian Inst. Press, Washington, DC, 2002, p. 22.

[15] Earl A. Thornton, "MIT, Jerome C. Hunsaker, and the Origins of Aeronautical Engineering," *Journal of the American Aviation Historical Society*, Winter 1998, p. 309.

[16] Gustave Eiffel, *La résistance de l'air et l'aviation: experiences effectuées au Laboratoire du Champ-de-Mars*, H. Dunod et Pinat, Paris, 1910.

[17] For information on Gustave Eiffel, see John Anderson, *A History of Aerodynamics and its Impact on Flying Machines*, Cambridge Univ. Press, Cambridge, England, U.K., 1997.

[18] For the quotation, see Jerome C. Hunasaker, "Europe's Facilities for Aeronautical Research, I," *Flying*, April 1914, p. 93.

[19] For the European trip, see Jerome C. Hunsaker, "Europe's Facilities for Aeronautical Research, II," *Flying*, May 1914.

[20] For the European trip, see A. F. Zahm, *Report on European Aeronautical Laboratories*, Smithsonian Miscellaneous Collections, Vol. 2, No. 3, Washington, DC, 1914.

[21] Like Hunsaker, Clark was an Annapolis graduate (class of 1907). He resigned from the Navy and enlisted in the Army, where he earned his Military Aviator wings in 1913. He would remain a major figure in the history of U.S. military aviation for the next two decades, serve as a member of the NACA, and develop the famous Clark Y airfoil. Following his retirement he would rise to the position of vice president and chief engineer of Consolidated Aircraft. Klemin, a graduate of the University of London, worked as Hunsaker's assistant before moving on to Columbia. In addition to his academic career, Klemin was active as a consultant and a leader of professional organizations, including both the Society of Automobile Engineers and the Institute of Aeronautical Sciences.

[22] Wayne Biddle, *Barons of the Sky*, Simon and Schuster, New York, 1991, pp. 82–83.

[23] Alex Roland, *Model Research*, NASA, 1985, p. 22.

[24] For a concise history of the NACA, see Roger Brillstein, *Orders of Magnitude: A History of the NACA and NASA, 1915–1990*, NASA, Washington, DC, 1989.

[25] C.V. Glines, "The Guggenheims: Aviation Visionaries," *Aviation History*, Nov. 1996.

[26] "A.S.M.E. Holds Aeronautic Session," *Aero and Hydro*, Vol. 8, No. 5, May 2, 1914, p. 58.

[27] *SAE in Aerospace: Reliability and Progress in the Air*, promotional brochure, author's collection, SAE, 2004.

[28] Grover C. Loening, *Our Wings Grow Faster*, Doubleday, Doran, New York, 1935, p. 72.

[29] Grover C. Loening, *Our Wings Grow Faster*, "The American Society of Aeronautic Engineers Established," *Aerial Age*, Vol. 1, No. 20, Aug. 2, 1915, p. 469.

[30] Grover C. Loening to Orville Wright, Aug. 12, 1915, Jerome Hunsaker Papers, National Air and Space Museum Archive.

[31] Grover C. Loening to Orville Wright, Aug. 14, 1915, Jerome Hunsaker Papers, National Air and Space Museum Archive.

[32] Orville Wright to Grover C. Loening, Aug. 25, 1915, Jerome Hunsaker Papers, National Air and Space Museum Archive.

[33] Hunsaker left the Navy in 1926 to take a position developing aircraft communication systems for Bell Laboratories. Two years later, he accepted a position as vice president of the newly organized Goodyear–Zeppelin Corporation. He returned to MIT in 1933 to teach the course he had developed two decades before and became the first head of the Department of Aeronautical Engineering when it was created in 1939. During World War II he chaired the NACA and coordinated U.S. Navy research. He continued to chair the department that he had created until 1956. One of the most distinguished engineers in the history of world aviation, he was awarded the Daniel Guggenheim Medal in 1933, the Wright Memorial Trophy in 1951, the Gold Medal of the Royal Aeronautical Society, and the NACA Distinguished Service Medal. He died in Boston on Sep. 10, 1984, at the age of 98 years and 1 month.

[34] Tom D. Crouch, interview with James Harford, tape in the AIAA collection, Aug. 4, 2004.

[35] J. Laurence Pritchard, "Lester Durand Gardner," copy in the Lester Durand Gardner biographical file, National Air and Space Museum Archive.

[36] Marvin W. McFarland, "Lester Durand Gardner: Elder Statesman of Aviation," *U.S. Air Services*, Oct. 1956.

[37] All of Gardner's surviving scrapbooks, including the Christmas card scrapbooks, are preserved in the Durand papers section of the AIAA History Collection, Manuscript Division, Library of Congress.

[38] AIAA History Collection, Manuscript Division, Library of Congress, scrapbook OV5.

[39] AIAA History Collection, Manuscript Division, Library of Congress, OV1.

[40] *Aviation and Aeronautical Engineering,* Vol. 1, No., 1, p. 7.

[41] Anon, *The Aeroplane*, Oct. 1916.

[42] Lester Durant Gardner, *Aviation*, Oct. 1919.

[43] Howard W. Blakeslee, "First Aerial Hitch-hiker Immortalized in Bronze," AP Feature Service, AIAA History Collection, Manuscript Division, Library of Congress, OV1.

Chapter 2

The Rocketeers: The American Rocket Society, 1930–1940

Interplanetary Dreamers, 1930–1931

It began in the early spring of 1930 at two tables in a corner of Nino and Nella's, a walk-down speakeasy in the basement of 450 West 22nd Street in the West Chelsea section of New York. The proprietor, Nino Bucalari, had a reputation as "one helluva chef," specializing in Italian dishes. His wife Nella's cousin, a wine steward on an Italian ocean liner, helped to fund the operation and supplied that rarest of Prohibition era commodities, good liquor.[1,2] The Bucalaris lived in an apartment just above the place. The Pendrays, Ed and Leatrice, rented the second-floor apartment, which they furnished with tables and chairs constructed ("in the cubist manner") from shipping crates. The downstairs establishment became the accustomed meeting place for their friends.

The Pendrays were westerners. Gawain Edward Pendray (1901–1987) was born in Nebraska and grew up on a Wyoming ranch. After graduating from the University of Wyoming in 1924, he moved to New York City where he earned an M.A. from Columbia University the following year. Pendray joined the staff of the New York *Herald Tribune* in 1925, starting as a reporter and moving on to successive positions as assistant city editor, picture editor, and science editor. He served as science editor of the *Literary Digest* from 1932 to 1936, when he shifted from journalism to public relations.[3]

A native of Colorado City, Texas, born on October 1, 1905, Leatrice May Gregory grew up in the small towns of Bosler and Rock Creek, Wyoming. Graduating from the University of Wyoming in 1927, she moved to New York, married Ed Pendray, whom she had met in college, and established herself as a newspaper columnist. Over the next two decades, her columns on entertainment, cooking, and beauty were syndicated in the women's sections of 300 newspapers.[4]

Like so many young people, the Pendrays were attracted by the excitement and sense of possibility in Jazz Age Manhattan. Ed spent a year writing and rewriting a novel about homesteading in Wyoming, all the while struggling with a series of part-time jobs to make ends meet. He taught English to foreigners, wrote jokes for *Life* and *Judge* magazines, produced a daily column for his hometown newspaper, the *Laramie* [Wyoming] *Republican-Boomerang*, and contributed an occasional piece to the *Saturday Evening Post*.

A quarter of a century later, a journalist asked Lee Pendray how her husband, with his literary tastes, had first become interested in rocketry and space flight. That, she responded, was a question that "... could well be answered by a psychiatrist." Ed had taken some engineering courses in college, she noted, and "had always had a remarkable understanding of science and the even more remarkable ability to interpret scientific subjects for the average reader."[5] As a youngster, he had devoured the novels of Jules Verne and H. G. Wells. His wife remembered that the occasional newspaper and magazine stories about the rocket experiments of Dr. Robert Hutchings Goddard, a Massachusetts physics professor, had also sparked his imagination.

The young journalist began to specialize in reporting science news. "I immersed myself in scientific books, living in libraries," he recalled. "I attended scientific meetings and managed to meet many prominent scientists."[6]

He also began to write science fiction yarns, "more for amusement and relaxation than anything else," Lee recalled. His stories found a home in the pulp magazines, where the pay was only a quarter of a cent a word, but there was a voracious appetite for new material. Printed on the cheapest paper and bound in garish and usually suggestive covers, these inexpensive publications had originally offered lurid tales of crime, Wild West adventure, aerial daring-do, and romance.

In the 1920s, however, entrepreneur Hugo Gernsback (1884–1967) had sent the pulps spinning in an entirely new direction. A native of Luxemburg who emigrated to the United States in 1904, Gernsback was fascinated by electricity and radio technology. He founded the Electric Importing Company, perhaps the world's first electrical supply house, in 1905 and began publishing *Modern Electrics*, a hobby magazine, in 1911.[7] Gernsback was convinced that technology would shape a utopian future for mankind. "What man wills, man can do," he explained." "Interplanetarian trips, space flyers, talking to Mars, ... death rays, gravity-nullifiers ... why not? If not today, well, then, tomorrow."[8]

Gernsback channeled his boundless optimism into his own wildly imaginative tales of the technological future, such as *Ralph 124C 41 +: A Romance of the Year 2660*, which he serialized in *Modern Electrics* in 1911. He originally referred to his stories as "scientifiction" and established *Amazing Stories*, a pulp magazine dedicated to the genre, in 1926. He followed up his initial success with the *Amazing Stories Annual* in 1927, *Amazing Stories Quarterly* in 1928, and *Science Wonder Stories* in 1929. Pendray's imagination was sparked by these optimistic tales of a future shaped by science and technology. He wrote another unpublished novel, *The Earth Tube*, and contributed a series of short stories to the Gernsback pulps, tales like, "A Burial in Space," "A Rescue From Jupiter," and its sequel, "Return to Earth."

Having established himself as a regular contributor to Gernsback publications, Ed Pendray and his wife began inviting David Lasser, managing editor of *Science Wonder Stories*, and associate editor Charles P. Mason to occasional story conferences at Nino and Nella's. Over time, the circle enlarged to include other members of the Gernsback fraternity. The talk at these informal gatherings always seemed to turn to the possibility of space flight, a subject that dominated their stories.

Like Pendray, these young dreamers were vaguely aware of the rocket experiments conducted by Robert H. Goddard of Clark University in Worcester, Massachusetts. In 1920, the Smithsonian Institution had published his *Method of Reaching Extreme Altitudes*, a short treatise demonstrating that a rocket was capable of propelling a vehicle through the vacuum of space. Three years later, Rumanian Hermann Oberth published *Die Rakete zu den Planetenraumen* (*The Rocket Into Interplanetary Space*). Densely packed with mathematical equations, the book nevertheless sparked a wave of enthusiasm for space flight in Germany and Austria. The leaders of the Soviet Union encouraged rocket research and elevated Konstantine Tsiokovsky, a Russian space theorist who had preceded both Goddard and Oberth, to the status of a national hero.

The space pioneers of the 1920s attracted considerable attention, not all of it welcome. Goddard fought a losing battle to maintain a measure of professional dignity. He received dozens of letters from individuals volunteering for the first flight to the moon. A Bronx promoter asked him to consider launching his "moon going rocket" from the Starlight Amusement Park, while a Hollywood publicist suggested that the first spaceship should carry a greeting from Mary Pickford, "America's Sweetheart." An editorial writer for the *New York Times* took Goddard to task for arguing that a rocket could travel into space where there was

The birthplace of the American Interplanetary Society, 450 West 22nd Street. (Courtesy AIAA.)

no air to push against. Goddard responded with a letter explaining Newton's three laws of motion. It was to no avail. For most common sense, down-to-earth folks, the notion of interplanetary travel was the very definition of the outrageous.

The small band of enthusiasts gathered around the two corner tables at Nino and Nella's found it difficult to understand why others did not share their excitement at the prospect of space flight. "To us," Pendray explained years later, "despite all evidence and scientific opinion to the contrary, a trip to the Moon seemed just around the corner." One evening in March 1930, as Nino began upending the chairs on the tables, the group adjourned to the Pendray's third-floor walk-up where the discussion continued, fueled by a gallon jug of red wine. "Somewhere around midnight," Pendray recalled, David Lasser suggested that they organize.

Why not? As individuals, they were a collection of struggling writers exploring the dream of interplanetary travel in science fiction stories aimed at adolescent readers. Together they might generate publicity for space flight and perhaps even raise money to support rocket experiments. They could publish a bulletin, sponsor public lectures, write letters to the editor, and issue press releases on society letterhead. Before the group broke up in the early hours of the morning, they discussed the points to be covered in a constitution and selected a bold name that expressed the scope of their dream: The American Interplanetary Society (AIS). "We felt," Pendray recalled, "the fervor of pioneers about to ride the Wave of the Future across the borders of a New Frontier."[1]

A month later, on April 4, 1930, 11 men and 1 woman who Ed Pendray regarded as the list of founders signed their names to a sheet of typing paper.[9] Decades later, Lee Pendray remembered them as they were then

- *David Lasser*, "the moving spirit" of the group, lost his slight tendency to stutter when "carried away by enthusiasm." The son of Russian immigrants, Lasser was born in Baltimore and grew up in Newark. He left school at 16 to work in his father's tailor shop and then enlisted in the Army when the United States entered World War I. Following his discharge, Lasser attended the Newark College of Engineering and then transferred to the Massachusetts Institute of Technology (MIT), where he graduated with a degree in Engineering Management in 1924. In spite of his training, Lee Pendray thought Lasser was more a poet than an engineer. Time would prove her correct. Following a few years in industry, he took a job as managing editor of *Science Wonder Stories*, which was much more to his taste.[10]
- *C. P. Mason* was a big man who loved food and wine and visited Nino and Nella's even when a meeting was not underway. A courtly pipe smoker, he drafted the constitution for the new group.

Four of the founding members of the American Interplanetary Society, with a guest: (left to right) G. Edward Pendray (ARS President for 1932), Lee Gregory Pendray, David Lasser (ARS President for 1930–1931), Sir Hubert Wilkens, and C. P. Mason. (Courtesy AIAA.)

G. Edward Pendray. (NASM Archive, Smithsonian Institution, 72-1076.)

- *C. W. Van Devander* was a wire service reporter who wrote science fiction under the nom de plume Peter Arnold. Boyish and slight, he exhibited "... the rather cynical streak of most young reporters."
- *Fletcher Pratt* wrote history as well as science fiction. He stood only five feet one or two inches tall and had scant sandy hair and blue eyes. Pratt had a talent for mathematics, collected chess sets, and devised strategy games in which toy soldiers and miniature ships were maneuvered on huge maps laid out on the floor. In addition to the books and games that crowded their apartment, Fletcher and Inga Pratt always seemed to have a number of cats on hand.
- *Laurence Manning,* who invented the game "Battleship," shared some of Pratt's enthusiasms and wrote his first science fiction story with his friend.[11] Lee Pendray recalled that he had a mathematical mind, a law degree, and "unlimited imagination." Rather than being "constrained" by a legal career, he operated a nursery for most of his life. Manning had a large collection of classical records and, during the period of his involvement with the rocketeers, was writing a symphony.
- Lawyer *Nathan Schachner* was one of the most popular authors in Gernsback's stable. Injured in an accident during an early rocket launch, he nevertheless remained an interested member for many years and always provided pro bono legal services for the group.
- *Adolph L. Fierst* was a rewrite man for Gernsback publications.
- *Dr. William Lemkin,* a Russian emigrant with a Ph.D. in chemistry, was the only professional scientist in the group.

John Shesta and Hugh Franklin Pierce standing behind the final ARS rocket engine test stand. (NASM Archive, Smithsonian Institution, 90-8379.)

- *Warren Fitzgerald* was president of the Scienceers, a group of New York City science fiction fans. Science fiction historian Sam Moskowitz identified Fitzgerald as an African American.[12]
- Nothing is known of *Everett Long,* the twelfth name on the list. It is interesting to note, however, that *Roy Giles,* a reporter and freelance writer whose name was not included, was later identified as a founding member of the AIS.[5,13]

Lasser presented the draft constitution when the group reconvened at the Pendray apartment on the evening of March 21, 1930 (Ref. 14). It contained the usual articles governing membership, the committee structure, bylaws, and the election and duties of officers. Lasser, Pendray, and Mason were elected president, vice-president, and secretary. Treasurer Manning would collect annual dues of $10 from regular members and $3 from associate members, who would enjoy all privileges except the right to vote. Individuals or organizations that assisted the society in one way or another could be recognized as contributing members. After operating for almost a year, the group would vote to incorporate under the laws of the state of New York on January 16, 1931.

The members would modify the organization in the years to come. The dues were readjusted in the spring of 1932. Active members who lived within 100 miles of New York City would pay $7.50 a year. Those who lived farther from the city would be assessed $5.00 (Ref. 15). An amendment adopted on April 29, 1938, created a new class of Junior members open to interested young people under the age of 18. The largest single organizational change came on April 26, 1935, when a seven-member board of directors was created to govern the Society.

The founders were quick to recruit friends and relatives into membership. Dr. Samuel Lichtenstein, Schachner's dentist brother-in-law, replaced Manning as treasurer and remained a key figure in the early years of the organization.[16] Hugo Gernsback was critically important in the early months. He arranged for a regular gathering place at the American Museum of Natural History, beginning with the third meeting on April 19, 1930, and opened the pages of *Science Wonder Stories* and his other magazines to news of the AIS.[17]

The short articles on the organization appearing in Gernsback publications attracted the attention of science fiction fans across the nation. Between April 1930 and March 1931, the AIS received 103 inquiries regarding membership. Thirty-nine individuals signed up as associate members, 18 of them from New York or the New Jersey suburbs. The others represented 18 states from New England to Florida and California. During the first year of operation, three individuals from Mexico, Canada, and Russia also signed up. One of the new members was listed as having joined while, "at sea." It can only have been Midshipman Robert Heinlein. The future science fiction writer had mailed his application while on a cruise aboard the aircraft carrier, USS *Lexington*. By the end of 1931, the society had enrolled 100 members.[18]

"I think that many young members were attracted to the AIS for the same reason I was," Bernard Smith recalled a quarter of a century later. "Existence in New York was fairly miserable, with hardly anyone feeling he could count on an income from one day to the next. To be transported to a whole new world, possibly away from the troubles of this one, was probably a subconscious motivation all along."[19] "It was a lousy planet," Smith noted on another occasion, and "the rocket ship was the only way to get off it."[20]

In addition to the young enthusiasts, the leaders of the AIS worked hard to attract the attention of men who could make their dreams come true. Lasser expected his cohorts to "ceaselessly hammer at the … public," and to enlist "converts, if not members, of all the technical men we meet; these men will be the centers for spreading our propaganda among others."[21]

G. Edward Pendray with a rocket tracking device. Photo taken for a Conoco Oil Company ad. The helmet is on display at the Stephen F. Udvar-Hazy Center at Dulles Airport. (NASM Archive, Smithsonian Institution, A-4562.)

AMERICAN INTERPLANETARY SOCIETY

302 WEST 22ND STREET, NEW YORK

This is to certify that

Miss Lee Gregory

IS AN Active MEMBER OF THE AMERICAN INTERPLANETARY SOCIETY FOR THE YEAR ENDING March 31, 1934.

Nathan Schachner
PRESIDENT

No. 156

C.P. Mason
SECRETARY

Lee Gregory's membership card. (Courtesy AIAA.)

The launch of ARS #2, "a considerable success." Bernard Smith is rushing back to the shelter, having risked life and limb to repair the launch mechanism with his clothes soaked in propellant. (Courtesy AIAA.)

Ultimately, the new members were attracted by the basic purposes of the AIS, which were spelled out in a single run-on sentence of the constitution:

> The objects of the Society shall be the promotion of interest in and experimentation toward interplanetary expeditions and travel; the mutual enlightenment of its members in matters bearing on the astronomical, physical and other problems pertinent to man's ultimate conquest of space; the stimulation by expenditure of funds and otherwise of American scientists toward a solution of the problems which at present bar the way toward travel among the planets; the collection, correlation and dissemination of facts, information, articles, books, pamphlets and other literature bearing on interplanetary travel and subjects relating thereto; the establishment of a library containing such literature for the information of members, scientists and others to whom the privilege may be granted by the Society; the raising of funds for research and experimentation, and such other activities as the Society may from time to time deem necessary or valuable in connection with the general aim of hastening the day when interplanetary travel shall become a reality.[22]

Pendray put the matter more concisely.
"We thought all that was necessary was to call people's attention to the subject, identify problems, and someone would put up the money and we'd go to the Moon!"[23] His press release announcing the formation of the group expressed that confidence. Pendray's own paper, the New York *Herald Tribune*, devoted almost a full column to the group. "Merely flying about through the air is already getting to be a somewhat dull business," the editor noted, "and who can fail to applaud the first mobilization of interest in the far more

breath-taking projects of 'astronautics'—the navigation of the solar system, if not ultimately of interstellar space."[24]

Although Lasser admitted that a New York *Evening Post* editorial on "Rocketeering," was at least "open-minded," he also noted that the attitude of most people toward the AIS was "... that of incredulousness, mixed with a broad dash of pity, toward those who seriously think that man can escape from the chains that bind him to earth." That was certainly the attitude of the *New York Times* editorial writer who commented that the purposes of the AIS were "too flighty and fantastic to be newsworthy." What was required, the editor suggested, was "an American Planetary Society, to emphasize our taking a greater concern in this planet's affairs."[25] Perhaps so, but even the skeptics of the *Times* regularly described the enthusiastic lectures offered at the monthly AIS meetings.[26]

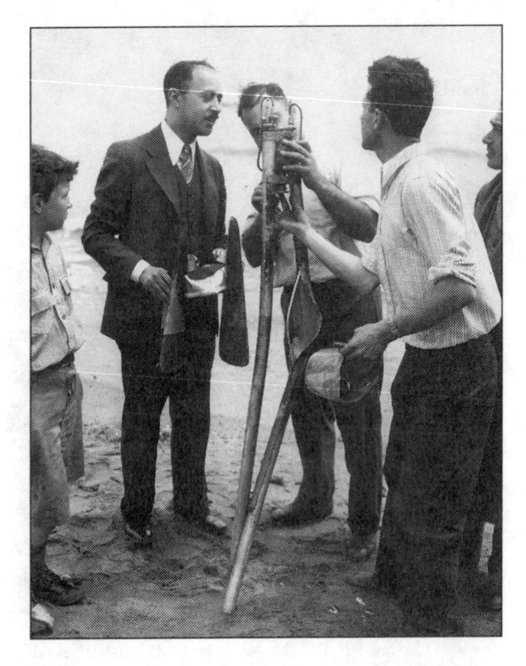

Pendray (third from left) and Bernard Smith (left) inspect the remains of ARS #2. (Courtesy AIAA.)

Lasser's discovery of a leather-bound treasure in a Second Avenue used book stall offered an opportunity for additional publicity. Published in 1640 by John Wilkins, Bishop of Chester, *The Discovery of a New World, or a Discourse Tending to Prove That There May be Another Habitable World in the Moon, and Concerning the Possibility of a Passage Thither,* was one of the first fictional accounts of a space voyage in the English language. Gernsback purchased the volume and arranged for a descendant of the author, famed arctic explorer Sir Hubert Wilkins, to present it to the young space enthusiasts at a public meeting on April 30. The event resulted in a fresh wavelet of newspaper interest and inspired dreams of a public event that would make an even bigger splash.

At the 11th regular meeting of the AIS on the evening of October 3, 1930, David Lasser announced that he had invited Robert Esnault-Pelterie to address a public meeting sponsored by the Society when he visited New York early the next year. One of the great pioneers of world aviation, REP, as he was universally known, had helped to found a French space flight society in December 1927. Three years, in 1930, later he published *L'Astronautique*, a solid overview of the subject. REP sent an inscribed copy of the book to the AIS in the spring of 1930, along with an application for membership and an offer to provide the group with the right to publish an American edition.

With REP's cable accepting the invitation in hand, Pendray persuaded a New York film distributor to loan the group a print of *Frau im Mond* (*Woman in the Moon*), German director Fritz Lang's new science fiction film about a trip to the moon. Surely such a spectacular double feature—a lecture by one of the world's most famous airmen together with a showing of an exciting new European film—would attract newspaper attention and a good crowd of potential members.

A post-war photo of John Shesta, ARS President for 1936. (Courtesy AIAA.)

In the interest of streamlining the program, Pendray ruthlessly edited the film, cutting the romance and melodrama and retaining only the dramatic blast-off, flight, and lunar scenes. Press releases went to every group and publication in sight. The rocketeers plastered walls, lamp poles, and subway cars with posters announcing the event. As the great day approached, American Museum of Natural History officials, concerned by a flood of inquiries about the program, insisted that the AIS arrange extra security for the meeting.

Lasser and the Pendrays met the boat bringing REP to New York on January 15, 1931. They escorted the great man to his hotel and invited him to dine with them at Nino and Nella's, where Pendray could put a good Italian dinner on his tab. Things went well that evening. REP enjoyed the quality of the wine available in Prohibition-era America and even outlined the remarks that he would make. Pendray took notes and would prepare a typed copy of the talk for REP.

Several days later, however, Pendray received a note from the great man: "Strong fever—the Dr. absolutely forbids lecturing on Tuesday." The young newsman doubted the severity of the ailment. He suspected that Esnault-Pelterie had simply lost confidence in the ability of his hosts to arrange the event and was afraid that he would face an empty auditorium.[5,27]

With only a few days remaining before the meeting, it was too late to announce a change in the program. REP's fears were unfounded. The 1,200-seat auditorium at the American Museum of Natural History was filled to capacity on the evening of January 27, 1931. Lasser explained that their guest was ill and introduced Pendray, who would read the prepared remarks. Pleased by the talk and the film, the crowd pushed forward at the end of the program. Many of them had either not heard Lasser's introduction or had misunderstood him and assumed that the balding Pendray with his elegant Van Dyke beard was the advertised speaker. Tired of trying to explain the truth, Pendray gave up and began signing REP's name. So many late-comers had waited outside the auditorium that Lasser and Pendray gave a second "performance" that evening.

In his first annual report to the members, Lasser announced plans to organize the world. He called for an International Interplanetary Commission that would link the work of space flight dreamers in all nations, complete with an international press service furnishing the newspapers of the world with the latest information on rocketry. Eventually, the project would inspire an international effort to travel into space. "I can," he concluded, "foresee the building of the first space ship only as a joint effort of a united earth." Nothing came of the effort, but his remarks epitomized the utopian idealism that motivated the founders of the AIS.[28]

Nothing was more important to the future of the organization than the early issues of what was originally known as the *Bulletin of the American Interplanetary Society*. The first issue,

(From left to right) Alfred Africano, Louis Goodman, Hugh F. Pierce (ARS President for 1940–1941), and Roy Healy (ARS President for 1942) take a lunch break during the tests at Midvale, New Jersey, 1937. (Courtesy AIAA.)

which appeared in June 1930, was a four-page mimeographed newsletter offering a mix of news and information. Fletcher Pratt introduced readers new to the field to the treatment of interplanetary travel in speculative fiction and to the work of the pioneers of rocketry and space flight, men like Robert Goddard and Hermann Oberth. There was news of the Society and notices of other New York events, such as Princeton Professor John Q. Stewart's lecture on rocketry at the Brooklyn Institute of Arts and Sciences. Finally, there was up-to-date news of developments in Europe, including the work of the Czech experimenter Ludvik Očenàšek and the death of Max Valier, a well-known German experimenter, in the explosion of a rocket motor on May 17, less than a month before the first issue of the *Bulletin* went to press.

Complimentary copies of the publication went to selected public and academic libraries and to at least 10 scientists, including both Goddard and his one-time graduate student, Dr. Percy Roope. At least a handful of respected professionals were also drawn into active membership, including Alexander Klemin, head of the Guggenheim School of Aeronautics at New York University; James Kimball, chief of the U.S. Weather Bureau in New York; and Dr. Clyde Fisher of the American Museum of Natural History.

Dr. Harold Horton Sheldon, a professor of physics at New York University (NYU), quickly emerged as a critically important member who could lend academic respectability to the organization. A Canadian educated at Queen's College, Brooklyn Polytechnic, and the University of Chicago, where he earned his Ph.D., Sheldon had joined the NYU faculty in 1927 and

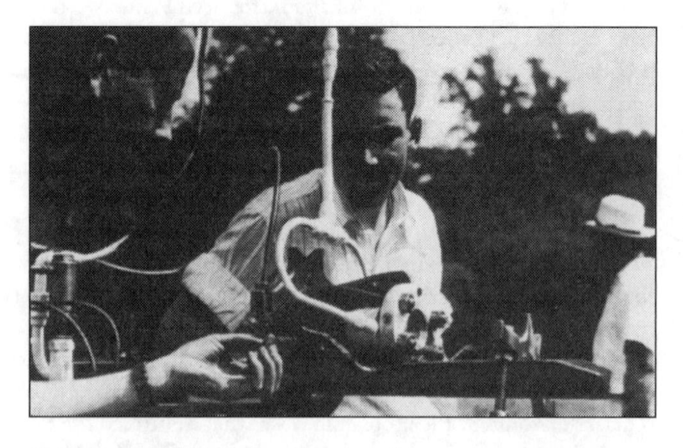

H. F. Pierce and MIT student Robertson Youngquist prepare to test an engine, 1941. (Courtesy AIAA.)

James Wyld with his regeneratively cooled rocket motor, Midvale, New Jersey, June 8, 1941. (Courtesy AIAA.)

doubled as science editor of the *New York Herald Tribune* from 1928 to 1931. The young professor had expressed his faith in space flight even before he met Pendray, who was also on the *Tribune* staff. "Einstein," Sheldon noted in 1929, "has presented us with the actual possibility of insulating ourselves from gravitation . . . ." That being the case, he continued, ". . . interplanetary journeys become a matter of simple accomplishment."[29]

With the *Bulletin* underway, Lasser wrote *The Conquest of Space*, the first English-language book on astronautics aimed at a popular audience. A foreword by Dr. Sheldon added an extra dash of credibility. When no publisher stepped forward, Lasser, Pendray, and Schachner pooled their resources and published 5,000 copies of the book in September 1931. A British publisher, Hurst & Blackett, brought out a new edition the following year.[30]

Arthur C. Clarke, perhaps the best known science fiction writer of the second half of the 20th century and one of the most important space advocates of the era, was a youngster when he saw a copy of Lasser's book in the window of a book shop a few dozen yards from his boyhood home in Summerset, England.

> "I saw it in the window, knew instinctively that I *had* to read it, and persuaded my good-natured Aunt Nellie . . . to buy it on the spot. And so I learned for the first time that space travel was not merely delightful fiction. *One day it could really happen* . . . my fate was sealed."[31]

Having spread their message through newspapers, journals, and books, the leaders of the AIS were quick to recognize the potential of new media. In the spring of 1932, Pendray simulcast his talk on "The Rockets of the Future," over both W2XAB, the Columbia Broadcasting System's pioneering television station, and W2XE, the company's long-distance short wave radio station. It was the 37th such talk in an aviation series sponsored by the *Brooklyn Eagle*.

Pendray replaced Lasser as president of the AIS in May 1932. For a time, the mailing address of the organization was that of the Milk Research Council, created to promote the nutritional value of dairy products, where Pendray was chief editor. Other changes were afoot. That month, the *Bulletin*, which had survived for two years as a no-frills, mimeographed newsletter, was reborn as *Astronautics*, an offset publication complete with photos. There were now 200 members spread across 21 states and 9 foreign countries.[32]

Lasser continued to edit the journal for another year, but the organization was moving in a new direction at a moment when his life was turning a very different corner. His idealism was not limited to the dream of a better world beyond the atmosphere. When Gernsback saw a news photo of his editor marching with a group of unemployed New Yorkers, he suggested, "... if you like the unemployed so much, why don't you join them."[33]

The young idealist did exactly that. In 1933 he founded the Workers Alliance of America, a union for the unemployed. He resigned in the 1940s because of increasing communist influence, but he had already been identified as a labor radical. When President Franklin D. Roosevelt appointed him to a position with the Works Progress Administration (WPA) in 1940, Congressional Republicans inserted a clause in the WPA appropriations bill barring the use of Federal funds to "... pay compensation to David Lasser." Congressman Martin Dies, chair of the House Un-American Activities Committee, explained that "... this fellow ... is not only a radical, but a crackpot, with mental delusions we can travel to the moon!"[34]

Although Lasser would hold several important government positions, including service as a labor advisor to Secretary of Commerce W. Averell Harriman, he spent most of his career in the labor movement, retiring in 1969 as assistant to the president of the International Union of Electrical Workers for economic and collective bargaining. In 1980, with Harriman's assistance, President James Carter cleared him of all charges of involvement with communist organizations.

In June 1982, on the occasion of an anniversary celebration, the Los Angeles Section of the American Institute of Aeronautics and Astronautics (AIAA) invited the 80-year old founder of its oldest predecessor organization to a celebratory awards dinner. In concluding his remarks that evening, Lasser proved that time had not tarnished his ideals. He imagined a day in the not too distant future when space tourists from all nations might have an opportunity to see the earth "as a lonely little ball in the vastness of space, fragile and helpless, slowly turning as it moved on its appointed circle around the sun." Attached to the "space bus" that carried them into orbit, he suggested, should be a banner "... inscribed with only a few words: For peace and democracy—Humanity United!."[35]

The Experimenters, 1931–1941

The AIS turned a corner in April 1931, when Ed and Lee Pendray splurged on a trip to Europe. The young couple combined visits to the usual tourist attractions with attempts to contact European rocket enthusiasts. Their most productive encounter was in Berlin, where they met with the members of the Verein Für Raumschiffahrt (VfR). The Society for Space Ship Travel (more popularly, the German Rocket Society) began when Max Valier, a German enthusiast who earned international fame with his rocket-propelled automobiles, boats, sleds, and gliders, argued the need for an organization to support additional rocket experiments. Attorney Johannes Winkler obtained a government charter and organized the initial meeting of nine men and a woman in The Golden Scepter, a Breslau tavern, on July 5, 1927.

With a small-scale wave of interest in space flight sweeping across Germany as a result of the work of Oberth, Valier, and popularizers like Berliner Willy Ley, the activity of the VfR generated considerable attention and attracted some 500 members during its first year. The society journal, *Die Rakete*, featured articles by leaders in the field and news from around the world. The publication was discontinued in the fall of 1929, however, and the money redirected toward an experimental program. Having built and tested some less than fully satisfactory liquid propellant rockets, the members of the VfR transformed an abandoned army garrison in the Berlin suburb of Reinickendorf into a "raketenflugplatz," or rocket flying field. Between April 1931 and May 1932, the members of the VfR conducted 87 rocket launches and some 270 static motor tests.

The Pendrays arrived in Berlin just as this period of extraordinary activity was getting underway. Their host, Willy Ley, a science journalist and a founder of the VfR, spoke little English. Ed Pendray admitted that "our German was nonexistent."[36] The enthusiasts found ways to communicate, however, and laid the foundation for a lifelong friendship. Ley introduced them to the members of the VfR team, including Wernher von Braun, "a young man in knickers," who would one day become the best known member of the American Rocket Society.[37] Pendray recalled:

> "They showed us everything, ... so freely that we got the feeling perhaps they thought we were rich Americans who might finance their experiments. Their repulsor rockets impressed us greatly. They were small—five or six feet high with gasoline-liquid oxygen powered motors the size of hen's eggs. They were crude ... but they worked! Until that moment I had never quite been able to visualize how a space ship would fly."[38]

The Americans had arrived too early to witness a flight, but they were treated to the test run of a engine. "And later, at a beer hall," Pendray recalled, "the language barrier was no barrier at all. There was much enthusiasm on both sides; much hand-waving, much sketching of plans on table-cloths, and much toasting each other's fortunes with beer steins."

Back in America, Pendray presented an enthusiastic report to his AIS colleagues, complete with the news that the members of the VfR had built and tested the world's first continuously firing liquid-propellant rocket in June 1930.[39] Most of those who attended Pendray's talk on the evening of May 1, 1931, were aware that Robert Goddard had also moved beyond theory to build and test rockets. In July 1929, newspapers reported that one of his flights had attracted the attention of Massachusetts police and fire authorities, who ordered him to cease experimenting with his noisy, dangerous contraptions. Pendray referred to that mysterious incident in his remarks, noting that Goddard, "... has carefully prevented the results of his work from becoming public," and suggesting that the 1929 rocket had been propelled by a series of explosions rather than continuous combustion.

When Pendray's remarks appeared in the next issue of the AIS *Bulletin*, Goddard lost no time in responding. "It may be said," he began, "that on July 17, 1929, a trial of a liquid-propelled rocket was made at Worcester, Massachusetts, the device functioning satisfactorily as regards the flow of liquid, the ascent of the rocket, and its rapid motion through the air." He concluded by noting that "my work with rocket motors, using liquids and continuous combustion, dates back to 1920."[40]

Even now, Goddard failed to tell the whole truth, neglecting to mention the first flight of a liquid-propellant rocket that he had conducted on March 16, 1926, from a Worcester cabbage patch owned by a distant relative, Aunt Effie Ward. Instead, he based his claim on the July 17, 1929, flight that had been reported in the newspapers. One can only assume that he was holding the date of his first launch in reserve, to be revealed in the event that a competitor claimed a flight prior to 1929.

True to form, Goddard accepted an honorary membership in the AIS but declined all invitations to address the group. "It happens," the professor noted, "that so many of my ideas and suggestions have been copied abroad without the acknowledgement usual in scientific circles that I have been forced to take this attitude." Goddard and the AIS would go their separate ways until World War II. Finally, in the last two years of his life, Goddard went so far as to accept a position on the Society's board of directors.

The AIS began to reinvent itself in the spring of 1931. The members had set out to publicize the dream of space flight and to interest competent scientists and engineers in the problem. "None of us," Pendray explained, "had conceived that we might ourselves become experimenters." The Germans had changed his mind. Those rockets, he explained to the members of the AIS, "were, after all, relatively easy to build."[38]

When Pendray first suggested that the AIS begin experimenting with rockets, "a storm of objection was raised." Some of the older members argued that the society was not organized or equipped to undertake such a program. Experimentation was "the work of lone, daring heroes," presumably men like Robert Goddard. Others protested that the effort would be far too expensive for the group. Pendray, his imagination fired by the vision of a rocket soaring aloft, characterized those who dropped out of the organization as "thin-skinned creatures ... who could not ... stand the thought of putting our sacred declarations of faith to the test of the proving stand and the launching rack."[41]

The process of redefining the central purpose of the organization began on April 3, 1931, when the members created a Research and Experiments Committee. Less than a month later, on May 1, the organizers of the new committee reported that they were encouraging "various eastern colleges" to undertake rocket research. More to the point, they had begun to consider the design of combustion chambers and had contacted various machine shops for estimates on the cost of a test article.

Pendray, filled with enthusiasm and ideas gathered from the Germans, became chairman of the committee. Ohio-born Hugh Franklin Pierce, whom one historian has described as "... an unpolished, hard-drinking ... New York subway ticket taker," emerged as another activist.[42] Along with collecting an interesting assortment of tattoos, Pierce had learned the bare rudiments of machining during his time in the Navy. He would do most of the rocket building in the small machine shop that he had established in the basement of his Bronx apartment building. The AIS members established a fund for experimentation on June 5, 1931, to support the effort. In view of the fact that they were attempting to turn fantasy into reality, a group of members agreed to contribute 10% of their science fiction earnings to the fund.

Pendray and Pierce published the preliminary design for a rocket motor test stand in the September issue of the *Bulletin* and unveiled the rocket itself before "a large and enthusiastic audience," at the American Museum of Natural History on the evening of February 18, 1932. Accompanied by a Dr. H. H. Sheldon, the pair transported their rocket to the Washington Square Campus of NYU, where they gave a lecture on space flight.

The central features of ARS #1 (see Appendix C) were the propellant and oxidizer tanks, a pair of aluminum pipes, each 5½-feet tall and 2-inches in diameter. The liquid oxygen was fed to the combustion chamber by the pressure generated by evaporation. The gasoline tank was pressurized by a cylinder of stored nitrogen. The motor, an aluminum casting, 6-inches long and 3-inches in diameter with walls 0.5-inch thick, fit into a light framework at the top end of the two tanks. The AIS team believed that placing the engine on the nose of the vehicle would increase stability. Four large sheet metal fins were attached to a frame at the bottom of the tanks.

At the outset, Pendray estimated the cost of the first rocket at perhaps $1,000. In fact, the final price tag was $49.40—$30.60 for the rocket and $18.89 for the test stand, propellants, batteries, and other supplies.[43] They had scoured the city for donations. George Slottman, vice-president of the Air Reduction Company, donated fifteen litres of liquid oxygen, a Dewar flask for transporting the stuff, a 3,000-pound nitrogen pressure tank, several feet of pressure tubing, and a pressure gauge. Pendray suggested that the AIS organize a subsidiary organization, the purpose of which would be to sing the praises of the Air Reduction

Company "... at all hours of the day and night, when not otherwise engaged." John O. Chesley, in charge of new developments at Alcoa, provided three duralumin blanks that Pierce could machine into combustion chambers. The Schrader Tire Valve Company donated specialized tire valves.[44,45]

In later years, Pendray was careful to point out that the AIS had not "gulled" their benefactors. "They were men with foresight who knew what they were doing. If what we were doing worked, and a rocket industry materialized, it would mean—as many years later it did—important business for their firms." Writing a quarter of a century later, Pendray noted that all three companies remained in the missile business and were still corporate members of the American Rocket Society (ARS).[45]

Improvisation also helped to keep the costs down. Pierce obtained some of the required valves by presenting himself as an interested manufacturer and asking for samples. The aluminum can that served as a water jacket for the motor was actually a malted milk shaker, a premium from a chocolate syrup company. Lee Pendray stopped at Macy's on her way home one evening to buy the $5 worth of pongee silk with which to craft the recovery parachute, based on a full-size parachute she had seen at an aviation show.

The next step was to locate a test site, no easy task for a group of city folk who did not want to repeat Goddard's experience with the local authorities. They finally settled on an empty farm field near Stockton, New Jersey, owned by David Lasser's uncle, "Ace" Hewitt. Pendray noted with some pride that the spot was not far from where George Washington had led his troops across the Delaware River to attack the British at Trenton. "We doubted that our efforts would prove so historic, but I dare say most of us secretly hoped so."[46]

Beginning in late August, the core members of the Research and Experiments Committee spent their weekends transforming a fallow field into their own version of the rakentenflugplatz. They dug long slit trenches with sandbag escarpments and built a wooden launch tower fourteen feet tall. They struggled with the usual assortment of problems. "Bugs in valves, bugs in fuel lines, bugs in ignition lanyards and bugs in fuses" one of them recalled.[47] Lee Pendray, a new mother who was managing a new home and turning out two newspaper columns a week, still found time to furnish the picnic lunches that were the high point of each expedition. None of them would ever forget the occasion on which she spread the tablecloth on a patch of poison ivy, a plant with which these city-bred rocketeers were not yet familiar.

Finally, on the afternoon of Friday, November 11, 1931, a small auto caravan headed west out of Manhattan, bound for their test site in the wilds of New Jersey. The Pendrays and Hugh Pierce were accompanied by Lasser, Schachner, Lemkin, Manning, and an Acme Newsreel photographer. There were two new members of the group as well, Alfred Best and Alfred Africano, who would emerge as one of the leading rocket designers of the group. Before any attempt was made to launch the thing, however, ARS #1 would be locked down on the launch stand and its motor test fired. The fledgling rocketeers would be happy if the small power plant worked as expected.

They spent a rainy night camped out on the concrete foundation for a garage that had never been built and rose in the morning to discover that their slit trenches were filled with water. Between draining the ditches, solving the final problems, and fueling the rocket with a garden sprinkling can, it was sunset by the time they were ready to throw the switch for the first time. There was a steady roar as a dagger of bluish flame issued from the nozzle. Ignoring the safety rules, the members of the team stood up in the trenches for a better view.

The scale on the test stand registered 60 pounds of thrust. That, they calculated rather optimistically, would have been enough to lift their rocket five miles into the air. The only real disappointment was that the newsreel photographer had not recorded the moment. Wet and cold, he was napping under a blanket on the back seat of one of the cars. Preparations for a flight test were underway the next morning when the rocket slipped and dropped a dozen feet to the ground. Twisted out of alignment, AIS #1 ended its short career.

The AIS experimenters dutifully recorded eight lessons learned from the experience:

1) Oxygen tank should be designed for easy filling
2) Rocket should be sturdy and resistant to damage
3) Fuel valves should be self-contained
4) Use an electrical ignition system
5) Employ permanent wiring to avoid the problems created by temporary connections
6) The rocket should have a streamlined body
7) Position the recovery parachute on the base, rather than on the nose
8) Ground equipment should be sturdy[48]

Bernard Smith, a 22-year-old newcomer, volunteered to build AIS #2 out of parts salvaged from the original rocket along with "Kansas bailing wire, old razor blades and cast-off pieces of metal—everything but Mrs. Pendray's kitchen sink."[49] The new rocket would feature balsa wood fins and valves scavenged from gas light fixtures. Smith disposed of the water jacket, punching holes through a makeshift nose cone to assist in cooling a rocket motor the size of a goose egg.

Smith was typical of the enthusiasts who had bounced around Depression-era New York before finding a spiritual home in the AIS. His terminal degree was a grammar school diploma. A prize-winning sculptor, he earned his living as a tinsmith, locksmith, and iron worker and was fascinated by mathematics. With his attention focused on the rocket, Smith had a difficult time keeping body and soul together. When things became too tough for their young volunteer, the Pendrays invited him to stay at their new home in suburban Crestwood, New York, where he could live and earn some money.[50]

While Bernard Smith worked away on the rocket, Laurence Manning, a horticulturist and member of Gernsback's stable of writers, and a colleague, Thomas W. Norton, constituted themselves a Committee on Biological Research, whose task it was to prepare for the day when a rocket did travel into space. On June 18, 1932, they spun a pair of guinea pigs to 600 rpm in a centrifuge measuring only two feet in diameter. Subjected to an acceleration force of 30 gravities, the test subjects did not survive.[51]

This time the members of the AIS team were determined to publicize their first successful launch and to stage the event closer to home. Daniel De V. Harned, a supporter with political connections, pulled some strings at City Hall and obtained permission to launch the new rocket from Great Kills Park, a stretch of semi-isolated Staten Island Beach, so long as two officials from the New York Fire Department's Bureau of Combustibles were present.

Laurence and Edith Manning's Staten Island home became the unofficial AIS headquarters and mess hall during the spring of 1933. Last-minute preparations underway on the morning of May 14, 1933, were captured by crews from both Acme and Universal Newsreels. The launch team—Pendray, Smith, and Manning as well as newcomers Alfred Best, Carl Aherns, John Shesta, and Alfred Africano—mounted AIS #2 on the launch rail and undertook the difficult and dangerous task of fueling the beast. Smith had designed a relatively simple launch procedure. Once the tanks were full and charged, they would place a burning rag beneath the nozzle. Smith would then yank a lanyard removing a stick and opening the valves to allow the fuel and oxidizer to flow.

Things did not go as planned. With everyone but himself safely ensconced in the slit trenches, Smith pulled the string, but the valves remained closed. Ignoring the obvious danger, Smith ran up to the rocket, reset the stick, lit the rag once again, and ran back to pull the string a second time. The rocket roared 250 feet into the air when the oxygen tank burst with a loud pop. Without a parachute, AIS #2 fell 400 feet down range into lower New York Bay. A pair of boys who had been watching from a rowboat returned it to the overjoyed rocketeers, who had finally seen a liquid-propellant rocket fly for the first time.

Franklin Pierce, serving as accident investigator, examined the sad remains of AIS #2 and found that sediment had blocked the oxygen flow, leading to uneven combustion and a catastrophic over pressurization of the tank. As Pendray noted, however, the rocketeers still regarded the launch as a great success. "It was the first liquid propellant rocket any of us had ever seen get off the ground," he explained. "Considering the state of the rocket art at the time, [it] was a very considerable triumph."[52]

Times were hard. Secretary and Treasurer Dr. Samuel Lichtenstein reported to those attending the third annual meeting on April 13, 1933, that cash on hand totaled $15.37. Even so, President Laurence Manning announced that the organization would support the work of *three* rocket building teams. Publication of *Astronautics* would be suspended for a few months in order to fund the experimental program. In view of the new focus, a change in the name of the organization seemed appropriate. On April 6, 1934, those attending the fourth annual meeting voted to rename themselves the American Rocket Society (ARS).[53]

The changing nature of the organization was also apparent in its shifting membership. With the exception of the Pendrays, the old science fiction crowd was drifting away. In their place, Pendray explained, a new generation of "... scientists and engineers, attracted by the relatively impressive flight of AIS #2, came running."[54]

Alfred Africano was one of the first to arrive. He had earned both a B.E. and an M.E. from the Stevens Institute of Technology and was working as an engineer with New York's Interborough Rapid Transit Company in 1932, when he read a *New York Sun* article on the AIS. Here, he decided, was an engineering problem that would enable him to stretch his skills.

Then there was John Shesta. Born Ivan Schestacovsky in St. Petersburg, Russia, in 1901, he had immigrated to the United States with his family during the chaos of World War I. Shesta attended Columbia University, earning degrees in both chemical and mechanical engineering. The brilliant young émigré, a steady consumer of Gernsback pulps who had experimented with black powder rockets as a boy, joined the AIS in 1933, when a classmate told him of the organization.

James Hart Wyld and Lovell Lawrence were the last of the major prewar figures to join the ARS. A native New Yorker, born on September 10, 1912, Jimmy Wyld taught himself to read at the age of four and took pride in having several times made his way through the 20 volumes of the *Book of Knowledge*. He was tutored, attended prep schools, earned a degree in mechanical engineering with highest honors from Princeton University in 1935, and won the Sayre Fellowship, which allowed him to pursue graduate work in electrical engineering.

Wyld was fascinated by the latest discoveries in science and technology and seriously considered writing a book on his hobby—magic. Given his broad interests, he sometimes expressed uncertainty as to his course in life. That indecision came to an end when he read Ed Pendray's article, "Men in Space," in the October 1934 issue of *New Outlook* magazine. The young man found Pendray in the telephone book, signed up with the organization, and "... began to spend much spare time on the problems involved in the development of high-power rockets and reaction-propelled devices."[55]

Although Lovell Lawrence had always been good with his hands, he would make his reputation as a manager. Forced to drop out of college in the depths of the Depression, he bounced from job to job until 1934, when he took a position with International Business Machines (IBM). Fascinated by the notion of space flight, he enlisted in the ARS in 1936 and served as both secretary and president of the organization before helping to found one of the first commercial rocket firms in the nation. Some notion of the importance of this new generation of ARS leaders is to be found in the fact that between 1936 and 1946 Alfred Africano would serve three terms as ARS president, James Wyld and H. Franklin Pierce two terms apiece, and John Shesta and Lovell Lawrence one year each.

The attempt to develop three rockets simultaneously produced mixed results. ARS #5, the vehicle designed by a team that included Franklin Pierce, Nathan Carver, and Nathan Schachner, was so radical that the rocket was never built, although it suggested ideas that were explored in later motors tested by the ARS.

Pendray, Smith, and Africano designed and built ARS #3, an innovative and sleek-looking craft with a fatal design flaw that allowed the liquid oxygen to boil off as fast as it went into the tank. Although the rocket never flew, the project was not a total waste. Dubbed "Ronald Rocket," ARS #3 would be displayed at lectures and demonstrations for the next six years. Impressed, an editorial writer for the *New York Tribune* noted that "... the public, whether it admires the rocket or not, cannot fail to admire the earnestness of these enthusiasts of interplanetary navigation, who, at the cost of considerable pains and effort have brought this model of the 'spaceship' of the future into being."[56]

Ronald Rocket disappeared while on display in the Science Pavilion at the 1939 New York World's Fair. Pendray remarked that many years later, after the launch of *Sputnik I*, several "long-memoried" ARS members joked that perhaps the Soviet success had begun with the kidnapping of the rocket that had never flown. The Russian Pavilion, they recalled, had been close by. Others disagreed, suggesting that, if anything, ARS #3 would have slowed the Soviet rocket effort.[54]

John Shesta had designed ARS #4 with the assistance of Carl Ahrens, Alfred Best, and Laurence Manning. With the gasoline tank sitting on top of the liquid oxygen tank, the sleek craft stood 7½-feet-tall and measured only three inches across. The small water-cooled motor, still mounted on the nose, featured four nozzles angled a few degrees outward to avoid heating the tanks. Four small spring-loaded helicopter blades were attached to the nose of the craft. During flight they would be folded in place. Released at altitude, the whirling blades would break the speed of the tail-first descent of ARS #4. Pendray remembered that Shesta, whose hobby was silver smithing, had produced "valves like jewels."

Static tested at the Staten Island site on June 10, 1934, the four-nozzled engine developed too little power to lift the rocket. After the installation of a new power plant, ARS #4 was transported back to Great Kills on September 9 and launched to an altitude of 382 feet. The rocket traveled some 1,585 feet over the ground and reached a speed of 1,000 feet per second, almost the speed of sound.

The *New York Times* was not impressed, labeling the rocket a "dud" and suggesting that "Mars and all other possible goals" could breathe easy for the time being. The members of the ARS were more optimistic. From their point of view, the test would have been perfect if only the autogiro recovery system had worked. In any case, the active career of ARS #4 was at an end.[57]

Pendray believed that the time had come to expand the research program. He developed a proposal for funding the development of a high-altitude rocket to conduct scientific research.

Perhaps inspired by Robert Goddard, who had emerged as one of the best-funded scientists of the decade, with support coming from the Smithsonian Institution, the Carnegie Institution, and the Guggenheim Foundation, Pendray hoped to raise $100,000 from a variety of government and philanthropic organizations.[58]

Although the idea did not get off the ground, it would resurface several times over the next few years. "The field is now ready and waiting for a Maecenas who will make the next steps in research possible," Pendray insisted in a letter to Martin Savell of the Institute of Public Relations. The time was ripe for a far-sighted philanthropist "... who will finally develop the rocket by financing the men who have brought it to its present level ..." Such a man, he was sure, "... will gain an international reputation for doing it." In the end, the governments of a world at war, not a high-minded philanthropist, would fund the first steps toward space.[59]

Determined to press forward with their own resources, the leaders of the ARS recognized the need for a change in strategy. "We decided that recklessly launching rockets without first developing durable and reliable components and materials might be fun and get us a headline or two," Pendray remarked, "but was in all other respects [a] fools business."[60]

For the moment, the group would stop flying rockets and focus their attention on developing the single most important element of the system, a reliable rocket motor. Shesta designed and built a sophisticated test stand that would record the exact time of a run, the pressure inside the combustion chamber, fuel pressure, and thrust up to 100 pounds. With the rocket blazing away toward the sky, a movie camera would record the instrument readings, providing detailed information on each run and enabling the young engineers to calculate the thermal efficiency of their motors and graph performance. It was proof that the ARS had adopted the values and methods of the engineering profession.

The members of the ARS would have to find a new place to test their rockets, however. The city fathers of New York would no longer tolerate such noisy and potentially dangerous tests on public land. The Pendrays, who had bought a home at 491 Westchester Avenue in suburban Crestwood, New York, agreed to dedicate their back yard to the cause. Located in a marshy area with few neighbors, it would do for the test site. The first series of firings were conducted on April 21, 1935, with a rocket motor of the ARS #3 design. Five tests were run using two separate nozzles, one four inches and the other a foot long, at pressures of both 150 and 300 pounds per square inch. The short nozzle operating at the higher pressure proved most efficient.

The ARS rocketeers reconvened at the Crestwood "proving ground" on June 2, 1936. This time the goal was to test a nichrome nozzle, the impact of a new injection system, the utility of alcohol as a rocket fuel, and several other variables. The new nozzle proved superior to its aluminum predecessors, which had burned through on every single run in April.

A third series of trials on August 25 provided the opportunity to test four new engines produced by John Shesta along with a water-jacketed motor of spun aluminum supplied by Willy Ley, who had fled Germany six months before and was now the Pendrays' house guest. Things did not go well. The first motor tested simply failed to ignite. Ley's water jacket failed during the fourth run, allowing the entire nozzle to burn off. On the final run, the hot gases burned through an asbestos gasket and pierced the sidewall of the combustion chamber with such force that it overturned the proving stand just as the motor exploded.[61]

Any hopes that the tests might go unnoticed were quickly dashed. The morning after the latest trials, the *New York Times* reported: "Motor For Rocket Explodes In Test." The reporter described "hissing blasts" that shattered windows, "prolonged explosions" heard three miles away, and "fragments of hot metal" flying around Ed and Lee's back yard. The noise, it seems, had "... alarmed the countryside for miles."[62]

Indeed, alarmed citizens had complained to the authorities. On the morning of August 26, officials of the Yonkers Bureau of Combustibles conducted an unsuccessful search for the cache of explosives that they were certain must be hidden in the Pendray basement. Concerned citizens gathered at the local American Legion Hall to demand action. The company insuring Ed Pendray's home threatened to cancel his fire and liability policy. Ed's solution was to sign the house over to Lee, who agreed to resign from the ARS. The group did, at least, reach an accommodation with the priest of the neighboring parish, who complained that the roar of the rocket motors disrupted the Sunday morning service. The ARS quickly agreed to restrict future weekend testing to Sunday afternoons.

Undaunted, the ARS Experimental Committee scheduled a fourth series of static tests for October 20, 1935. The crowds were now threatening to spread beyond the confines of the Pendray backyard. A number of ARS members were on hand to witness the proceedings, along with an assortment of curious wives, neighbors, and girlfriends. The invited guests included a representative of the Esso Company, suppliers of the gasoline. The crowd seems also to have included every 12-year-old Buck Rogers fan for five miles around.

Maintaining safety standards was difficult given the growing number of visitors and the general record of ARS motors. Disaster struck on October 20, when a motor exploded, sending shrapnel whistling through the air, thudding into trees and sandbags, and shattering Miss Ramona Jennings' elbow. The guest of a member, she had stepped from behind a protective tree to take a photo. The victim was wrapped in a blanket with a tourniquet on her elbow and rushed to the New Rochelle Hospital, where doctors repaired her compound fracture.

Miss Jennings would take a year to recover. The members of the ARS paid her medical and living expenses and thanked their lucky stars that she did not sue.[63] Chastened, the leadership of the ARS called a halt to its program of rocket motor tests during the course of two meetings held on February 20 and April 9, 1936. The organization would, however, offer advice and lend support and equipment "... to any such endeavors as the experimental committee approved."[64]

Without the excitement of "spectacular developments" supplied by rocket launches and tests, the organization began to stagnate. On April 29, 1938, the secretary reported a "falling off" of membership. In the past, he suggested, it might have been the result of improving economic conditions and the fact that more members, or potential members, now had jobs. There was no doubt, however, "that a certain proportion of individuals join up because of the spectacular possibilities in the development of the rocket, then became disappointed because there have been no rocket shots such as they had anticipated, and therefore drop out."[65] At the annual meeting on April 21, 1939, Secretary Max Krauss reported that the ARS had 106 members on the books, a total gain of only 6 members over 1931. In 1940, on its 10th anniversary, the ARS had less than $500 on hand.

The ARS library, which had gotten off to such a well-publicized start, was also in sad shape. "The library has been so much talked about," Pendray explained to Roy Healy in the summer of 1940, "that one would think it a really large and important collection." The awful truth, he reported, "... is that it at present contains exactly 31 items, 21 of which are paperbound books and pamphlets."[66]

In the fall of 1938, just when it seemed that things could not get much worse, Ed Pendray stepped down as editor of *Astronautics* and resigned from the board of directors. His public relations business was booming and he had begun to think that perhaps it was time for younger members take the reins of ARS leadership. "I can best serve the Society, in future," he explained to Alfred Africano, "by making way for those who have more time, better engineering training, and perhaps larger funds of energy and money."[67]

Africano responded immediately, noting that his friend had "... kept the ARS going during most of the years of its existence."[68] He reminded Pendray that, in spite of the accident and other problems, there was reason for optimism. The French Astronomical Society had awarded Africano and the ARS the 1935 REP-Hirsch Prize. Established by banker André-Louis Hirsch and Robert Esnault-Pelterie, who may have been feeling a bit guilty for his earlier lack of confidence in the Americans, the prize was given annually for the most significant contribution of the year to astronautics. Africano was honored for his paper, "The Design of a Stratospheric Rocket." The ARS shared the prize for its report on the rocket experiments of 1932–1934.

Whether at the insistence of his friends or as the result of a new burst of enthusiasm for his old passion, Pendray decided to rethink his resignation. A master publicist, he was intent on portraying the ARS as a serious organization. A reporter for the *New York Daily Mirror* noted that Pendray "laughed at Buck Rogers." The one-time science fiction writer now insisted that "any fairy tales you read about rockets hovering in space or being steered in and out of nests of stars is all bilge."

"Let's get this straight," he explained, "neither you, nor I will ever go to the moon or any other planet in any rocket. Nor will our children. Our grandchildren might—we can't tell now." Why then, the reporter asked, fiddle with rockets? Simply because the rocket offered the only hope of one day going into space, Pendray explained. Moreover, long before they boosted the first human being into space, rockets could be used for meteorological research, to carry high-speed mail, and perhaps even to power aircraft. Far from being wild-eyed dreamers, he insisted, the members of the ARS were serious fellows working hard to build an exciting future.[69]

Pendray also led the ARS effort to assist industrial designer Raymond Loewy develop a "Rocketport of Tomorrow" display for the Chrysler exhibit at the 1939 New York World's Fair, where Ronald Rocket, ARS #3, would be the star attraction. In October 1939, *Life* magazine published an article on the ARS. The following spring, the Conoco Oil Company ran a full-page ad featuring the organization in *The Saturday Evening Post*. Titled, "Up From Here to the Stratosphere," the ad showed Ed and Lee Pendray and Franklin Pierce, outfitted in their World War I vintage safety helmets, gazing up into the sky as though watching their latest rocket, surely fueled with Conoco products, disappear into the blue.[70]

Officially, the leaders of the ARS kept their distance from active rocket experiments for three years, although individual members continued to punch holes in the sky with rockets of their own design, often using borrowed ARS equipment. Three times—on September 12, 1937; September 10, 1939; and November 19, 1939—Franklin Pierce used solid-propellant skyrockets to explore the stability of various body shapes and tailfin configurations. On the final occasion, 2-, 4-, and 6-pound rockets were supplied by the Unexcelled Fireworks Company. The largest model reached an altitude of more than 800 feet. While inviting ARS members to attend Pierce's 1939 trials at the Charles Westendarp farm near Mountainville, New Jersey, Pendray cautioned them to "... make no mention of rockets, as we do not wish any outsiders present."[71]

The memory of Miss Jennings' fractured elbow was fading by 1938, however. The leaders of the ARS consulted a number of lawyers, all of whom suggested that under the corporation laws of the State of New York, the members of the organization would be absolved of all responsibility in such a case. On April 14, 1939, the Board of Directors voted to resume active rocket experiments.[72]

The ARS rocketeers had no doubt as to the goal of a new round of experiments. Three years of experience had underscored the need to solve the problem of overheating, which had led to the destruction of one ARS rocket motor after another. "After a few seconds of operation," John Shesta explained, "chambers and nozzles were so badly scarred by melting and erosion as to be unusable after one run. High melting point materials such as nichrome and stainless steel were tried and gave better results but were still unsatisfactory."[73]

Eugen Sänger, a brilliant Austrian engineer who had published in *Astronautics*, first suggested the possibility of a regeneratively cooled rocket motor in which the liquid fuel would be circulated around the combustion chamber, cooling the motor while preheating the fuel for more efficient burning. Inspired by the Sänger article, James Wyld set off along the same path.[74]

Wyld was a brilliant, 25-year old engineer who had donated a portion of his Princeton graduate fellowship to help defray Miss Jennings' medical expenses. After completing his graduate studies, Wyld worked as a "gypsy engineer," moving through a series of short-term jobs that kept body and soul together. He published his design for a simplified version of Sänger's power plant in *Astronautics*. At the time, he was sharing a Greenwich Village apartment with a pair of electrical engineers. The trio of "violent gadgeteers" transformed their small pantry into a workshop where Wyld began to build his motor in 1937.

John Shesta was also working toward a resumption of active rocket motor tests. By the fall of 1938, he had completed work on ARS test stand #2. Painted a bright red, the stand was fully equipped with pneumatic controls for the fuel and oxidizer lines, a nitrogen pressure system, a water system to cool the motor after a run, a clock, and instruments to record the thrust and the pressure in the chamber and the tanks. Franklin Pierce broke the new stand in with tests of a small liquid-propellant power plant undertaken in the foundation of an abandoned building in New Rochelle, New York, on October 22, 1938.

Change was in the air. Professional engineers were finally beginning to pay attention. Lester Durand Garner, secretary of the Institute of the Aeronautical Sciences (IAS), the leading organization for technical professionals in aviation, attended the October trials with Professor Alexander Klemin, head of the Guggenheim Aeronautical Laboratory at NYU. Impressed, both of them immediately signed up as members.[75] The lightweight news items and historical pieces that dominated the early issues of the AIS *Bulletin* were giving way to the solid professional offerings, like James Wyld's, "Fundamental Equations of Rocket Motion," that now filled the pages of *Astronautics*.

More important, the ARS rocket experiments were approaching their climax. The next series of tests was scheduled for December 10. After a preliminary run with the Pierce motor, James Wyld's regeneratively cooled power plant was bolted in place for the first time. The progenitor of a long line of historic motors, it stood only eight inches tall and featured a combustion chamber measuring two inches in diameter. On the first try, the motor failed to ignite. While Wyld made some adjustments, Robert C. Truax, an Annapolis midshipman who was already beginning to emerge as a leading rocket engineer, bolted his own regeneratively cooled power plant on to the test stand. After a few seconds of steady roar, the fuel vaporized in the cooling jacket, allowing the combustion chamber to burn through.

With a smaller metering nozzle fitted to the fuel line providing a leaner mixture, the Wyld motor was ready for a second try. It blazed away for almost 14 seconds without destroying itself, producing 90 pounds of thrust until the liquid air and alcohol were exhausted. As Shesta noted, however, the disassembled motor did show evidence of erosion and even incipient melting of the aluminum inner sleeve. Still, it was the most impressive performance that any ARS motor had turned in to date.[76]

It was the last anyone would see of the promising rocketeer for awhile. Wyld resigned as president of the ARS Board of Directors and temporarily ceased experimenting in February 1939, when he went to work as a member of the engineering team operating the high-speed wind tunnel at the National Advisory Committee for Aeronautics (NACA) research facility at Langley Field, Virginia. Convinced that his destiny lay with liquid-propellant rocketry, however, he returned to New York in the summer of 1940 and went back to work fine-tuning his regeneratively cooled motor, substituting a monel inner sleeve for the aluminum original.

Shesta had spent the hiatus modifying ARS test stand #2 to support longer runs and more provide accurate data. The first round of tests with the modified stand took place at a new test site near Midvale, New Jersey, on June 8, 1941. Wyld's motor produced between 80 and 85 pounds of thrust for 26 seconds. A second experimental motor produced by Nathan Carver and Charles Piecewiez, a unique design in which combustion occurred outside the motor, fired for only eight seconds.

They were back at the New Jersey site on June 22. The first motor tested that day, constructed by Alfred Africano, was a wrought iron pipe with a refractory ceramic lining intended to prevent the chamber from burning through. While the lining was spewed out through the nozzle during the run, the rocket developed so much thrust that the test stand could not provide an accurate measure. Further testing seemed warranted.

Robert Youngquist, a promising representative of the MIT rocket club, an affiliate of the ARS, supplied the next test motor, which employed a liquid oxygen spray to cool the lower section of the motor and nozzle. The little power plant produced 35 pounds of thrust for 13 seconds before exploding. Finally, Franklin Pierce tested an "historic" motor design of the sort that had powered ARS #1 and #2, surrounded by a sheet metal water jacket. It fired for a record 48 seconds, producing 35 pounds of thrust.

James Wyld brought his rocket motor back to Midvale on August 1, 1941. Running on liquid oxygen rather than the liquid air employed as an oxidizer on earlier runs, the little power plant ran for 21.5, 23, and 45 seconds, producing up to 135 pounds of thrust. As Roy Healy reported in *Astronautics*: "The test proved conclusively that a reliable motor for aerological sounding rockets has at last been designed, built and tested."[77]

Shesta immediately set to work on a small gyroscopically stabilized rocket to be powered by Wyld's motor. That fall, Lovell Lawrence, an IBM employee and a member of the Wyld/Shesta circle, was dispatched to Washington, DC, to supervise the installation of some IBM equipment. While in the nation's capital he was able to interest the U.S. Navy Bureau of Aeronautics in the Wyld motor. Lieutenant Charles F. Fischer traveled to New Jersey in late November to witness a test firing. Suitably impressed, he explained that although the Navy was not interested in flying rockets, it might have another use for a rocket motor. He asked for a report outlining the details of the little power plant but noted that the government would not contract with individuals.

Wyld was preparing the report at Shesta's home on Sunday, December 7, 1941, when the dance music to which they were listening was interrupted by an announcement that Pearl Harbor had been bombed. Eleven days later, Lovell Lawrence (President), Hugh Franklin Pierce (Vice-President), Wyld (Secretary), and Shesta (Treasurer) incorporated Reaction Motors Incorporated (RMI). They were the only stockholders, each of them controlling 125 shares, and the only employees.

The first American company founded to produce liquid-propellant rockets, RMI was founded on a shoestring. The group set up shop in a tumble-down North Arlington, New Jersey, garage owned by Shesta's brother-in-law. With the first Navy contract in hand early in

1942, they resumed testing with the ARS test stand, which the board of directors voted to loan to RMI on April 16, 1942. They shipped the first motor to the Navy only 180 days after signing their contract.

The early years were a struggle to overcome problems, both technical and bureaucratic. The company graduated from the garage to a converted nightclub in Pompton Plains, New Jersey. By 1943, they were testing rocket motors developing 3,000 pounds of thrust, enough to boost heavily laden flying boats into the air. Two years later, Wyld and his colleagues were hard at work on the 6000CR, or XLR-11, as the Air Force designated it. It was a 4-chambered power plant developing 6,000 pounds of static thrust that would push the Bell X-1 faster than the speed of sound for the first time in October 1947 and propel the Douglas D-588-2 to twice that speed soon thereafter.[78]

Ironically, the first real breakthrough marked the end of the experimental era in the history of the ARS. The August 1, 1941, tests were the last rocket trials conducted under the auspices of the organization. Thanks to the ARS, rockets were no longer a rather dangerous hobby pursued by eccentric amateurs. Having worked to convince Depression-era Americans that the ultimate human destiny lay out there among the stars, the members of the ARS had laid the foundation for a new American industry that could make that dream come true. With that task complete, they set about to reinvent their organization one last time.

References

[1] Howard Eisenberg, "Paul Reveres in Space Suits: The American Rocket Society," manuscript in the G. Edward Pendray Papers, Princeton Univ. Library.

[2] F. H. Winter, notes on a telephone interview with G. Edward Pendray, Oct. 8, 1980, Pendray biographical file, National Air and Space Museum Archive.

[3] In 1936, Ed Pendray accepted a post as assistant for public relations and education to the president of the Westinghouse Electric and Manufacturing Company. His achievements in that post ranged from the creation of the annual Westinghouse Science Talent Search to responsibility for the famous time capsule buried at the 1939 World's Fair. Pendray founded his own public relations firm in 1945, Pendray & Company, retiring as senior partner in 1971. *Public Relations News* named him Public Relations Professional of the Year for 1964. His best known book, *The Coming Age of Rocket Power*, appeared in 1947. He sat on the board of the Guggenheim Foundation and was a fellow of the American Association for the Advancement of Science. During these years his involvement with the Guggenheim Foundation linked his professional expertise to his lifelong fascination with rocketry. He remarried following Leatrice Pendray's death in 1971. Eighty-six year old G. Edward Pendray died of complications resulting from Parkinson's disease on September 15, 1987. The key figure in the foundation of the earliest predecessor organization, he was a Fellow of the AIAA.

[4] Leatrice Pendray continued to work as a newspaper columnist and free lance writer until 1945, when she helped her husband launch Pendray & Company, a firm in which she became a partner in 1959. The couple moved from New York City to Crestwood, in Westchester County, NewYork, in 1932. They lived in Pittsburgh from 1942 to 1944 to be closer to Westinghouse headquarters. They returned to Westchester County in 1944 and remained there until 1967, when they moved to 50-C Emerson Lane, Rossmor, Jamesburg, New York. The Pendray's were the parents of three daughters. Lea died in Princeton, New Jersey, on October 7, 1971.

[5] Lee Pendray to Milton Lehman, Nov. 1, 1955, the Papers of G. Edward Pendray, Mudd Library, Princeton Univ. Libraries, Box 15, folder "L. Pendray." Most of the information on the founders of the ARS is drawn from this letter.

[6] Lucy Greenbaum, "Quest for Fiction Data Started Pendray On Road to Fame As A Scientist," *White Plains NY Times*, Feb. 26, 1937.

[7] Huqo Gernsback flourished as a publisher. His empire included some fifty periodicals in fields ranging from electricity, electronics, and radio to science fiction, economics, and sex. Although *Life* magazine once referred to him as "the Barnum of the space age," Gernsback was highly regarded in the electronics field. He was granted 80 patents, introduced a generation of youngsters to radio technology, and sparked their imaginations. As John Pierce, director of electronic research at Bell Laboratories and a pioneer of the communications satellite, noted: "It is hard to overestimate Gernsback's effect on young people who later became members of the technical and scientific community." Gernsback was married three times. He willed his mortal remains to the Cornell University Medical School. The Hugo, the premiere award offered by American science fiction, is named in his honor, as is a crater on the moon.

[8] Hugo Gernsback, "Science Wonder Stories," *Science Wonder Stories*, Vol. 1, No. 1, June 1929, p. 1.

[9] G. Edward Pendray to David Lasser, Aug. 5, 1955, G. Edward Pendray Papers, Princeton Univ. Libraries.

[10] "David Lasser, Founding President, Dies At 94," *AIAA Bulletin*, July 1996, p. B5.

[11] F. C. Durant, Interview with G. Edward Pendray, manuscript in Pendray biographical file, National Air and Space Museum archives.

[12] Sam Moskowitz to Frank Winter, Hugo Gernsback biographical file, National Air and Space Museum Archive, Nov. 22, 1974.

[13] Details on the Founders of the ARS can be found in Frank H. Winter, *Prelude to the Space Age*, Smithsonian Inst. Press, Washington, DC, p. 73.

[14] The precise date of the founding of the AIS, the earliest predecessor of the AIAA, is not clear. In later years, Pendray repeatedly dated the first meeting, when the dozen charter members added their names to a piece of typing paper, as April 4. The official minutes of the meetings, however, establish a date of March 21 for the first meeting. Because the group discussed a draft constitution that evening, it is clear that the late night gathering when Lasser got things underway must have occurred at an earlier date. The evidence suggests that the first gathering was in early March.

[15] Constitution and by laws of the American Rocket Society, Amendment, Oct. 1, 1932, in "Minutes of the American Rocket Society, 1930–1948," AIAA Collection.

[16] With the exception of the Pendrays, most of the founding members of the AIS left the organization within a few years.

C. W. *Van Devander* served as press secretary to New York Governor W. Averell Harriman and was a national newspaper columnist.

Fletcher Pratt (1897–1956) earned national fame as the author of popular books on naval and military history.

Nathan Schachner (1895–1955) published a series of highly regarded biographies of Thomas Jefferson, Alexander Hamilton, and Aaron Burr following World War II.

[17] Science fiction historian Sam Moskowitz thought that Hugo Gernsback had masterminded the organization of the AIS through his employee, David Lasser. Sam Moskowitz to Frank Winter, letter in the National Air and Space Museum Archive, Nov. 22, 1974.

[18] "Report of the Secretary [AIS] 1930–1931," G. Edward Pendray Papers, Princeton Univ. Library.
On Heinlein see Winter, *Prelude to the Space Age*, p. 76.

[19] Eisenberg, "Paul Reveres in Spacesuits," p. 10.

[20] Winter, *Prelude to the Space Age*, p. 77.

[21] David Lasser, "President's Annual Report," *Bulletin of the American Interplanetary Society*, March–April 1931, p. 1.

[22] Minutes of the American Rocket Society, 1930–1948, volume in the collection of the American Institute of Aeronautics and Astronautics.

[23] F. C. Durant, III, interview with G. Edward Pendray, in G. Edward Pendray file, National Air and Space Museum Archive.

[24] "Editorial," *New York Herald Tribune*, Jan. 31, 1931; reprinted in the *Bulletin of the American Interplanetary Society*, March–April 1931, p. 4.

[25] "Editorial," *New York Times*, quoted in G. Edward Pendray, "The ARS at 31," *Astronautics*, Nov. 1961, p. 71.

[26] See for example, "Predicts 3,000 Mile Speed," *New York Times*, May 16, 1931, and "Predicts 'Hitching' Rockets to Stars," *New York Times*, March 26, 1932; see also Eisenberg, "Paul Reveres in Spacesuits," p. 6.

[27] Eisenberg, "Paul Reveres in Spacesuits," p. 8.

[28] "Plan Ship to the Stars," *New York Times*, April 4, 1931.

[29] "Dr. H. H. Sheldon, A Physicist, Dies," *New York Times*, Dec. 24, 1964, p. 19.

[30] Winter, *Prelude to the Space Age*, p. 80.

[31] Neil McAleere, *Odyssey: The Authorized Biography of Arthur C. Clarke*, Victor Gollancz, London, 1993, p. 22.

[32] "Report of the President of the American Interplanetary Society," *Astronautics*, May 1932, p. 7.

[33] Frank Winter, "ARS Founders—Where Are They Now?" *Astronautics & Aeronautics*, May 1981, p. 88.

[34] http://orpheus.uscd.edu//specco//testing/html/mss0322d.htm, cited by the author on Jan. 23, 2004.

[35] Allan P. Bangs, "In Appreciation," *Astronautics & Aeronautics*, Sept. 1996, p. B11.

[36] G. Edward Pendray, "Early Rocket Developments of the American Rocket Society," *First Steps Toward Space: Proceedings of the First and Second History Symposia of the International Academy of Astronautics at Belgrade, Yugoslavia, 26 September 1967, and New York, U.S.A., 16 October 1968*, edited by Frederick C. Durant, III and George S. James, Smithsonian Inst. Press, Washington, DC, 1974, p. 142.

[37] Durant, "Interview with G. Edward Pendray," p. 3.

[38] Eisenberg, "Paul Reveres in Spacesuits," p. 13.

[39] G. Edward Pendray, "The German Rockets," *Bulletin of the American Interplanetary Society*, No. 9, May 1931, p. 5.

[40] Robert H. Goddard to G. Edward Pendray, *Bulletin of the American Interplanetary Society* (No. 10, June–July 1931), pp. 9–10.

[41] G. Edward Pendray, "Why Shoot Rockets?" *Journal of the British Interplanetary Society*, Vol. 2, Oct. 1935, p. 9.

[42] Frank Winter, "Bringing Up Betsy," *Air & Space*, Dec. 1988/Jan. 1989, p. 78.

[43] Although the $49.40 was far less than the $1,000 estimated cost of constructing the rocket, it should be noted that the cost amounts to $666.94 in 2003 dollars, as per the Consumer Price Index. See http://www.eh.net/hmit/compare/result.php?use%5B%5D=DOLLAR&use%5B%5D=GDPDEFLATION&use%5B%5D=UNSKILLED&use%5B%5D=GDPCP&use%5B%5D=NOMINALGDP&amount2=49.40&year2=1932&year_result=&amount=&year_source=, cited June 9, 2005.

[44] Frank Winter, *Prelude to the Space Age*, p. 78.

[45] Eisenberg, "Paul Reveres in Space Suits," p. 15.

[46] Eisenberg, "Paul Reveres in Space Suits," p. 16.

[47] Eisenberg, "Paul Reveres in Space Suits," p. 17.

[48] Anthony Springer, "Early Experimental Programs of the American Rocket Society, 1930–1941, AIAA paper 2000-3279, 2000.

[49] Bernard Smith, quoted in Eisenberg, "Paul Reveres in Space Suits," p. 19.

[50] Bernard Smith as told to Frederick I. Ordway, "Some Vignettes From An Early Rocketeer's Diary: A Memoir," *History of Rocketry and Astronautics: Proceedings of the Seventeenth History Symposium of the International Academy of Astronautics, Budapest, Hungary*, Univelt, San Diego, CA, 1983, edited by John L. Sloop, Smith would eventually move to California, where he became a member of the rocket team at the California Institute of

Technology. During World War II he finally earned his own degree in physics and spent a career as a leading figure in American rocketry, including leadership of the U.S. Navy Bureau of Naval Weapons.

[51] The members of the IAS were not the only would-be space travelers to sacrifice small animals in the cause of science. In the spring and summer of 1930, Wernher von Braun and Constantine Generales conducted similar centrifuge tests with mice at their Swiss boarding school.

[52] Pendray in 1932 report, AIAA collection.

[53] All of the annual meetings in this period were held at the American Museum of Natural History.

[54] Eisenberg, "Paul Reveres in Space Suits," p. 21.

[55] James Wyld to David Fallon, June 12, 1938, Wyld Biographical File, National Air and Space Museum Archive; for more information on James Wyld, see Frank Winter, "Bringing Up Betsy," p. 79.

[56] "Editorial," *New York Tribune*, Feb. 18, 1932.

[57] "Planetary Rocket Remains on Earth," *New York Times*, June 11, 1934.

[58] Winter, *Prelude to the Space Age*, p. 82.

[59] G. E. Pendray to Morton Savell, June 4, 1939, Pendray Papers.

[60] Eisenberg, "Paul Reveres in Space Suits," p. 23.

[61] "Report on the Rocket Motor Tests of August 25th," *Astronautics* March, 1936, p. 3.

[62] "Motor For Rocket Explodes in Test," *New York Times*, Aug. 26, 1935.

[63] "Woman Injured in Rocket Blast," *New York Times*, Oct. 22, 1935.

[64] The decision to halt officially sanctioned rocket experiments was the result of motions passed on February 20 and April 9, 1936.

[65] Minutes of the ARS Board of Directors, April 29, 1938, AIAA Headquarters Files.

[66] G. E. Pendray to Roy Healy, Aug. 5, 1940, Box 10, Papers of G. E. Pendray, Princeton University Library. It should be noted, however, that those 31 items constituted one of the finest collections of works on astronautics in the nation, including a complete run of *Die Rakete*, the VfR publication, and virtually everything available on the subject in English, French, and German.

[67] G. E. Pendray to Alfred Africano, Oct. 10, 1938, Box 10, G. E. Pendray Papers, Princeton Univ. Library.

[68] Alfred Africano to G. E. Pendray, Pendray Papers, Oct. 16, 1938.

[69] Anon, "Only Human," *The [New York] Daily Mirror*, May 19, 1938.

[70] *The Saturday Evening Post*, June 8, 1940.

[71] Mimeographed invitation, Box 10, Pendray Papers.

[72] Minutes of the Meeting of the ARS Board of Directors, April 14, 1939, AIAA Headquarters Files.

[73] John Shesta, "Reaction Motors Incorporated—First Large Scale American Rocket Company: A Memoir," Preprint, IAF-78-A-5, Shesta File, National Air and Space Museum Archive.

[74] Eugen Sänger, "The Rocket Combustion Motor," *Astronautics*, Oct. 1936, pp. 2–8.

[75] Klemin's enthusiasm was so great that in 1936 he designed a small, rocket-powered model airplane used in an unsuccessful mail delivery stunt.

[76] Shesta, "Reaction Motors Incorporated," p. 1.

[77] Roy Healy, "Wyld Motor Retested," *Astronautics*, No. 50, Oct. 1941, p. 8.

[78] The details of the early history of RMI are drawn from Shesta, "Reaction Motors Incorporated."

Chapter 3—
Organizing a Profession: The Institute of the Aeronautical Sciences, 1928–1935

First Steps

Lester Gardner's love of aviation went well beyond the usual enthusiast's desire to see airplanes fly ever higher, faster, and farther. He loved the very notion of flying and shared J. Lawrence Pritchard's deep fascination with the history, legend, and lore of the air. A booster and an organization man to the bone, he saw the need for a society that would meet the needs of engineers, scientists, and other professionals employed in aviation while at the same time preserving and communicating the traditions of flight.

Gardner took the first step toward creating such an organization in June 1928, when he called together a group of like-minded friends "... to discuss the practicability of forming an aeronautical association ... ."

> That came up by the fact that several aviators had gone to England and addressed them, and had been able to confer with the entire aeronautical profession in England as such, and when they went to Germany they could go to the corresponding organization, and the same in France; but when the corresponding French, English, and German aviators came here there was no representative organization through whom they could meet the American aeronautical profession.[1]

Gardner even had a model in mind for his new organization. "It was obvious," he wrote many years later, "that when an independent aeronautical society should be organized in the United States, the Royal Aeronautical Society could provide a general pattern of organization and operation."[2]

He could think of no better or more effective ally in attempting to establish such an organization than Jerry Hunsaker. As Massachusetts Institute of Technology (MIT) grads and fellow movers and shakers in aviation, they knew one another well. Since the first of the year, the two had been discussing investment arrangements that might be useful in shoring up underfinanced aviation companies. That spring, Gardner broached the subject of an aeronautical society.

The notion of an organization that would facilitate contact between the many and varied technical professionals employed in aeronautics intrigued Hunsaker. He suggested:

> What the United States needed, in the opinion of those who dreamed of making flight safe, swift and economical was an organization to bring together those who had knowledge in various special areas to share, and who knew what new knowledge was needed. It was agreed that we did not need an Aero Club or Society of enthusiasts but rather an institute of qualified people. The Royal Aeronautical Society of London presented a good example of what could be done ... .[3]

It is useful to note at the outset that Gardner and Hunsaker were describing slightly different organizations. The Major had in mind a sort of club for gentlemen engineers, and

Lester Durand Gardner (1876–1956). (NASM Archive, Smithsonian Institution, A-44244.)

his friend saw a need for a professional society that would enable elite specialists to interact with colleagues in other disciplines. Together, they would craft an organization that was a little of both.

Gardner took the first step, contacting Colonel the Master of Sempill (later Lord Sempill), president of the Royal Aeronautical Society (RAeS), seeking advice and information on the operation of the organization. Secretary J. Laurence Pritchard responded on July 1, 1928, emphasizing that the RAeS encompassed both science and engineering. He explained:

> Normally, with well established industries there are two bodies concerned, one the scientific body and the other the engineering body. It is felt, however, that in the young state of aeronautics the time has not yet come for the industry to be able to support two such organizations, and the functions of both are those [thus] combined in the Royal Aeronautical Society ....[4]

For that reason, he continued, the RAeS had two distinct classes of membership. Fellows were usually focused on research, while Members were "practical aeronautical engineers, navigators, pilots and the like, with good technical experience." Both Fellows and Members, he emphasized, had "to be well qualified technically before they are accepted by the Council." The organization also offered Associate Fellowships and Memberships to individuals "... who have good practical experience, but who are not so highly qualified." Finally, there were classes of membership for students and for those who had no technical qualifications. Clearly, the RAeS was intent on separating the professional wheat from the chaff.

Pritchard went on to describe the system of branches, "the object of which is to encourage interest in the technical side of aviation." The fee for branch membership was nominal, and affiliation carried no technical status. The branches did sponsor lectures and other programs for those who were not regular members or who could not travel to London, however. He also outlined the publication, information, and library programs and explained the Joint Committee through which the RAeS, the Royal Aero Club, the Air League of the British Empire, and the Society of British Aircraft Constructors were able to work together to define their relationships with one another.

Gardner passed Pritchard's letter along to Hunsaker, who responded on August 9, expressing interest but suggesting that they not take any action until the fall, when the individuals whom they wished to involve would have returned from their vacations. By the time they did raise the issue with colleagues, the stock market had collapsed. The flush times were over, and most engineers were far more interested in keeping their jobs than in launching a new professional society. In any case, as Gardner and Hunsaker later noted, "... there was opposition to the plan by the officers of the older engineering societies ...."[5]

It is by no means clear what led the pair to resurrect the idea in the fall of 1932. Perhaps they believed that conditions in the industry were more promising, or perhaps Gardner could simply no longer contain himself. Whatever the reason, they met on September 26, 1932, "and decided to revive the idea of a scientific society." Over the next week they discussed the matter with George Pegram, chairman of the department of physics at Columbia University, and a select group of aeronautical colleagues.

Gardner later remarked that they had "... decided to limit the canvas to men who had been honored with the grade of fellowship or associate fellowship in the Royal Aeronautical Society."[6] Opening the discussion with such distinguished figures, Hunsaker explained, signaled their desire to avoid creating an organization with "... too inclusive a membership."[7]

Their list included Grover Loening and Lawrence Bell, both of whom ran their own aircraft companies, and Theodore Paul Wright, chief engineer at Curtiss Aeroplane and Motor. Charles Lawrance had pioneered the development of the radial engine in America. Samuel Stewart Bradley was general manager of the Manufacturers Aircraft Association and director of the Aeronautical Chamber of Commerce, where Gardner was serving his year as president. Earl D. Osborn, who had take over the reins of *Aviation* from Gardner, had recently hired National Advisory Committee for Aeronautics (NACA) veteran Edward Pearson Warner as editor of the magazine. Edward Eugene Aldrin had shaped aeronautical engineering in the Air Corps, and, like Wright and Warner, was a graduate of the aeronautical engineering program at MIT.

Charles Lawrance arranged for a planning group of New Yorkers that included Hunsaker, Gardner, Osborn, Aldrin, and himself to meet at New York's Yale Club on October 3, 1932. Gardner was instructed to prepare and file incorporation papers for an organization to be known as the Institute of American Aeronautics. Any detailed discussion of plans and purposes would wait until a core of incorporators could be recruited. Two days later, Gardner met with Warner, who had been out of town for the first gathering. Warner approved, so long as it was "to be a scientific society and not an engineering group" and suggested an alternate name, the Institute of the Aeronautical Sciences (IAS).

Rather than paying a lawyer to draft the articles of incorporation, Gardner contacted Harold P. Wiseman, secretary of the Institute of Radio Engineers (IRE), and received permission to copy key sections of their articles, simply substituting the word "aeronautical" for the word "radio." The Secretary of the State of New York rejected the title Institute of American Aeronautics on October 6, pointing out that the name was too close to that of the American Aeronautical Corporation, a Delaware firm that had recently registered in New York. The IAS however, did not present a problem.[8]

Hunsaker, Gardner, Loening, and Lawrance met again at the Yale Club on October 10 to sign the papers of incorporation. In order to satisfy a legal technicality, Gardner, Bradley, Lawrance, Loening, and Warner met to sign a fresh copy on October 14. Because five incorporators were required, Bradley signed the final papers for Hunsaker, who was out of town on business. Hunsaker's name was later added to the list of incorporators, as was that of Virginius E. Clark and a number of others. State Supreme Court Judge Louis A. Valente pronounced the papers complete and ready for filing.[9]

Lester and Margaret Gardner drove out of New York City at 6:00 a.m. on October 15, headed for the state capital at Albany. By 11:00 a.m., Alexander Rice McKim, an MIT alumnus and state official, had squired them into the office of Mr. Fletcher of the State Bureau of Corporations, who approved the articles of incorporation with a few changes.[10] In addition, Fletcher explained that because the organization related to "art and science," the Commissioner of Education would have to approve. Dr. Ernest E. Cole, Deputy and Acting Commissioner of Education, provided that paperwork. Before the end of the workday, Garner had paid the required $40 fee and filed the articles of incorporation. Less than a month after he and Hunsaker had decided to set things in motion, the IAS was on its way.

Launching the Institute

In the fall of 1932, Jerry Hunsaker was in the process of shifting from his position as a consultant on aeronautical communications issues with Bell Laboratories to a new role as vice-president of the Goodyear–Zeppelin Corporation. Lester Gardner, who had the clearest vision of the shape of the organization in any case, was spending full time on the project, as he would for the next 15 years.

It is impossible to overestimate Gardner's role in the foundation of the IAS. As his successor, S. Paul Johnston, would note many years later, the secretary's success in attracting the handful of elite movers and shakers in American aviation as founders of the organization was critical to the ultimate success of the group. Without those names and the companies they represented, "... he would never have been able to work the miracle of providing for its [the IAS] financial support."[11]

"The number of shirt-fronts that he drenched with his tears, the number of desks he pounded to splinters, the number of doorbells he rang in those days he alone can tell," Johnston continued. "We hope that he leaves us a personal record of that critical period when the life of the infant Institute hung in the balance." Gardner's home, 251 West 101st Street, was the organization's first mailing address and work place. Margaret Gardner supplied the coffee and doughnuts that kept her husband's crew of volunteers going on late evenings when they were assembling material for council meetings, preparing press releases, or answering correspondence.

Gardner finished the first draft of a constitution on October 12. "I had before me the Constitutions of the S.A.E. and the Institute of Radio Engineers and several other membership corporations ...," he explained in a memorandum to Hunsaker, Loening, Lawrance, and Warner. The secretary of the IRE had warned him not to attempt to develop a detailed constitution intended to cover all contingencies. "The Constitution adopted at first should be very simple and throw all the power into the Council until it, through By Laws, found out what should go into the Constitution."[12]

Gardner worked with that advice in mind and sent the document to Hunsaker, Lawrance, Loening, and Warner on October 12, suggesting that they meet in his office at the Aeronautical Chamber of Commerce, 10 East 40th Street, New York, at 3:30 p.m. on Monday, October 17. He also suggested a slate of temporary officers who would hold office until the first meeting of the Institute. Hunsaker would serve as President, with Loening and Warner as vice presidents. Lawrance would be Treasurer, and Gardner himself volunteered to function as Secretary.

His first choice for members of the governing council were George Mead, vice-president of Pratt and Whitney; Holden C. Richardson, chief engineer at the Naval Aircraft Factory; George Lewis, director of research for the NACA; Leslie MacDill, a force to be reckoned with in the Air Corps; and Charles Lindbergh, who needed no introduction. As alternates he suggested aircraft manufacturers Donald W. Douglas, Igor Sikorsky, and Giuseppe Bellanca; Ralph Upson of Goodyear; Boeing's Claire Egtvedt; William B. Mayo, in charge of all aeronautical activities for Henry Ford; Clarence Young, director of aeronautics with the

Department of Commerce; famed test pilot and MIT Ph.D., James Doolittle; Commander Walter Webster of the Naval Aircraft Factory; and Edward Aldrin.[13]

Gardner intended that all of these men would contribute names to a much larger list of suitable individuals who would be invited to join the IAS. With that accomplished, the group could plan its first national meeting, at which time regular elections could be held and the constitution and bylaws approved.

On October 19, the Secretary mailed a copy of the corrected draft of the constitution ("as I understand it to be finally adopted") to Hunsaker, Warner, Loening, and Lawrance. He promised that a draft of the bylaws and a suggested list of possible members would follow and asked that the members of the Council who were in the city attend a weekly meeting that would convene at the Yale Club promptly at 12:30 p.m. each Monday. At the first two Council meetings (October 17 and an "adjourned" meeting on October 24), the temporary constitution and bylaws were adopted and Gardner's suggested slate of officers was approved, with some modifications. In the end, the temporary Council included Aldrin; Doolittle; Hunsaker; Lawrance; Loening; Mead; Warner; Virginius Clark; T. P. Wright; Clark Millikan of the California Institute of Technology; Joseph S. Ames, chairman of the NACA; James H. Kimball, a weather bureau official who had assisted long-distance fliers; and Carl-Gustav Rossby, an MIT meteorologist.[14]

The birth of the Institute created a flurry of interest in the press. The *New York Times* announced, "Hunsaker Heads Group Formed to Promote Scientific Research." Four days later the paper published a second note, remarking that: "The announcement of its list of officers should go far to establish confidence in the soundness of the new Institute of Aeronautical Sciences . . . ." The New York *Herald Tribune* published its own hopeful notice.[15]

The aeronautical press was also enthusiastic about the IAS. Earl Findlay, editor of *U.S. Air Services*, noted:

> It differs from other aero organizations in our country in that its aim is to bring together beneath its banner those individuals primarily who are devoted to aeronautics as a science and who are themselves capable of making individual contributions thereto. In its devotion to this emprise [sic], no disrespect toward other societies is entertained or implied. Rather does it evidence an appreciation for a need which other organizations are not designed to fill. Its project is ambitious, but it is one that every genuine supporter of the industry and the art can whole-heatedly endorse. It is a hopeful harbinger of a new era of national acculturation—the springtime of a new ideal in aeronautical letters.[16]

Gardner could not have paid for that sort of a birth announcement! Unfortunately, not all of the responses were so favorable. Howard Coffin, who was to the Society of Automotive Engineers (SAE) as Gardner was to the IAS, ". . . asked some pertinent questions as to why we got up a new Society when the SAE organization is available." Hunsaker responded by assuring Coffin that the SAE ". . . admirably serves the automotive part of aeronautics, especially standards work, and any engineer needs to belong to it." At the same time, he pointed out, the American Society of Mechanical Engineers (ASME) was also very active in aeronautics. Hunsaker himself also belonged to the American Physical Society, the Geographical Society, ". . . and would like to be in touch with the meteorologists, radio and electrical engineers, metallurgists, [and] chemists . . . ."

What was missing, he argued, ". . . is just what the Institute of the Aeronautical Sciences is designed to provide: a coordination of specialists in these many fields of science and

Edward P. Warner. (NASM Archive, Smithsonian Institution, 95–8754.)

engineering devoting themselves to aeronautical developments. Such men belong to their own professional societies, but in them are not in contact with others who are able to criticize and evaluate their work from other points of view."[17]

It was the strongest argument that could be made. It probably did not satisfy Coffin, however, who could respond that the SAE was also an umbrella organization that brought disparate specialists together to discuss common issues. Although Hunsaker would not have admitted it, the truth was that the aeronautical types felt that the time had come for them to stand alone.

On October 27, Gardner sent a memorandum to each Council member, suggesting that they create a loose-leaf binder of materials on the new organization. "They will receive from time to time perforated sheets, consecutively numbered," he noted, "which would keep them abreast of all decisions and actions.[18]

"The Institute will bring together in one organization those persons who are interested in all the sciences which have to do with aeronautics in its broadest meaning," the secretary announced in the first item intended for those binders. Perhaps with memories of the old SAE in mind, Gardner explained that the founders of the IAS were taking action in order to "... avoid the complications that might arise from the formation of a similar group, possibly under unfortunate guidance ..." Unlike the visionary language of the American Rocket Society (ARS) pioneers, the founders of the IAS couched their Statement of Purpose in barely comprehensible bureaucratese and made it broad enough to cover any direction in which the new organization might move:

> To advance the art and science of aeronautics; to publish works of literature, science and art for such purpose; to do all and every act necessary, suitable and proper for the accomplishment of any of the purposes or the attainment of any of the objects or for the furtherance of any of the powers hereinbefore set forth, either alone or in association with other corporations, firms or individuals, and to do every other act or acts, thing or things, incidental or appurtenant to or growing out of or connected with the aforesaid science and art, or powers or any part or parts thereof, provided the same are not inconsistent with the laws under which this corporation is organized or prohibited by the laws of the State of New York.[19]

The first IAS publication, a pocket-sized leaflet issued early in 1933, included a copy of the constitution, bylaws, and a membership list and provided a much clearer and more accurate description of the goals of the founders.

> The progress of aeronautics requires the intensive application of many sciences. The organizers of the Institute believe that the future development of aircraft will come

> from scientific research, experiment and design. The Institute will bring together for the discussion and evaluation of ideas, scientific workers engaged in many specialized fields and engineers immediately concerned with the application of new knowledge to aeronautical development.

That came close to expressing the core of the founder's vision for their organization, as expressed in Jerry Hunsaker's letter to Howard Coffin. The IAS would be an umbrella organization in which the entire range of scientific and technical professionals involved in aviation could meet and share information. "As practically every scientist or engineer already belongs to his own special professional society, in order to bring together all specialists contributing to aeronautical development," Gardner continued, "the dues of the Institute have been made nominal." Dues would be on a sliding scale, from $5 for the highest ranking Fellows, to $3 for run-of-the-mill Technical Members.[20]

No subject of discussion was as contentious as the qualifications and grades of membership. Initially, individuals invited to join the IAS could qualify for any one of seven membership categories:

Honorary Fellows were to be persons of eminence in aeronautics selected by the Council.

Benefactors were to be persons who were "interested and active" in aeronautics who had contributed not less than $1,000 to the organization.

Fellows would be elected from the roster of either scientific or engineering members with at least six months in the prior grade. Fellows would be persons who had attained a position of "distinction" and had made "notable and valuable contributions" to flight science or engineering. No more than 40 Fellows were to be elected the first year and no more than 10 each year thereafter.

Scientific Members were to be persons of "recognized standing" in one of the aeronautical sciences.

Engineering Members were to be persons of "recognized standing" in aeronautical engineering.

Industrial Members would be persons who had achieved "recognized standing" in "the application or development" of aeronautics.

Junior Members would be individuals under the age of 25 who were students or working in flight science or engineering.

Scientific Members should be at least 30 years of age and engaged "for a substantial period of time in research, teaching, or the application of science to aeronautics." Preferably, the work in question "should not be of an immediate commercial application." Publications, academic position, or membership in other scientific societies might be regarded as proof of "recognized standing."

Engineering members were expected to be at least 30 and to have had charge of "important aeronautical engineering work." Alternatively, they could qualify on the basis of five years of work as an aeronautical engineer. Aeronautical designs created, publications, or membership in other engineering societies would be proof of "recognized standing." Persons involved in the business side of aviation would qualify as Industrial members.

By the end of October, the members of the Council had begun to submit lists of the names of colleagues whom they regarded as potential members, along with some indication as to why the individual was regarded as suitable. Gardner and his volunteers collected the names and sent them back out to the Council members, who would vote yea or nay on each name. Two negative votes would bar a nominee from membership. By November 2, Gardner had heard from a number of Council members and expressed real concern in a letter to "Dear Jerry."

It had become clear to him that there was a wide divergence of opinion on the Council as to the definition of what constituted "recognized standing." In particular, Clark Millikan, one of the best known professors of aeronautical engineering in the nation, and Virginius Clark had indicated that they would hold the applicants to a very high standard. Both men argued that even Engineer members "should have scientific standing as well as an engineering background." Gardner feared "... that so many [applicants] will receive two unfavorable votes that everyone will be unhappy." He realized that if too many well-known and well-liked men were blackballed, his organization would die before it was properly born.[21]

As a way around the problem, he suggested that they take the advice of the RAeS and establish an Associate grade for both Scientific and Engineering members. Virtually any graduate engineer with 5 or 10 years of service in industry would qualify for that grade. "I was surprised to learn from Mr. Weston of the Institute for Radio Engineers," he continued, "that 65 percent of their membership consists of Associate members." In the case of the IRE, 5,400 of the 6,000 members were Associates. Those Associate memberships, Weston explained, were the financial backbone of the organization. With Associate memberships in place, Gardner argued, "... we can take in almost everybody from a trained draughtsman to the top."

Ultimately, the Council approved the addition of two more membership categories.

Pilot Members were to be those aviators whose work related to science or engineering, such as test pilots or "persons of distinction who are pilots but who would not properly be regarded as Industrial, Engineering or Scientific members."

Technical Members were individuals interested in the advancement of the aeronautical sciences and usually involved in the industry in some capacity, who did not qualify under any other category.

Hunsaker, Gardner, Warner, Loening, and Lawrance, the usual crowd of New Yorkers, were on hand for the "third adjourned first meeting" of the Council on December 12, 1932. Gardner noted that the Council had responded to seven mailed ballots to date, six of which involved the selection of members. Four hundred and twenty invitations had been sent to individuals approved by the Council. One hundred and forty of those individuals had signed up. Gardner warned the members not to share the list of those who had not responded, lest individuals who had not been invited to join should become upset.

The alterations to the constitution and bylaws involving membership categories and qualifications were proposed and passed. Hunsaker, Warner, and Gardner were appointed a subcommittee to identify from 50 to 100 distinguished foreign practitioners of the aeronautical arts and sciences who could be invited to joint the IAS. Kimball, Richardson, Lewis, and new members Glenn Martin, Ralph Upson, and Starr Truscott were named to a committee that would nominate members of the Council for the coming year.

For the first time, Gardner offered a financial report. Seven hundred dollars, representing the dues of the 140 members, had been deposited with the Banker's Trust Company. The secretary had personally paid all of the IAS bills to date for stationary, mailing costs, incorporation fees, and the like. Treasurer Lawrance was instructed to reimburse the secretary and to authorize $50 as a petty cash fund on which Gardner could draw to meet incidental expenses.

Less than a month later, on January 7, 1933, (at the fourth adjourned first meeting) Gardner reported that the membership had climbed to 303. Orville Wright had agreed to be named the first Honorary Fellow of the Institute. The IAS now included 128 engineering members, 74 scientific members, 72 industrial members, and 29 pilots. The account at Banker's Trust stood at $621.26, with all bills paid. The secretary was careful to note, however, that

114 of the new members had not yet paid their dues, totaling $570. Total funds on hand and receivable, he was proud to report, amounted to $1,191.26.[22]

Sherman Mills Fairchild, c. 1928. (NASM Archive, Smithsonian Institution, 95–8033.)

No Women Need Apply

One thing was clear. The Founders of the IAS were consciously creating an "Old Boy's Club." The Council had rejected the nomination of Amelia Earhart as a pilot member. There were only eight pilot members on the original list of founders. They included Major General Benjamin D. Foulois, Chief of the Air Corps; Assistant Chief Brigadier General Oscar Westover; World War I ace Raymond Brooks; pioneer aviator Beckwith Havens; and pilots in the employ of Curtiss, Pratt & Whitney, and Transcontinental & Western Airlines. None of them, not even Foulois, was as well known as Earhart, who had flown into the hearts of all Americans that spring when she had become the first woman, and the second pilot, to solo the Atlantic.

The decision was not limited to Amelia Earhart. The Council decided that no women were to be admitted into membership. It is true that women were rare in aeronautical engineering. The case of Miss Isabel C. Ebel, an MIT graduate in aeronautical engineering who was pursuing graduate work in the field at New York University (NYU), was so rare that the *New York Times* headlined a story on her: "One Co-ed Studies With 3,000 N.Y.U. Men; Seeks Degree in Aeronautical Engineering." Still, there were qualified women. Pearl Young, a trained physicist serving as chief editor of NACA technical publications, would have met the most stringent criteria.[23]

Gardner was sufficiently concerned about the matter to take a second ballot. He wanted to be certain that the members were serious about excluding half the human race, including a pilot whose fame was second only to that of Charles Lindbergh. The outcome was the same. Clearly, the members of the Council regarded gender as a key element in defining a distinguished professional. Their decision would stand until January 1939, when Miss M. Elsa Gardner, an aeronautical engineer employed as editor of the *Technical Data Digest* at Wright Field in Dayton, Ohio, was invited into membership.[24]

The Founder's Meeting

Gardner mailed an invitation to the first annual meeting of the IAS to everyone who had joined as of December 8. The meeting would be held at Columbia University's Pupin Physics Laboratory, Broadway at 120th Street, on January 26, 1933. Professor George B. Pegram had arranged for the venue. A founding member of the IAS, Pegram served for many years as chairman of the physics department and dean of the graduate school at Columbia. As a leader of the American Physical Society (APS), he had played a key role in helping Hunsaker, also a member of the APS, grapple with membership and other issues faced by the founders of the IAS.

The Pupin Laboratory was an especially appropriate site for the founder's meeting. The building was named for Michael Pupin, who had taught physics at Columbia since 1889 and who had been appointed an emeritus professor in 1931. Pupin had served as president of both the IRE and the American Institute of Electrical Engineers and had been a member of the NACA from the time of its foundation in 1915 until 1922.

As Pegram pointed out in his welcoming remarks to the founder's meeting, the IAS was adding to an already distinguished tradition. Columbia had hosted the first meetings of the American Chemical Society, the American Mathematical Society, and the APS. This would be a long and happy relationship. With the exception of the 1945 annual meeting, which was cancelled as a wartime conservation measure, the IAS would return to the Pupin Physics Laboratory for the technical sessions of its annual meeting every year until 1947, by which time the gathering had simply outgrown Columbia. On December 22, 1954, IAS members dedicated a bronze plaque on the entrance wall of the Pupin Physics Laboratory, identifying the spot where the Institute had begun and where it had met for so many years.[25]

The date of the founder's meeting was not firmly set until December 20, 1932, when Gardner informed the members of the Council that Professor Auguste Piccard had cabled his agreement to address the members of the IAS on the afternoon of January 26 and to be the guest of the Institute at a reception and dinner to be held in the Columbia faculty club that evening. "Please disregard all other dates," he continued, "as no meeting will be held before this."

Piccard was the greatest scientific explorer/adventurer of the day. A native of Switzerland and a professor of physics at the University of Brussels, he combined an interest in ballooning with a professional desire to study conditions in the upper atmosphere. Piccard designed and built the first sealed, pressurized balloon gondola, which enabled scientists to reach record heights in relative safety. On May 27, 1931, Piccard and Paul Kipfer launched from Augsburg, Germany, and reached an altitude of 51,775 feet, returning with information on conditions aloft as well as data on cosmic rays. On a flight from Zurich on August 18, 1932, Piccard and Max Cosyns ascended to 53,152 feet.[26]

Gardner recognized that Piccard, with his frizzy hair and huge eyes peering out from behind thick glasses, had captured the attention of the nation's newspaper readers. Here was just the fellow to boost attendance at the founder's meeting and attract the press. He paid William Feakins, Piccard's American agent, a $150 fee, half the cost of luring the scientist to New York. Columbia's Institute of Arts and Sciences would pay the other half and offer a free evening lecture to the public at the MacMillan Academic Theater. Gardner assured the members of the board that although Piccard's evening talk would be aimed at a popular audience, his address to the members of the IAS would be rigorous and SCIENTIFIC. Still, he would make tickets to the evening talk ("The World from Ten Miles Up," complete with movies) available to the members for just $1 apiece. He hoped that money raised in that fashion would help to defray the cost of the founder's meeting.

With less than a month to go before the meeting, it was obviously too late to invite members to prepare papers to read in formal sessions of the "scientific symposium." President Hunsaker had a solution to the problem. "My own idea," he suggested, "is to avoid all formal papers this year and confine ourselves to a somewhat managed discussion of what seems to be ahead for 1933 in the way of new knowledge, what is needed and the prospects of getting it."[27] They would handpick leaders in key fields, who would offer remarks designed to raise questions and prompt a discussion. Following that approach, the planners drafted a schedule of events that would begin with welcoming remarks at 11:00 a.m., and conclude with Professor Piccard's evening lecture.

Gardner, Hunsaker, and the members of the Council can only have been pleased as 132 of the leading figures in American aviation filed into the Pupin Physics Building at the appointed time. The manufacturers were well represented, from corporate giants like George Mead, director of United Aircraft and Transport, to smaller producers such as Clayton Bruckner of WACO. LeRoy Grumman, the head of a relatively young company, rubbed shoulders with pioneers like Glenn Martin, Igor Sikorsky, and Grover Loening. The academics were out in full force: Theodore von Kármán and Clark Millikan from California Institute of Technology; Hunsaker, who would soon be returning to MIT for good; and the ubiquitous Alexander Klemin.

There were government officials from the NACA, Department of Commerce, Weather Bureau, and military services. Propulsion specialists met and mixed with instrument designers, navigation experts, structures men, and accessory manufacturers. A swarm of well-known aviators—including Jimmy Doolittle, Richard Byrd, Clarence Chamberlain, and Bernt Balchen—rounded out the meeting. They were, the *New York Post* commented, "... the aristocracy of aviation."[28]

Professor Pegram, standing in for Columbia President Nicholas Murray Butler, who had a cold, welcomed the members of the IAS. Hunsaker responded, underscoring what the members of the Council hoped would be the nature and purpose of the new organization.

We call ourselves the Institute of Aeronautical Science.

> We are in a way self chosen. Perhaps that is necessary; every man chooses his life's work, and so having responded to a call to the aeronautical profession, why here we are. We have the responsibility for founding a group that represents the profession as we decided to define it. We are responsible for making flight swift and safe and economical; we are professional because we are interested more in knowledge than in a product. The cynic says we are paid for what we know and not for what we do; I think that is one definition of a professional man. Then we are devoted to the perfection of a new means of transportation which must, as such new means always have, have a pronounced effect upon the social organization, political ideas, dress, food, and even morals of our civilization.[29]

With the preliminaries out of the way, James Kimball of the Weather Bureau began the technical sessions at 11:15 a.m. by chairing a discussion of problems and research opportunities in aerology. Next, Jimmy Doolittle presented a short paper and managed a discussion of "scientific flight," followed by an overview of aeronautical research in 1933 from the NACA's George Lewis. After an informal luncheon at the Faculty Club, E. P. Warner introduced, "The Application of Science to Design." C. Fayette Taylor of MIT preceded Piccard's lecture with a discussion of "Progress in Power Plant Fundamentals." The First Annual Meeting of the IAS began at 4:00 p.m., and was followed by a reception and dinner at the faculty club and the evening lecture.

With few exceptions, things went very smoothly. A *New York Times* reporter did note one "amusing incident." At one point during the proceedings, Commander Byrd began to light up his pipe only to notice a horrified look on Auguste Piccard's face. Someone leaned over to Byrd and explained that tobacco smoke made their guest of honor deathly ill. Byrd spent the rest of the day resolutely puffing on an empty pipe.

Edwin Eugene Aldrin, Sr. (NASM Archive, Smithsonian Institution, 89–13625.)

As Gardner had hoped, the press turned out in full force. They could not get enough of Piccard. One astonished Midwestern newsman watched in awe as the professor demonstrated his "two-fisted" approach to the use of a blackboard. "With his left hand he drew pictures of astronomical zones and rough diagrams marking 'highways' in the heavens. Simultaneously, with his right hand he solved an obviously difficult equation in physical chemistry." All the while, his twin brother Jean, who was living in America, stood beside him and translated Piccard's remarks.[30]

Reporters were also intrigued by reports of startling new technology. George Lewis had mentioned a new rotary wing craft designed by Philadelphian H. H. Platt in his session. Those remarks had sparked a flurry of stories in which reporters predicted a bright future for what they described as a "Paddle Airplane," or a "Side Wheel Plane," a craft that would "rise and descend almost vertically and eliminate some of the dangers of air navigation." Although the reporters found it difficult to find just the right words to describe this new sort of craft, the serious discussion of the current state of research and development in the field and prediction of a bright future for what would soon become known as a helicopter were indications of the real value that an organization like the IAS could play.[31–33]

What began as a one-day founder's meeting in 1933 was expanded to a two-day annual meeting by 1935 and to three days in 1937. An annual Honors Night banquet and awards event grew out of the first dinner given in honor of Piccard. It would remain the highlight of the annual meetings until 1937, when the banquet and presentation of awards was shifted to December 17 so that the first IAS Wright Brothers Lecture, offered by Cambridge University Professor B. Melville Jones, "Boundary Layer Experiments in Flight," could be presented on the anniversary of the first flight. The first three Honors Nights were staged at the Columbia Faculty Club. Beginning in December 1937, the event was shifted to the Hotel Biltmore. From that point on, the high point of the annual meeting was held at a major New York hotel, from the Waldorf-Astoria to the Astor. Beginning in 1947, the technical sessions were held at the same hotel.

On the occasion of the 25th anniversary of the IAS, Clark Millikan (president, 1937) remembered the atmosphere of the early annual meetings:

> Those who participated with some regularity in the Columbia gatherings will always have nostalgic memories of them ... the long subway rides out to the University with the high probability, for non-New Yorkers, of finding oneself far in the Bronx and on the wrong train; the frigid walks to and from the Physics building across the snowy campus; the academic atmosphere of the lecture halls in which the papers were presented ...."[34]

He remembered the high jinks, as well. Luis de Florez entertained those attending the informal "smokers," usually ". . . with the able assistance of such fellow conspirators as Casey Jones and Ed Aldrin." With a deadpan expression, he introduced his fellow engineers to such Rube Goldberg devices as the "Compound, Regenerative Air Regulation Disintegrator and Decoder." On another occasion he proposed a new mathematical symbol, "the Multi-Bellied Beta," "for use by engineers when they reach a point in a formula where they wish to insert a sound resembling a Bronx cheer."[35]

The founders were enormously pleased with the first meeting. The turnout had been nothing less than extraordinary. More than one-quarter of the total membership had traveled from across the nation to attend. The last-minute program planning had caused some light-hearted concern. "I am one stage further than Major Doolittle," Ed Warner had remarked at the outset of his session. "He said he hadn't heard until last night that he was supposed to prepare a paper; I haven't heard it yet!" Still, the presenters had been the best men in the field and the discussions had been very lively.

The first business meeting also went well. With Hunsaker presiding, the slate of Council members proposed by the nominating committee (Kimball, Lewis, Martin, Richardson, Starr Truscott, and Ralph Upson) was accepted. The members of the new Council then extended the terms of the acting officers who had superintended the foundation of the organization for another year.

Secretary Gardner then took the floor to explain the alterations required to the constitution and bylaws that the members had received in the mail. Those documents were unanimously accepted, as was Lawrance's treasurer's report. Following a general discussion of the nature of the organization, Hunsaker brought the business meeting to a close. "It is like getting married," he remarked. "Now we are all bound together."[36]

The Journal

President Hunsaker called the next full Council meeting into session at 8:00 p.m. on May 4, 1933, on board a steamer returning up the Chesapeake Bay to Washington following a NACA meeting at Langley Field in Hampton, Virginia. Gardner reported that the first technical publication of the IAS would be the printed proceedings of the Founder's Meeting. He had held the project up for a few weeks until prospective foreign members had an opportunity to join, thus increasing his pool of potential advertisers. He was pleased to announce that advertising had paid all costs of publishing and binding the volume, which also included a new membership list.

Next, Hunsaker mentioned that several members had suggested the need for a technical journal. The founders had originally assumed that the initial IAS publishing program would be limited to a newsletter. Following "a lengthy discussion," however, the members of the Council instructed the president to study the issue and report back with a plan. It would be "advantageous," they advised, if an arrangement could be negotiated with the APS, which might be willing to handle "the mechanical part of the publication of a Quarterly."[37]

Hunsaker succeeded in enlisting the support of the APS. Their publication staff would handle the copyediting of articles that the IAS had approved for content. The journal would be published right along with the APS publications. The larger print runs would represent a savings for both organizations.

President Hunsaker agreed to serve as editor. Although he was willing to handle the business of soliciting articles as well as receiving, vetting, and responding to authors for the first few issues, an editorial committee would be required in the near future. He even agreed that the funds of the Guggenheim Laboratory at MIT could underwrite the venture, if required. "The expense can be absorbed in our overhead," he offered, "and justified by the educational value of the quarterly to all students of aeronautics, including our own."[38]

Gardner distributed 4,000 circulars announcing the first issue and mailed 900 sample copies. As a result, he received 230 subscriptions, many from foreign members, who were asked to pay for their subscriptions above and beyond their dues. Combined with the normal membership list, this gave the journal a total of fewer than 800 paid subscribers.

The secretary convinced the Council to sell advertising, "which not only pays for the cost of publication, but makes a profit in good years." Even before the announcement of the first issue, Gardner was able to convince 19 aviation companies to advertise in the journal at rates that ran from $75 for a full page to $25 for a quarter-page. Indeed, advertising revenue for the first issue eventually amounted to $1,500, three times the cost of publication.[39]

Volume 1, Number 1 of the *Journal of the Aeronautical Sciences* was dated January 1934. Hunsaker was determined that the publication would be an important element in creating a distinguished intellectual tradition for the IAS. The first issue could scarcely have set a higher standard. Theodore von Kármán wrote on turbulence and skin friction. The head of the Guggenheim Aeronautical Laboratory of the California Institute of Technology, he was rapidly replacing Ludwig Prandtl, his own major professor, as the world's leading aerodynamicist. Max Munk, another Prandtl protégé, explored the aerodynamics of slotted wings, and C. G. Rossby, who had helped to pioneer air mass analysis, wrote on the latest trends in meteorological research. Athelstan Spilhaus, a future president of the University of Minnesota, described the fine points of gyroscopic instrument design. John Stack, the NACA's leading aerodynamicist, offered his thoughts on the impact of compressibility on high-speed flight.

The journal was launched as a quarterly. It became a bimonthly publication beginning with Volume 2, Number 1, in January 1935. Just one year later, beginning with Volume 3, Number 3, the journal was transformed into a monthly. Between October 1934 and May 1935, Alexander Klemin and Clark Millikan served as editorial supervisors of the publication while Hunsaker was working with a Federal aviation commission. Beginning with the issue of October 1939 (Volume 6, Number 12), Arnold M. Kuethe took up his duties as Associate Editor, a paid position on the IAS staff. Hunsaker remained the editor of record for more than seven years, until the fall of 1941, when Hugh Dryden of the NACA became head of the editorial committee and editor of record.

Other IAS publications included the annual membership list and copy of the constitution and bylaws. Early on, the Institute also made available off-prints of papers offered at meetings. From the outset, each issue of *The Journal of the Aeronautical Sciences* included a section entitled, "News From the Institute," which announced meetings, information on the latest happenings at headquarters, obituaries, and a notice of new publications of interest to members. Beginning in November 1940, that information was included in a new publication, *Aeronautical Review*, which was initially presented as a separate section of the journal, complete with its own cover. By January 1944, this arrangement had been discontinued and *Aeronautical Engineering Review* appeared as a completely separate publication.

Skyport!

For the first year of its existence, the IAS made its headquarters in Gardner's home or in one of a number of offices from the Aeronautical Chamber of Commerce to the

Manufacturers Aircraft Association at 300 Madison Avenue. Twenty years later, S. Paul Johnston, Gardner's successor, noted that: "Some of the staff still remember long nights spent in inadequate offices working on borrowed furniture to get out meeting announcements or to put together an issue of the then embryonic *Journal*."[11]

Theodore P. Wright, IAS President for 1938. (NASM Archive, Smithsonian Institution, 00181594.)

Space would no longer be a problem after the late summer of 1933, when the IAS took up residence in the most prestigious and well-publicized skyscraper in the world. There were not enough superlatives for the Depression-era publicists who touted the wonders of Rockefeller Center, a complex that would eventually grow to include 14 buildings and external spaces stretching along Fifth Avenue between 48th and 52nd Streets.

It was "... the largest building project ever undertaken by private capital," a circumlocution that evaded huge government construction efforts from the pyramids to the Panama Canal. As was ever the case in such situations, the writers of press releases never grew tired of searching for stunning comparisons. The number of people going to work in Rockefeller Center each day would be greater than the number of Mexicans in Guadalahara or Scots in Dundee. The complex would house more people than the island of Tasmania or the Republic of Turkestan. "If all fourteen buildings were piled on top of one another, they would reach 3,890 feet, almost ¾ of a mile." Nor were the publicists above stretching the truth a bit. Six decades after the completion of the 66-story RCA Building, reference books and architectural historians still frequently describe it as having 70 stories, the figure provided in early press releases.[40]

What had begun in 1928 as a means of funding a new home for the Metropolitan Opera became the business dream of John D. Rockefeller, Jr., with the opera nowhere in sight. "Junior," as he was universally known behind his back, had forged ahead with work on this city within a city in spite of the fact that the family had lost half of its fortune in the stock market crash.

Rockefeller Center generated considerable controversy. Critic Lewis Mumford argued that by attracting so many people into a relatively small space, the project promoted "super-congestion" and represented a "failure to recognize civic responsibilities." Architect Frank Lloyd Wright remarked that the plan would produce "the latest atrocity committed upon a people already about to revolt." Americans, from native New Yorkers to tourists and millions of people who never visited the big city, soon decided otherwise. In a remarkably short period of time, the complex came to symbolize the energy, excitement, and sophistication of a city that captured and held the attention of the world.

The project marked a coming of age for 25-year-old Nelson Rockefeller, who hopscotched over his older brother into a key position in the enterprise, with responsibility for filling the thousands of offices with paying tenants and superintending the effort to decorate the complex with works of art. His battle with Mexican artist Diego Rivera, who had included a portrait of Soviet leader V. I. Lenin in the large mural commissioned for the lobby of the RCA building, catapulted the artist, the patron, and the place into the national memory.

It was broadcasting that put Rockefeller Center on the map. NBC established its corporate offices in the RCA Building, the central structure of the complex, and broadcast its first program from the Center in November 1933. The company ultimately constructed 35 broadcast studios in the building, the largest of which could seat an audience of 1,400. They were engineering wonders, rooms suspended within rooms, floating on fabric-covered springs to ensure soundproofing. Because these unconventional structures could not support the weight of a normal steel, stone, and glass building, the western side of the RCA Building, where the studios were located, was only 16 stories tall. That section of the building contained more than 1,500 miles of wire and cable and was designed for the future. Two stories of the studio area were intended for television and were unused until that technology came of age.

From the outset, two shows were normally broadcast at any one time, one on the NBC "Red" network and one on the "Blue." In addition, any number of rehearsal studios would be in use. The opportunity to join the studio audience of a favorite radio program became a major attraction for tourists visiting New York. During the first year of operation alone, more than three-quarters of a million people visited Rockefeller Center to take advantage of that opportunity Across the nation, millions of Americans gathered around their radios became familiar with Radio City, the Rainbow Room, and Rockefeller Center.

Edwin Aldrin, the manager of aviation activities for Standard Oil of New Jersey, a corporation with which the Rockefellers had some familiarity, first broached the possibility of establishing the IAS headquarters in the RCA Building.[41] Presumably, Gardner was quick to follow up on the idea. In the end, Nelson Rockefeller and Lawrence Kirkland, who were in command of the all-important leasing effort, decided to give the IAS free office space for one year.

Why was the Rockefeller organization so generous with the new Institute? The strategy called for attracting clusters of corporations into the building. With the National Broadcasting Company as their largest and most visible tenant, the entertainment business was a natural. Indeed, Rockefeller Center soon became the East Coast home of major movie studios and talent agencies. Petroleum companies quickly followed suit. Rockefeller and Kirkland must have regarded aviation corporations as another group of potential renters. Perhaps a bit of generosity to the IAS would serve to attract business from that quarter.

If that was the thinking, it worked. Over the fall and winter of 1933 and into the spring of 1934, other aviation companies moved into the RCA Building, from manufacturers like Sperry, Curtiss-Wright, and Fairchild to smaller operations such as the Casey Jones School of Aeronautics and the Information Division of the Aeronautical Chamber of Commerce, which claimed to maintain the largest aeronautical library in the nation.

The IAS signed its lease on October 1, 1933. Gardner officially welcomed the members to their new headquarters on the 54th floor (5431) of the RCA Building at 30 Rockefeller Plaza, on November 20, 1933. Dubbed the Skyport, the suite served as an office area and club room for members. Gardner took extraordinary pride in his new quarters. "From the RCA Building you can see fifty miles," he explained to readers of the new *Journal of the Aerospace Sciences*. "You can look north to Westchester, east to Long Island, south to the Atlantic and west to New Jersey .... At night the multicolored lights of the city and its suburbs offer a picturesque spectacle. The neon lights of Times Square, the strings of lights in Central Park, the illuminated windows of skyscrapers and the streams of moving lights on highways and trestles provide an unforgettable panorama."[42]

The Skyport was furnished in a manner appropriate to the setting. Grover Loening covered the decorating costs and donated the furnishings, with the exception of 13 stylish aluminum chairs presented by the Aluminum Company of America. Loening's gift, valued at $2,000, earned him a place as the first Benefactor of the Institute. Rueben Fleet, head of

Consolidated Aircraft, provided a typewriter and an aircraft model. T. P. Wright donated models of two Curtiss airplanes. Luis de Florez presented a radio receiver, and photographer Edward Steichen and the RCA Corporation donated a large photo montage originally intended for the Rockefeller Center theater. Not to be outdone, Paul Litchfield of Goodyear–Zeppelin provided a photomural of the airship USS *Macon*.

Dr. Clark Blanchard Millikan, IAS President for 1937. (NASM Archive, Smithsonian Institution, 00169168.)

Gifts for the Skyport would continue to pour in during the days, weeks, and months that followed. Model airplanes soon dotted the office area, along with photos of aircraft that the members had designed, built, and flown. Gardner was especially fond of special aeronautical instruments presented to the IAS. There was a "Stormoguide," for example, a gift of the Taylor Instrument Company. His favorite, however, was a special hand-held altimeter provided by instrument manufacturer Paul Kollsman. Gardner carried it when squiring visitors upstairs on the elevator. He could show them both their absolute altitude and the rate of climb as they ascended to the 54th floor.[43]

The RCA Building would remain the home of the IAS for the next 12 years, but the organization would move around a bit inside the complex. The need for additional office space by 1937 led to a shift to new quarters on the 51st floor. Three years later, with the creation of the aeronautical archive and the requirement for collections storage and display area, the IAS took over half of the 15th floor on the west side of the building.[44]

First Steps

Lester Gardner would soon have company in his office in the sky. Not long after the move into new quarters, the Council approved the hiring of the first permanent, full-time staff members to assist the secretary with IAS operations: C. E. Sinclair, who would succeed Gardner as secretary, and George R. Forman, who would become assistant treasurer and editor of the *Aeronautical Engineering Review*, the first nontechnical IAS serial publication.

One of the first projects undertaken in the new office was the selection of an Institute seal. Forty-six entries were submitted in response to the Council's announcement of a competition. The winner combined two very similar sketches, one by George Lewis of the NACA and the other by W. S. Simpson of the U.S. Navy Bureau of Aeronautics. The result showed a symmetrical, tear-drop airfoil with laminar flow lines over the top and bottom, circled by the name of the organization. It would remain the logo of the Institute until the merger with the ARS.

The first IAS technical meeting outside the annual meeting was held in the RCA Building at 1730 hrs on March 29, 1934. Elmer Sperry addressed those present on "The Development

Grover Loening, a founding member of the IAS, with Amelia Earhart, who was not. (NASM Archive, Smithsonian Institution, 78–16309.)

of Blind Flying." W. Lawrence Le Page spoke on April 26 on "Rotary Wing Aircraft." In addition to these monthly meetings, the IAS sponsored a luncheon every Wednesday at a restaurant near Rockefeller Center, where "old and new friends . . . discuss . . . aeronautical matters of current interest." Then there were special excursions. On June 7, 1934, 200 members toured the aircraft carrier USS *Lexington*.[45]

The rich assortment of meetings and get-togethers available to members living in New York underscored the single most important problem faced by the early IAS. Although the East Coast was the center of finance and government in the United States, the center of gravity of the aviation industry was shifting to the West Coast.

The answer to that problem came on December 4, 1934, when 180 members and their guests gathered at the Athenaeum, the California Institute of Technology's "attractive social center," to organize the Pacific Coast Section of the IAS. The organizing committee was a "Who's Who" of West Coast aviation: C. L. Egtvedt of Boeing; Donald Hall, who had designed the *Spirit of St. Louis*; Lockheed's Hall Hibbard; John K. Northrop; Gerard Vultee; Arthur Raymond, who headed the Douglas team that produced the DC-3; Baldwin Woods of the University of California; and Clark Millikan of California Institute of Technology, who had spearheaded the creation of the Section. Donald Douglas presided over the meeting, which featured papers by such comers as Clarence "Kelly" Johnson of Lockheed and F. R. Collbohm, who would one day head the team of Douglas engineers who created the Rand Corporation.

Lester Gardner, who flew out for the occasion, congratulated the organizers on having attracted a larger crowd than had attended the IAS Founder's Meeting. "As the membership in the Institution grows," he remarked, "other sections will be organized in centers where there are a sufficient number of members to permit the holding of well attended local meetings." The Sections, he realized, would be one of the keys to the future of the organization. Although national leadership was essential for success, the regional Sections would provide a framework for local meetings, speakers, awards, and a means of linking individual members into the organization.

Following the Founder's Meeting, Hunsaker had sent a letter to the membership warning them that the second year was the most difficult time in the history of any organization. In the case of the IAS, the second and third years would be the making of the organization. By the time of the second annual meeting on January 31, 1934, members could lounge on the new furniture and enjoy the view from the Skyport, their prestigious new headquarters.

On the occasion of the third annual meeting on January 29–30, 1935, Charles Lawrance, second president of the IAS, could only shake his head in wonder. Originally, he explained, ". . . it was expected that the activities of the Institute would consist only of an annual meeting and brief bulletins to the members. No rental for a headquarters, no Journal, no Membership Roster, and no permanent staff were contemplated." Now, less than three years after the charter had been granted, all of those things were in place. The founders had achieved a success well beyond anything they had hoped for.

References

[1] Lester D. Gardner, Opening Remarks at the Business Meeting of the IAS, Jan. 26, 1933, Binder, AIAA files.

[2] J. C. Hunsaker and Lester D. Gardner, "Background and Incorporation of the Institute of Aeronautical Sciences," Hugh L. Dryden Papers, Box 253, Folder 1952, Manuscript Division, Johns Hopkins Univ. Libraries.

[3] Jerome Hunsaker, Jr. to Tom Crouch, Jan. 7, 1981, with an attachment by J. Hunsaker, Sr., "Institute of Aeronautical Sciences."

[4] J. Laurence Pritchard to Lester D. Gardner, July 1, 1928, AIAA files.

[5] Hunsaker and Gardner, "Background and Incorporation," p. 7.

[6] Lester Durand Gardner, "Events Leading up to the Formation of The Institute of Aeronautical Sciences," AIAA files, 1952.

[7] Jerome Hunsaker, Jr. to Tom Crouch, Jan. 7, 1981; Hunsaker and Gardner said that they met with six colleagues that week: Charles Lawrance, Edwin Aldrin, Grover Loening, Earl D. Osborn, Samuel Bradley, and Lawrence Bell. They could not reach Edward P. Warner, who was out of town.

[8] The first-hand accounts of the first series of meetings are slightly contradictory. The account provided is an amalgam of Hunsaker, Jr. to Crouch, Jan. 17, 1981, "Institute of Aeronautical Sciences"; Hunsaker and Gardner, "Background and Incorporation of the Institute of the Aeronautical Sciences"; Gardner, "Events leading up to the formation of the Institute of Aeronautical Sciences," AIAA files. All correspondence relating to the incorporation of the IAS is held in AIAA files.

[9] The first certificate of incorporation signed by Hunsaker, Loening, Gardner, and Lawrance survives in the AIAA files. The final certificate has not been located. The full list of names added to the articles is not known.

[10] The incorporators established a board of directors numbering between 5 and 21 members. Fletcher indicated that the board should have 12 members.

[11] S. Paul Johnston, "Lester Durand Gardner ... 80 Years Young," unidentified source, Lester Gardner biographical file, NASM Archive, p. 17.

[12] Lester Durand Gardner memorandum to Jerome Hunsaker, Charles Lawrance, Grover Loening, and Edward Warner, Oct. 12, 1932, binder of materials sent to each council member, AIAA Files.

[13] Lester Durand Gardner memorandum to Jerome Hunsaker, Charles Lawrance, Grover Loening, and Edward Warner, Oct. 12, 1932, binder of materials sent to each Council member, AIAA Files.

[14] IAS circular, 1932, Binder, AIAA files.

[15] "Air Institute Founded," *New York Times*, Oct. 18, 1932; *New York Times*, Oct. 23, 1932; *New York Herald Tribune*, Oct. 23, 1932.

[16] Earl Findley, "The Institute of Aeronautical Sciences," *U.S. Air Services*, Jan. 1933.

[17] Jerome Hunsaker to Howard Coffin, Dec. 21, 1932, Binder, AIAA files.

[18] Lester Durand Gardner to Members of the Council, Oct. 27, 1932, AIAA Files.

[19] Binder, General Information, Oct. 27, 1932, AIAA files, p. 2.

[20] *Constitution, By-Laws and Membership List, Institute of the Aeronautical Sciences*, copy in binder of all original materials sent to council members, AIAA files, 1933.

[21] Lester Gardner to Jerome Hunsaker, Nov. 2, 1932, AIAA files.

[22] All facts and figures are from minutes of the meetings of the IAS Council, Binder, AIAA files.

[23] "One Co-ed Studies With 3,000 N.Y.U. Men; Seeks Degree in Aeronautical Engineering", *New York Times*, Nov. 4, 1933, p. 17.

[24] "Aeronautical Institute Honors Field Engineer," *Dayton News*, Feb. 2, 1939.

[25] "IAS Plaque Unveiled at Columbia University," *Aeronautical Engineering Review*, March 1955, p. 23.

[26] Piccard would ultimately make a total of 27 balloon flights, reaching altitudes of up to 72,177 feet. He then turned his attention to exploring the ocean depths. He developed the bathyscaphe, which enabled scientists to reach the bottom of the ocean. His son, Jacques, employed his father's invention to become one of the first two men to reach the deepest point in the ocean. His grandson, Bertrand, was one of the first two men to circumnavigate the Earth on a nonstop balloon flight.

[27] J. C. Hunsaker, undated memorandum to members of the Council, Binder, AIAA files.

[28] *New York Post*, June 23, 1933.

[29] J. C. Hunsaker remarks, Transcript of the Founder's Meeting, Binder, AIAA files.

[30] "Piccard Surpasses Boldest Utopian in Forecasting 'World of Wonders,'" *Clinton [Wisconsin] Observer*, Feb. 3, 1933.

[31] Untitled article, *Greenville, S. C. News*, Jan. 30, 1933.

[32] "Side Wheel Plane Up At Meeting," *New York Evening Post*, Jan. 25, 1933.

[33] "Paddle Airplane Made by American," *New York Times*, Jan. 27, 1933.

[34] Clark Millikan, "Reflections on our First Quarter Century," *Aeronautical Engineering Review*, March 1957, p. 32.

[35] Millikan, "Reflections," p. 33.

[36] J. C. Hunsaker, Transcript of the First Business Meeting, Binder, AIAA files, p. 21.

[37] "Minutes of the Adjourned April Meeting of the Council of the Institute of the Aeronautical Sciences, Inc." Binder, AIAA files.

[38] J. C. Hunsaker to the IAS Council, Feb. 26, 1934, Binder, AIAA Files.

[39] Undated memorandum to the Board, "Prepared by Dr. J. C. Hunsaker and Lester D. Gardner," AIAA files.

[40] The material on Rockefeller Center is take from Daniel Okrent, *Great Fortune: The Epic of Rockefeller Center*, Viking, New York, 2003.

[41] *History of the Institute of Aeronautical Sciences, Tenth Anniversary, IAS, 1932–1942*, IAS, New York, 1942, p. 5.

[42] "The Skyport," *Journal of the Institute of the Aeronautical Sciences*, Jan. 1934, pp. 50–51.

[43] "Indoor Aviator," IAS scrapbooks, AIAA History Collection, Manuscript Division, Library of Congress.

[44] *History of the Institute of the Aeronautical Sciences, Tenth Anniversary, 1932–1942*, IAS, New York, 1942, p. 5.

[45] "Meetings of the I.Ae.S.," *Journal of the Aeronautical Sciences*, July 1934.

Chapter 4—
Coming of Age: The Institute of the Aeronautical Sciences, 1935–1945

Speaking Up

The Federal Aviation Commission held it first meeting on July 11, 1934. Appointed by President Franklin D. Roosevelt under the terms of Section 20 of the Air Mail Act of 1934, the panel was to "make an immediate study and survey" and to recommend "a broad policy covering all phases of aviation and the relation of the United States thereto." The six-man commission, headed by J. Carroll Cone, Director of Air Regulation for the Bureau of Air Commerce, included a newspaper editor, a lawyer, and a labor relations expert. Hunsaker and Edward P. Warner, far and away the most experienced aeronautical authorities on the panel, were quick to seek advice and testimony from their fellow members of the Institute of the Aeronautical Sciences (IAS).[1]

No one was more pleased than Charles Lawrance, who had succeeded Hunsaker as president of the IAS. "This is the first time to my knowledge," he remarked to those attending the third annual meeting on January 30, 1935, "that aeronautical scientists and engineers ... have had an opportunity to present to any government agency a statement on national aviation policy." Speaking with the voice of the organization, the IAS leadership offered six recommendations, from a call for the government to give preference to companies that maintained "adequate engineering and technical personnel," to support for university research and the work of the National Advisory Committee for Aeronautics (NACA). The Institute recommended that the military services recognize the importance of relying on officers with technical training and suggested that, in general, Congress and the White House recognize the critical need to protect funds for research and development.[2]

It was the first time that the IAS had spoken for the practitioners of flight science and technology. This was not something that would happen a great deal in the decades to come. Although the postwar American Rocket Society (ARS) would work hard to influence government policy and the executive secretary of the IAS would participate in various postwar government studies and panels, the leaders of the IAS were generally reluctant to speak out on public issues that might have an impact on the business arrangements of their corporate members. None of the founding members mentioned the potential of the organization to effect public policy. That function of the organization lay 40 years in the future.

Honors and Awards

The honors and awards program was underway, as well. The grade of Honorary Fellow was reserved for the most distinguished professionals. As noted, Orville Wright was the first person to be so honored, even before the Founder's Meeting. As late as 1942, only 18 individuals had been made Honorary Fellows. By the end of the first year of operation, 23 members had been elected Fellows of the IAS.

Not long after the Founder's Meeting, the members of the Council circulated an invitation to any member who might be willing to endow an IAS award. Sylvanus Albert Reed, one of the first group of IAS Fellows and a pioneer in the development of aluminum propellers, was the first to respond. He bequeathed $10,000 to the organization to support an annual prize of $250 recognizing "a notable contribution to the aeronautical sciences resulting from experimental or theoretical investigations, the beneficial influence of which on the development of practical aeronautics is apparent." Announced in December 1934 just before the second annual meeting, the first Sylvanus Albert Reed Award went to meteorologists C. G. Rossby and H. C. Willett.

The Lawrence Sperry Award, named in honor of the young man who had pioneered the autopilot and gyroscopically stabilized flight instruments, was endowed in April 1936 by Elmer A. Sperry, Jr.; Edward G. Sperry; and Helen Sperry Lea, his brothers and sister. The $250 annual prize would honor, "a notable contribution made by a young man to the advancement of aeronautics." The selection would be made by a board appointed by the donors. Initial members included James Doolittle; Glenn L. Martin; Elmer Sperry, Jr.; Gardner; Loening; and Lawrance. William C. Rockefeller was the first recipient. Clarence "Kelly" Johnson, the second honoree, would emerge as one of the great aircraft designers of the century.

The Institute established the Octave Chanute Award in 1939. Named for a leading late 19th century engineer and flying pioneer, the prize was awarded for "a notable contribution by a pilot to the aeronautical sciences." The annual recipient was to be chosen by a committee that included the vice president of the Institute, the Chief of the Army Air Corps, the Chief of the Navy Bureau of Aeronautics, the Director of the Bureau of Standards, and the Director of Research for the NACA. Early winners included Boeing test pilot Edmund T. Allen in 1939, Howard Hughes in 1940, and NACA engineer/test pilot Melvin Gough in 1941.

What would become the Loesy Atmospheric Sciences Award and the Jeffries Aerospace Medicine and Life Sciences Research Award were both established in 1940. The first, offered "in recognition of outstanding contributions to the science of meteorology as applied to aeronautics," was named in honor of Robert M. Loesy, a pilot and meteorological advisor to the Chief of the Air Corps, who died during a bombing raid on Dombas, Norway, on April 21, 1940, while serving as Assistant Military Attaché to the U.S. legation to Norway and Sweden. The first Jeffries Awards, named for physician John Jeffries, the first American to make a free flight in a balloon, went to Dr. Louis Bauer in 1940 and Major Harry G. Armstrong in 1941, pioneers of aerospace medicine in the United States.

The Bane Award was named in honor of Colonel Thurman H. Bane, aeronautical engineering pioneer and first commander of the U.S. Army Air Service Engineering Division during World War I. Established in 1943 and administered by the IAS, it was presented annually until 1953 to an officer or civilian employed by the Army Air Forces Matériel Command (later the Air Research and Development Command of the United States Air Force) for an outstanding achievement in aeronautical development for that year.[3]

The Reed, Sperry, Chanute, Loesy, Jeffries, and Bane Awards constituted the list of IAS prizes for the next two decades. The Institute had other ways to honor its members, however. Since Sir Horace Darwin had offered the first Wilbur Wright Memorial Lecture in 1913, the Royal Aeronautical Society (RAeS) had honored the inventors of the airplane by inviting a distinguished contributor to flight technology to offer a lecture each year. The IAS followed suit on December 17, 1937, when Professor B. Melvill Jones gave the first Wright Brothers Lecture to members of the Institute. The success of the first lecture inspired Edmund C. Lynch to endow the IAS with a $10,000 Vernon Lynch Fund in 1938 to support the annual Wright Brothers Lecture. The lecturer would be selected by those Americans who had been invited to present the Wright Lecture to the members of the RAeS.

Dr. Hugh Dryden, Chief of the Mechanics and Sound Division of the National Bureau of Standards and soon to become a major figure in the administration of the NACA, gave the second Wright Brothers Lecture, "Turbulence and the Boundary Layer," at Columbia University on the evening of December 17, 1938. As a means of commemorating the 35th anniversary of the first flights at Kitty Hawk, IAS Sections and Branches held simultaneous meetings across the nation at which selected members read Dryden's address at the same time he was delivering it in New York. The California Institute of Technology's Clark B. Milliken presented the third lecture in 1939, by which time the annual Wright Brothers Lecture had become a major event of the IAS year.

The members of the Institute also cooperated with other organizations in selecting aerospace contributors for a variety of honors. Beginning in 1938, the Institute was invited to join representatives of the American Society of Mechanical Engineers (ASME) and the Society of Automotive Engineers (SAE) on the Board of Award for the Guggenheim Medal presented annually by the trustees of the Daniel Guggenheim Medal Fund "for notable achievement in the advancement of aeronautics."

The IAS also participated in selecting the recipients of the Musick Memorial Trophy. Established by citizens of Aukland, New Zealand, the award honored the memory of Pan American pilot Captain Edwin C. Musick and six crew members of the *Samoan Clipper* who were lost on a flight from the United States to New Zealand on January 12, 1938. The Institute and the RAeS awarded the trophy to citizens of the United States and of the British Commonwealth of Nations who had made important "contributions to safety in the air with special regard to transoceanic flying." The winners received a small copy of the large trophy, which they held for a year. The Musick Trophy was not presented during World War II, when it was displayed in the IAS headquarters. Last presented in 1957, the trophy has been on permanent display in Aukland since that time.

Over the decades, the IAS would also participate with other organizations in selecting the recipients of the Robert J. Collier Trophy, the Elmer A. Sperry Award, and the Wright Brothers Trophy. In the fall of 1935, following the creation of the Student membership grade and the organization of the first Student Branches, the IAS instituted the first student awards. Each year, two members of each Student Branch received a certificate and an honorarium consisting of two years of prepaid membership at the entry level Technical grade following graduation. The Student Scholastic Award was presented to the student with the highest academic standing in each Branch. The Student Branch Lecture Award went to the member who had presented the best original paper at a Student Branch meeting in any given year.

In 1940, not long after the inauguration of the government-sponsored Civilian Pilot Training Program, the IAS administered the Shell Aviation Scholarships and Awards program. Funded with a $15,000 gift from the Shell Oil Company, the competition involved various tests of flying proficiency in which flight students from more than 500 schools participated. The IAS created a network of regional judges and a National Board of Award. There were prizes for top regional competitors and scholarships of $1,000, $750, and $500 for the national winners. The schools and flight instructors who had produced student pilots with the highest records of proficiency received trophies.

Spreading Our Wings

The IAS celebrated its fifth anniversary on October 15, 1937. The Institute was growing slowly. That year the roster included 1,795 members scattered across 33 states and U.S. territories and 25 foreign nations. New York led the states with 392 members; California was second with 213; Michigan, with 168 members, was the only other state with more than

100 names on the roster. England led the foreign nations, with 59 IAS members; Germany had 38; France had 29; Italy had 19; and Japan had 10.

The membership was up from the 1,305 names on the roster as of January 1, 1936. At that time the breakdown of members had been as follows:

IAS Membership by Grade[4]

	Foreign	American	Total
Honorary Fellows	1	4	5
Fellows	0	47	47
Scientific Members	94	118	212
Engineering Members	77	303	380
Technical Members	0	132	132
Pilot Members	7	70	77
Industrial Members	8	148	156
Student Members	0	298	298
Total	187	1,120	1,307

Five years later, in 1941, new members (except for Honorary Fellows, Honorary Members, Benefactors, and Student Members) paid a $5 initiation fee. Dues were $20 for Fellows, $15 for Associate Fellows, $12.50 for Members and Industrial Members, and $7.50 for Technical Members.

The IAS established a Corporate membership grade in 1936. "The larger aviation companies preferred to make contributions to a scientific and engineering society and support the publication of the *Journal* directly, rather than through advertising," an Institute spokesman explained. Annual dues for "manufacturers of airplanes, airships, or aircraft engines" would be $200, "or 1/20th of one percent of their gross annual business, which ever is larger." In no case, the IAS leadership was quick to note, would the dues exceed $2,500 for any one company.[5] Corporate members were "... entitled to deduct their contributions in determining their net income for tax purposes." Moreover, "Corporate Members engaged in government work ... may include as part of the cost of performing such contract or subcontract ... [a] portion of the membership dues paid by them to the Institute."[6]

It was apparently an appealing offer. Seventeen companies signed up before the official announcement of the new membership grade. They included aircraft manufacturers (Curtiss–Wright, Douglas, Martin, Fairchild, Northrop, and United Aircraft), parts specialists (Edo, Sperry Gyroscope, Kollsman, Goodyear Tire and Rubber, and Summerill Tubing), and oil companies (Socony Vacuum and Ethyl Gasoline). By February 1937, the number of Corporate members had climbed to 26 companies.

In 1939, the Institute asked its members to subscribe to a Code of Ethics. The Council and the Advisory Board were quick to note that they "assumed that a member is a person in whom qualities of high character are inherent, such as patriotism, integrity, devotion to high ideals of personal honor and of professional honesty, a broad spirit of fairness to all concerned, and that these qualities fundamentally control all of his judgments and decisions." Just in case, however, the IAS offered a nine-point code, ranging from an admonition "to advertise only in a dignified manner," to promises not to claim false credit or to "suppress information bearing vitally upon the safety of a structure or of its operation," and concluding with a caution to maintain the confidentiality of information "obtained by him as to the business affairs, and technical methods or processes of a client or employer."[7]

"As no special effort has been made to secure new members," the IAS annual report to the members noted in 1935, "the growth in membership indicates that the Institute is now regarded as part of an aeronautical specialist's professional equipment."[8] Perhaps so, but the creation of a regional and local framework, beginning with the organization of the Pacific Coast Section on December 4, 1934, was critically important to the success of the organization.

In celebration of the fifth anniversary of the Institute, the first meetings of Branches were held in New York; Washington, DC; and Philadelphia. Members in those cities and in Los Angeles, where the relationship was regularized, signed Branch Charters on October 15, 1937. That year, Professor B. Melville Jones, who offered the first Wright Brothers Lecture in New York, traveled to Los Angeles, where he repeated his talk for members of the local Section meeting in Pasadena on December 21.

The following year, T. P. Wright, sixth president of the Institute, made a tour of the new Sections. By the 10th anniversary in 1942, Sections had been organized in aeronautical centers across the nation. The original West Coast Branch had spawned three Sections—Los Angeles, San Diego, and Seattle—each of which represented an obvious center of aircraft manufacturing. Philadelphia was an early focal point for rotary wing developments, and military bases in Dayton, Ohio, and San Francisco had large enough contingents of military officers, engineers, and scientists to support a Section. South Bend, Indiana, represented Notre Dame University.

During the first decades, headquarters seems to have treated local Sections and Student Branches almost interchangeably. South Bend, for example, obviously a Student Branch, was initially listed as a Section. By the fall of 1944, the pattern of local organizations was falling into place. IAS membership had grown from 3,000 in 1940 to more than 10,000 members in 1944, enrolled in nine active Sections: Buffalo, Cleveland, Detroit, New York, Philadelphia, Texas, San Diego, Los Angeles, and Seattle. At the same time, 862 student members were enrolled in 34 active Student Branches. Between January and October 1944, the Sections held 24 local meetings, offering their members a total of 23 lectures and 6 films, and the Student Branches met 75 times, sponsoring 53 lectures and 42 films.[9]

The organization of the Institute was evolving as well. In January 1934, the constitution had been amended to create a 20-person advisory board, the members of which would be elected annually. By 1936, nine standing committees were responsible for

- *Executive operations* (Glenn Martin, Sherman Fairchild, Charles Lawrance, Edwin Aldrin)
- *Finance* (Sherman Fairchild, Charles Lawrance, Grover Loening)
- *Nominations* (G. W. Lewis, Alexander Klemin, Glenn Martin, Holden C. Richardson, R. H. Upson)
- *Meetings* (G. W. Lewis, Raymond Haskell, J. C. Hunsaker, Clark Milliken, Arthur Nutt, James Taylor, D. W. Tomlinson)
- *Annual Dinner* (Luis de Florez, George Post, James Taylor)
- *Admissions* (V. E. Clark, C. H. Biddlecombe, Jerome Lederer, John Sanborn, Daniel Sayre)
- *Student Affairs* (Clark Millikan, John Akerman, Peter Altman, William G. Brown, George Haskins, Bradley Jones, Richard Smith, E. A. Stalker)
- *Reed Award* (J. C. Hunsaker, Lester Gardner, Charles Lawrance)
- *Editorial and Professional* Activities (J. C. Hunsaker).

The original attempt to survey new developments and prospects in the entire field during the early annual meetings was giving way to recognition of the importance of specialization.

Dr. Sylvanus Albert Reed, who provided a $10,000 endowment for the first IAS award, for "a notable contribution to the aeronautical sciences resulting from experimental or theoretical investigations, the beneficial influence of which on the development of aeronautics is apparent." (Courtesy AIAA.)

By 1936, the IAS had established technical committees covering Aerodynamics; Aeronautical Design, Materials, and Structures; Education; Instruments; Metallurgy; Meteorology; Physics and Radio; and Power Plants and Fuels. The fifth annual meeting, held January 27–29, 1937, was made up largely of simultaneous technical sessions offering presentations by members in standard areas of interest suggested by the new committee structure.

The number of meetings was expanding as well. In addition to the annual meeting, the Wright Brothers Lecture, and the informal gatherings for New Yorkers sponsored by headquarters, a Pacific Coast Annual Meeting was inaugurated in 1938. The following year the meeting was redesignated the Annual Summer Meeting to underscore the fact that it was a national IAS meeting and not a regional gathering.[10]

Then there were the Section and Student Branch Meetings. By 1936, the IAS was also cosponsoring both local and national meetings with other organizations, including the American Association for the Advancement of Science (AAAS), the ASME, the SAE, the American Society of Metals, the Soaring Society of America, and others.

Specialist meetings focused on specific segments of the industry and specific technical problems were also scheduled for the first time. The Philadelphia Section sponsored a meeting on rotary wing flight on September 23–24, 1938. The event was a great success, attracting some 200 attendees. Just a month later, the Institute staged an Air Transport meeting at Chicago's Hotel Morrison. Earlier sessions on new commercial aircraft and revolutionary new instruments, systems, and procedures for blind flying had sparked so much interest at the annual meeting that the Council decided to stage this special event. Attendees included the Air Attaches from Germany, Britain, and Poland and representatives of France and Amtrog, the Soviet trade organization that conducted Stalin's intelligence operations in the United States.[11]

The International Congress of the Aeronautical Sciences, which the IAS planned to sponsor at the New York World's Fair on September 11–16, 1939, was a major effort to attract the international community to the United States. The biggest names in the Institute and in American aviation agreed to serve on the General Arrangements, Meetings, Finance, or Program Committees planning the event. Invitations went out to the "governments, universities, and aeronautical organizations throughout the world to send delegates . . ." The meeting would follow the pattern of the IAS annual meeting, offering technical sessions in a wide rage of areas including aerodynamics, airplane design, structures, instruments, engines, meteorology, and air transport.

Columbia University offered to house up to 100 foreign delegates in residence halls during the week of the conference. Each nation and organization represented at the conference would be asked to present a written tribute to the Wright brothers at a formal banquet on the evening of September 15. The event would culminate on September 16, which would be "International Aviation Day" at the Fair.

It was not to be. In the fall of 1939, the Council of the IAS announced the postponement of the International Congress until 1941. "It was the unanimous opinion that the engineers and scientific specialists in the United States and countries which were planning to send delegates were so occupied with problems of design and production that they will find it difficult to participate." Not to mention the fact that Germany had invaded Poland on September 1, 1939.

For the American aircraft industry, the war began long before December 7, 1941. Expansion was the order of the day, as companies competed to meet the needs of embattled Allied nations and to supply aircraft to an American Army and Navy struggling to prepare for the day when the United States might also be drawn into the conflict. As the nation approached a crisis and the IAS approached its 10th anniversary, the members of the Council agreed that the time was ripe to consolidate the gains made by the Institute over the past decades and to remind members of the value of the organization.

In April 1941, just nine months before the Japanese struck Pearl Harbor, James Doolittle and Lester Gardner, the president and executive vice-president of the Institute, set out on a tour of the nation. Over the next six weeks they would travel more than 18,000 miles, visiting 20 cities and speaking to 31 IAS Sections and Branches. Most of the trip was made via commercial airlines, although Doolittle, an official of Shell Oil, arranged to fly three legs of the trip in company airplanes. They often spoke to a noon gathering in one city and then flew on to an evening presentation in another town. Before they were through, they had spoken to an estimated 4,000 members, students, and guests of the Institute.

Doolittle usually began with an account of his early experiences as an Army test pilot, emphasizing the distance that aviation had traveled in less than two decades. Gardner would then focus on the role of the Institute and the advantages that it offered, especially to young engineers. They usually concluded the meeting with a showing of a new film, "Conquest of the Air," a 71-minute English documentary on the history of flight produced by the Alexander Korda and featuring Lawrence Olivier in a cameo role as Vincenzo Lundardi, an 18th-century balloonist. Back in New York on May 16, Doolittle and Gardner were confident that their extended pep talk had reminded members from coast to coast of the important role that the Institute had played in promoting the role of scientific and engineering professionals in aviation.[12]

The Library and the Aeronautical Index

Lester Gardner believed that meeting the information needs of the members, many of whom had no access to a good technical library, was one of the most important services the IAS could provide. He took the first step in that direction at the time of the move into the Skyport, when he donated his personal library of several hundred volumes on aviation, many of them rare and historically significant, along with his collection of aeronautical pamphlets and photographs.

The secretary explained to members in the fall of 1935 that although the Institute did "not wish to house a large collection of books," it was important to build "a representative library." The Council would not appropriate money for books until 1938, so Gardner invited interested parties to follow his example and send lists of any books that they were willing to donate to the Institute. "Scientific and engineering books are particularly desired," he noted, promising that "each book will be marked with the name of the donor and the gift acknowledged in the *Journal*."[13]

Gift books began to arrive one or two volumes at a time. As promised, Gardner dutifully acknowledged each donation in the *Journal of the Aeronautical Sciences*. He also provided foreign aeronautical organizations with subscriptions to the *Journal*, and received their publications in return. By September 1935, issues of 44 aeronautical journals from the United States, Argentina, Austria, Australia, England, France, Germany, Italy, Spain, and the Soviet Union were added to the IAS library collection each month.

As the library grew, Gardner realized that having access to books and journals was the least of the problems facing an engineer in search of information. The real difficulty was how to find what you were looking for in a rapidly expanding literature. It was a problem as old as flight itself.

The traditional answer was to prepare a bibliography. French balloonist Gaston Tissandier had published his classic, *Bibliographie aéronautique: catalogue de livres d'histoire, de science, de voyage et de fantasie, traitant de la navigation aérienne ou des aérostats*, in 1887.[14] During the early 20th century, as library shelves began to groan under the weight of books, articles, journals, and papers covering aspects of aviation, professional librarians stepped into the breach.

The New York Public Library issued "A List of Works ... Relating to Aeronautics" in a 1908 edition of its *Bulletin*. Paul Brockett of the Smithsonian Institution launched a much more substantial effort. Brockett had begun work as a Smithsonian messenger in 1886 and, with additional training and experience, rose to the position of assistant librarian of the Institution. During the course of his own flying machine experiments, Samuel Pierpont Langley, who had led the Smithsonian from 1887 until his death in 1906, had amassed one of the world's most complete aeronautical libraries. Charles Doolittle Wolcott, his friend and successor, asked Brockett to prepare a bibliography based on existing works like Tissandier's and the material in the Langley library.

Published in 1910 as Volume 55 of the Smithsonian Miscellaneous Collections, the *Bibliography of Aeronautics* contained 13,487 individual entries covering flight-related items appearing prior to 1909. A second volume covered the years 1909 to 1916, and a third, published in 1923, covered works that had appeared since 1917. The Aeronautical Chamber of Commerce (ACC) carried on the work with the *Library Bulletin* published from 1922 to 1935.

As past president and an active supporter of the ACC, Lester Gardner was fully aware of the indexing project sponsored by that organization and regarded it as a critically important function that the IAS could perform more effectively that the ACC. The secretary announced a new initiative to the members at the Third Annual Meeting on January 30, 1935. "There have been many indexes of books and magazines," he began, "and there have been digests of technical papers and bibliographies."[15] None of those aids was as complete as they might be, however, and none included such key graphic items as photographs, drawings, and maps. With that in mind, Gardner approached William Zelcer, New York City's Deputy Commissioner of the Department of Plants and Structures for Aviation, suggesting that he sponsor such an index as a project that could be undertaken by unemployed engineers and aeronautical professionals. Zelcer liked the idea, and the Emergency Relief Bureau of New York agreed to fund the effort.

Gardner would act as technical director of the project. Dr. Merle S. Ward, of the Works Division of the Emergency Relief Bureau, would supervise the "aeronautical specialists, filing clerks and typists" who would record bibliographic information on thousands of 3 × 5 index

IAS President Charles L. Lawrence (center) presents the first Reed Award to meteorologists Dr. H. C. Willett (left) and Professor C. G. Rossby (right), December 1934. (Courtesy AIAA.)

cards. "Practically every airplane or aircraft engine will be found indexed by country, manufacturer and type," he explained. "Such subjects as accidents, aerology, aerobatics, airports, airships, armaments, aviation, balloons, gliders, maps, as well as hundreds of others, will all have their special files." There would be files within files. "Separate headings under the subdivision airplane parts, for example, include fuselage, landing gear, propellers, cockpit, wings, ailerons, slots, wheels, floats, pontoons, dope, chairs, interior furnishings, and cowling."[16]

The founding members of the IAS chipped in to support the project. Sherman Fairchild donated filing cabinets, each of which could hold 80,000 index cards. Luis de Florez, a wealthy Massachusetts Institute of Technology (MIT) alumnus, donated $250, enabling the IAS library to acquire the annual volumes of the *Engineering Index*, 1926–1935. This series indexed publications in all fields of engineering. Each volume ran to 1,200 pages and included 40,000 entries drawn from 2,000 publications.[17] Harry F. Guggenheim provided the library with a complete set of *Les Fisches Aeronautiques*, a bimonthly classified digest and index of aeronautical materials published by the French Centre de Documentation Aéronautique Internationale. The French project was supported by a grant from the Daniel Guggenheim Foundation for the Promotion of Aeronautics.

The press was enthusiastic. Gardner convinced his friend Lowell Thomas to feature the Aeronautical Index project on his radio program.[18] Reginald Cleveland, the *New York Times* aviation specialist, was quick to congratulate the Institute. "Until this work was begun," he noted, "it was practically impossible to locate aeronautical information without a long search of many books, magazines, and newspapers. When this index is complete, it will be possible to find a picture of almost anything that has happened in aviation, or of any aeronautical product."[19]

The number of completed cards had reached 15,000 by May 1935. Gardner predicted that the number would double over the next month, "... so that the index will take its place as one of the largest and most complete indexes of aeronautical literature in the world." His words were prophetic. The Aeronautical Index soon took over the original office space in Rockefeller Center, necessitating the first move to the new quarters on the 51st floor.

By July 1935, the index employed 25 people who had completed 90,000 cards. The goal was to reach 200,000 by the end of the year. That month the Relief Bureau released a complex "Organization and Work Chart" for what was officially known as Project 89-Pb-1592-X. The tasks outlined included typing, clipping, filing, and indexing. In addition to the collection of bibliographic information on books, articles, pictures, and biographies, the Index project was also producing a chronology of aeronautics, "which will be comprehensive and cover all the principal activities in aeronautics from legendary times." IAS members were asked to furnish "chronologies of their specialties," which would go into the compilation of a master document.

The Federal Works Progress Administration (WPA) took over direction of the Index and most similar locally supported relief projects in the fall of 1935. The New Dealers allocated $34,650 "... to be expended in ... expanding the work already underway." There were now 120,000 cards on file, with 5,000 more being generated each week. Gardner reported that the Index had reached a total of more than 400,000 cards by January 1937. In the end, the project produced some two million cards, employed more than 100 people, and cost $150,000.

Although the work of keeping the Index updated would continue on a smaller scale, the basic project was approaching completion after three years of effort. The first specialized bibliographies covering Propellers, Stratospheric Flight, Rocket Propulsion, Women in Aviation, and Airports were ready for distribution in the fall of 1937.[20]

The complete Aeronautical Index was issued in 1938. It consisted of 28 volumes made up of 50 separate bibliographies covering the entire range of aviation topics. More than 300 complete sets of the mimeographed bibliography, some 16,000 volumes, were distributed to universities, research facilities, and libraries across the nation. In addition, the Institute, in cooperation with various state and local WPA offices, also produced a series of important free-standing bibliographies of aeronautical articles in key 19th and early 20th century publications, including the *Scientific American* and *Harper's Weekly*.

The secretary had worked hard to interest IAS members in his pet project. In the end, he seems to have recognized that it was a losing battle. "I fear few of our members appreciate the magnitude of this work or its great importance," he complained as the project was winding down in February 1939. That did nothing to reduce Gardner's determination to establish the Institute's reputation as the source of aeronautical information, however.[21]

The Kollsman Library and the Pacific Aeronautical Library

Gardner was pleased to announce yet another great library coup at the annual summer meeting of the IAS in Pasadena in August 1940. IAS founding member Paul Kollsman had recently sold his aircraft instrument firm to the Square D Company of Detroit, Michigan. Now F. W. Mangin, president of Square D, joined Kollsman in providing the IAS with a $50,000 gift, thought to be the largest presented to a scientific or technical society up to that time. The money would go to found a Kollsman lending library, a considerable expansion of the existing operation.

With librarian Robert R. Dexter as director, the Kollsman Library initially offered a collection of 2,000 books on aviation for mail order loan, with another 10,000 volumes in the IAS reference library. In addition, Gardner had reached an agreement with Dr. Harrison W. Carver, director of the Engineering Society Library, that gave IAS members access to a mail order lending library of 100,000 volumes on all aspects of engineering.[22,23]

It was only the beginning. The Kollsman Library would continue to grow over the years. By the fall of 1941, donations from members including Richard Fairey, Ed Aldrin, Grover Loening, Ed Warner, and others had raised the collection of books available for loan to more than 5,000 volumes.

As use of the Kollsman Library increased during the first year of operation, it became apparent that the West Coast, with its concentration of aircraft companies, had special information needs. In order to reduce the time taken to ship materials and to provide a convenient local center that engineers and other researchers could visit in person, the IAS established the 12,000-volume Pacific Aeronautical Library at 6715 Hollywood Boulevard, Los Angeles, in October 1941. The Paul Kollsman Fund Committee, which administered the original gift, contributed $10,000 to the project. West Coast aviation companies also supported the project. Cal Tech's Ernest W. Robischon, who would remain a familiar name to IAS members for many years to come, was named librarian.[24]

In order to extend the reach of its bicoastal and lending library system to members and general researchers, the IAS launched a new publication, the *Aeronautical Reader's Guide*, in September 1939. Issued quarterly ("... priced at $1 for the four issues in each volume"), it offered notices and abstracts of new books and articles in the field, all categorized for easy reference.[25] "Probably no single industry today is so plentifully supplied with printed works of all kinds as aviation," Lester Gardner, now editor of the new *Aeronautical Review*, noted. "That is not only a sign of public interest, but a creator of it. Day by day, the printing press helps develop the greater industry of tomorrow." The *Aeronautical Reader's Guide* was designed to help engineers remain abreast of that literature and of the latest developments in their field.[26]

Establishing a Past for the Technology of Tomorrow

By 1940, Lester Gardner had established the IAS as the most authoritative source of information on American aviation. He was determined to go beyond that to create the national's finest archive of materials related to the history of aeronautics—a collection of papers and objects that would celebrate the history, legend, lore, and traditions of flight.

Few achievements can match the impact of flight on the human imagination. Evidence of the age-old dream of taking to the air is to be found in world myth, religion, and literature. Flight symbolized those capabilities to which we most aspired—the capacity to soar over obstacles in our path, the ability to exercise control over our destiny, the opportunity to taste the absolute freedom of the skies. From the beginning, we placed our gods in the sky and made flight, the one gift we had been denied, an attribute of divinity. "The natural function of a wing," Plato explained, "is to carry that which is heavy up to the place where dwells the race of gods."

For millennia, flight was regarded as the very definition of the impossible. "If God had wanted us to fly," it was said, "he would have given us wings." When human beings finally did take to the sky on wings of their own design, the excitement was almost overwhelming. It happened twice. A wave of enthusiasm swept over Europe and America in 1783, when hot air and gas balloons carried the first human beings aloft. That excitement was reflected in popular culture. Beverages and dances commemorated the first balloon flights. Clothing and hat styles were inspired by the colorful craft rising above the rooftops of Paris and London. Balloon motifs decorated items of furniture, jewelry, ceramics, boxes, wallpaper, fans, upholstery fabrics, and dozens of other items.

When Wilbur and Orville Wright made the world's first powered, sustained, and controlled flights in a heavier-than-air machine just 120 years later, most people found it difficult to believe that the age-old dream of winged flight had been achieved. Artist and aviation enthusiast Gutzon Borglum described the reaction of the spectators gathered at Fort Myer, Virginia, in the fall of 1908 to watch Orville Wright's first public flights. "The crowd stood open mouthed," he noted, "with murmurs of wonder and an occasional toot from [an] automobile horn; then as he passed over us everybody let go in an uproar of shouting and handclapping. The miracle had happened! .... Nothing seemed impossible."[27]

Lester Gardner was sure that the collection, preservation, and display of records, objects, and images able to communicate the sense of excitement and bright promise that was so much a part of the tradition of flight was a worthy goal for the IAS. He was the latest in a long line of collectors, stretching back to the late 18th century, who had delighted in searching out items connected to the great events of the air age and the people who had written their names large in the history of flight. Before he was done, a great many of those collections would find a home at the IAS.

John Cuthbert, an English aeronautical enthusiast, was one of Gardner's earliest predecessors. He began his collection in 1819 when he acquired samples of fabric from several famous balloons. Over the next few decades, he filled a series of scrapbooks with a wealth of newspaper clippings, pamphlets, prints, and manuscripts on aeronautical topics. He sought to assist his heroes as well, heading a committee created in 1837 to assist the widow of Robert Cocking, an aeronautical showman who died in a parachuting accident. J. Fillenham, a friend of both Cocking and Cuthbert, was another English collector of note, specializing in aeronautical autographs, posters, and handbills.[28]

By the late 1870s, Gaston Tissandier (1843–1899) had moved beyond the antiquarian impulse of men like Cuthbert and Fillenham to build the world's first great collection of aeronautica and to lay a solid foundation of bibliographic and historical scholarship in the field. A chemist trained at the Lycee Bonaparte and the Conservatoire des Arts et Metiers, Tissandier was appointed director of the Laboratoire de Chimie de Union Nationale at the age of 21. Fascinated by the prospect of exploring the upper atmosphere with balloons, he completed more than 50 major ascents during the course of an active aeronautical career that stretched from 1866 to 1886 and earned a dual reputation as one of the great balloonists of the age and a major contributor to the technology of buoyant flight.

Having first come to public attention with a series of flights in the world's largest free balloon, Tissandier emerged as one of the aeronautical heroes of the Franco-Prussian War in 1870, earning the Legion d'Honeur for his service as a balloonist during the Siege of Paris and with the Armèe de Loire. On August 17, 1875, he piloted the balloon *Zenith* to an altitude of more than 26,240 feet and was the only member of the three-man crew to survive the failure of an experimental oxygen apparatus in the bitter cold of the substratosphere. In the fall of 1883 Gaston and his architect/brother Albert Tissandier cruised back and forth above the Paris skyline for an hour at a time in their electrically powered airship. The craft, with an average speed of 3 mph, was scarcely practical, but it did mark Tissandier as a leader in aircraft design.

The founder/editor of the journal *La Nature* and the prolific author of books and articles on a wide range of scientific and technical subjects, Tissandier enjoyed a reputation for meticulous research that matched his record of fearlessness in the air. By the late 1870s, with the assistance of the world's leading rare book dealers, he began to build an extraordinary collection of items relating to the history of flight. In the absence of trustworthy secondary sources of information on the subject, Tissandier built a solid foundation of original materials on which to base his own account of the ancient dream of flight and its realization—from the myths and legends of the distant past, through the invention and early history of the balloon, down to the events of his own time.

Hugh Latimer Dryden, Wright Brothers Lecturer, 1938, ARS President for 1943. (NASM Archive, Smithsonian Institution, 78-3945.)

It would be difficult to overstate the importance of Tissandier's collecting activity. The titles contained in his own library provided the starting point for his classic *Bibliography aeronautique: catalogue de livres d'histoire, de science, de voyage et de fantasie, traitant de la navigation aerienne ou des aerostats* (1887), the first useful guide to the literature of flight. In addition, Tissandier collected a wide variety of engravings, news clippings, coins, and objects related to the history of aeronautics. The manuscript treasures of his collection range from the letter in which Benjamin Franklin informed Sir Joseph Banks, President of the Royal Society, that human beings had flown for the first time to dozens of letters and other documents tracing the career of the pioneer aeronaut Jean Pierre Blanchard.

The collection eventually filled the walls, bookcases, and closets of Tissandier's fashionable apartment on the Rue de Chateaudun. "In the drawing room," noted a visiting English reporter, "are to be remarked a series of drawings, representing various episodes of the terrible ascension of 1875, which nearly cost Mr. Tissandier his life." Tissandier's cabinet of aeronautical relics and curiosities contained some 3,000 objects. "In huge portfolios stored in the dining room are hundreds of engravings, colored pictures, posters, and handbills of the period of the discovery [of the balloon], among which are many most curious and interesting in character."[29]

Tissandier's son Paul inherited both the collection and the desire to build on and improve what his father had begun. Wilbur Wright, who taught the younger Tissandier to fly, described him as a "splendid fellow ... in every respect and very trustworthy." He was, as Wilbur explained to his father, the "son of Gaston Tissandier, the most celebrated of French balloonists," and possessed of "a magnificent apartment and a wonderful collection of aeronautical books and curiosities."[30]

The rapid rise in the number of collectors and exhibitions of engravings, printed aeronautica, and aeronautical antiques during the early years of the 20th century served to educate and inspire a public already fascinated by the arrival of the Air Age. A special Musèe Retrospectif Aerostation, prepared for the Paris Exposition Universelle Internationale de 1900, placed contemporary European interest in airships within an historical context and set the stage for the great exhibitions of aeronautica that would follow over the next half century.[31] The historical material included as part of two large Aero Shows held in New York City in 1906 and the impressive exhibit of 18th century balloon materials offered at Frankfurt by the Internationalen Luftschiffahrt-Ausstellung in 1909 were typical of these early attempts to provide a past for the technology of the future.[32]

Patriotism, coupled with the extraordinary attention focused on aviation during the world war, further underscored interest in the history of flight and the collection and public display of aeronautica. Kathleen Burn Moore, Countess of Drogheda, drew on the riches of her own remarkable collection and those of her friends to mount such an exhibition of "paintings and prints of the earliest and latest types of aircraft" at London's Grosvenor Gallery in 1917. H. G. Wells wrote the introduction for the catalog of the exhibition, the proceeds of which were donated to the Flying Services Fund of the RAeS and the Irish Hospital Supply Depots for the British Red Cross.

Wells was especially intrigued by the lessons in the history of flight technology embodied in the exhibition. Although the focus was on the balloons and airships of the 18th and 19th century, the airplane was clearly the vehicle of the future. He remarked:

> But there can be no doubt in the minds of those who are fairly well informed in these matters that the picture of a defeated and wrecked Zeppelin, with which this series is rounded off, does also round off and close the first great chapter in the story of man's adventure in the air.
> When, a hundred years hence, the Lady Drogheda of that day opens her revival of this show—I hope for some quite other cause than the Red Cross—I doubt if there will be very much to add to the balloon and airship series. It will be pictures of multiplanes, helicoptres, and every sort of great aeroplane, that will make the bulk of the matter added to what we have here to-day.[33]

The Golden Age of aeronautical collecting and exhibiting that had begun in the decade prior to the world war gathered momentum and reached a climax during the quarter of a century following the Armistice. The heroic figures of the first air war gave way to a new breed of men and women who earned fame and extraordinary public acclaim by constantly pushing the limits of the possible—crossing oceans; linking continents; and flying higher, faster, and farther.

Lieutenant Colonel C. Lockwood Marsh and J. E. Hodgson, private collectors and aeronautical historians of note, played a major role in building and publicizing one of the world's great institutional collections of aeronautica at the RAeS.[34,35] Interest in the history of flight was deeply rooted in the RAeS. Christopher Hatton Turnor (1840–1914), a founding member of RAeS Council (1866), was the collector/author of *Astra Castra*, one of the earliest treatments of the history of flight in English.

Hodgson, who served as honorary librarian of the RAeS following World War I, was responsible for planning an extraordinary display of historical treasures for the International Aero Exhibition in Olympia on July 16–27, 1929. Charles Dollfus, the greatest of the French historians of flight and founder of the Musèe de l'Air, cooperated with Hodgson in gathering material for this most impressive of all the interwar exhibition of the historic relics of flight. "Never before has there been gathered together such a remarkable selection of historical material giving so complete a record of man's attempts to fly in Great Britain from the earliest legendary days to 1914," the editor of the *Journal of the Royal Aeronautical Society* reported with pride. Rudyard Kipling was among those who congratulated Hodgkins on the success of a project so "vital" to stimulating the public imagination.[36]

Lester Gardner had studied the historical and archival activities of the RAeS with great interest. By 1939, the IAS was firmly established as the representative professional organization for aeronautical engineers and scientists in America. Anxious to build on the success of the Aeronautical Index, he now dreamed of creating one of the world's finest aeronautical archives/research centers/museums at the IAS, a place where the material and graphic heritage of aeronautics could be preserved and maintained as a resource for the use of aeronautical professionals and the public alike. It followed the pattern established by the RAeS and fit Gardner's vision of the IAS as an elite club for gentlemen engineers.

A series of 1940 photos shows the IAS office suite in Rockefeller Center looking more like a research library or museum than the work area of a well-managed professional society. There are bookcases everywhere. Every wall is covered with photos, historic prints, maps, and paintings. Model airplanes and other aeronautical memorabilia crowd table tops and desks. The stream of gifts flowing into headquarters was becoming a flood. A few paragraphs expressing gratitude for recent "Gifts to the Institution" was now a regular feature of the "Institution Notes" column of the *Journal*.

The acknowledgments offered in the April 1938 issue are typical. E. E. Aldrin presented three bookcases filled with books. Ernest Dichman donated "several hundred valuable pamphlets and reports." A. J. Hamon provided a set of Francis Trvelyan Miller's new compilation *The World in the Air* and a copy of George Johnson's *Peru From the Air*. Dr. Lyman J. Briggs, director of the National Bureau of Standards, gave Gardner a sealed glass tube containing air from the stratosphere obtained by the crew of the U.S. Army Air Corps/National Geographic Society balloon *Explorer II* when they reached their record altitude of 72,395 feet over South Dakota on November 11, 1935.

Then there were gifts commemorating the achievements of corporate members. T. P. Wright, then president of the IAS and a Curtiss–Wright executive, donated a rib from the Curtiss NC-4, the first aircraft to fly the Atlantic. Glenn L. Martin presented a large photomural of one of his seaplanes, and G. E. Woods Humphrey, managing director of Britain's Imperial Airways, offered an enormous metal scale model of the two-aircraft Short–Mayo composite with which he hoped to bridge the Atlantic. Not to be outdone, United Airlines provided a model of the DC-4E in company livery. Pan American officials offered something a bit more exotic, a decorative stand supporting a glass ball used as a float by Japanese fishermen, which had been found by company personnel based on Midway Island.

Gardner unveiled a real prize in July 1938. Late in 1937 Orville Wright had presented the IAS with an irregularly shaped (3.5 × 6 inch) piece of original fabric and a splinter of wood from the airframe of the world's first airplane. The secretary had forged a warm friendship with the inventor of the airplane since 1932, when Mr. Wright agreed to accept the first Honorary IAS Fellowship. "The more I realize what an important relic you sent the Institute," Gardner responded in January 1938, "the more strongly do I feel that we should preserve it with the utmost care, as it is probably the only piece of the original plane which any society has." He informed his benefactor that he planned to "... have a flat aluminum case made to hold it so that it can be bolted to the wall and kept behind a locked door."[37]

Gardner published a photo of the finished display in the July issue of the *Journal*. "The relic was such a valuable addition to the collection at the Institute," he announced to the membership, "that Glenn L. Martin agreed to make an aluminum case to protect it." That case, he concluded, "is probably the finest example of aluminum workmanship ever made." The aluminum "case" was mounted on a wooden board. A key unlocked a door that opened to reveal the precious objects within. Mounted in a place of honor on the Institute wall, Gardner's prize acquisition was nothing more or less than an Air Age reliquary.[38]

The serious collecting effort began in the spring of 1939, when Gardner traveled to Europe with Dr. George W. Lewis, then IAS president and chief of research for the NACA. Lewis gave the annual Wright Brothers Lecture to the members of the RAeS on May 23, 1939, after which the pair set off on a Grand Tour of aeronautical Europe. In Paris, the members of the Aero Club de France hosted a special reception in their honor. Lewis went on to Berlin and Gardner moved on to Turin, where he attended an international gathering of aeronautical editors. True to form, Gardner then flew on to Athens, Amsterdam, Berlin, and Oslo. Everywhere he went, he was treated to dinner and offered a chance to renew his acquaintance with the movers and shakers of European aviation.

For Gardner, a conversation with Hart O. Berg in Paris had been the most important moment of the trip. A native of Philadelphia and educated in Belgium, Berg had been the manager of European operations for Charles Flint and Company, an investment firm that had sold an entire fleet of ships to Brazil, sold submarines to Russia, and marketed the Wright airplane on the continent during the years 1907 to 1912. He had been a great favorite and friend of the Wright brothers, who, in turn, offered Berg's wife the honor of being one of the first women to fly as an airplane passenger.

Berg asked Gardner and Lewis for advice as to the disposition of his large collection of letters, business documents, photos, and clippings relating to the Wright brothers. Gardner jumped at the chance to acquire the material for the IAS. There were a few issues to be resolved, however. Apparently, Berg had previously offered the collection to Mrs. Barton K. Yount, wife of an Air Corps Brigadier General who served as Assistant Military Attaché for Air in Paris and Madrid in the early 1930s. When Mrs. Yount agreed to step aside, Berg asked his old friend Luis de Florez, a founding member of the IAS, to arrange the papers before presentation to Gardner. The IAS Council accepted the gift on July 20, 1939, expressed their thanks to Berg, and voted him in as a Benefactor.[39]

The Aeronautical Archive

The acquisition of the Berg Papers set the stage, attracting other donations to the IAS. The same month that he announced the Berg gift, Gardner also thanked New Yorker Henry Ramsey for turning over his collection of 25 years worth of clippings about aviation. The following month, Jimmy Doolittle offered up two boxes of old clippings, including some early notices of the Wright brothers. The decision of William Armistead Moale Burden (IAS President, 1949) to loan his collection of 10,000 aeronautical books, magazines, photographs, clippings, and reports to the IAS, however, was the real turning point.

A decade before, while still a Harvard undergraduate, Burden had been so fascinated by aeronautics and so intrigued by the business opportunities available in the aviation industry that he subscribed to both *The Journal of the Royal Aeronautical Society* and *Aviation*. Following graduation in 1928, he rode the post-Lindbergh boom to prosperity as a financial analyst specializing in aviation securities. Determined to understand the fundamentals of the industry to which he had tied his future, Burden joined the IAS and continued to build his collection as an information resource. "The books, magazines, and press clippings overflowed my tiny office downtown and I moved the file home," he recalled. "I was one of the very few in finance who was making such a thorough collection."[40]

Over the next decade, Burden's "thorough collection" continued to grow larger, richer, and deeper. In 1939, Lester Gardner congratulated Burden on having created "the finest private aeronautical library in the world and probably second only in its scope to the collections of the Library of Congress."[41] In fairness to other private collectors and to such great aeronautical libraries as that of the RAeS, however, it should be noted that Gardner was not a disinterested commentator. In the late summer of 1939, having acquired the Hart Berg papers, Gardner convinced William Burden to loan his priceless treasures to the IAS.

William Armistead Moale Burden, the "thorough collector," benefactor and IAS President for 1949. (Courtesy AIAA.)

Meeting on September 29, 1939, the IAS Council agreed to the conditions of the Burden loan, which were spelled out in an eight-page legal agreement signed on January 29, 1940, and sealed by the exchange of dollar bills.[42] The initial loan was to run for five years, with the possibility of renewal upon mutual agreement. Gardner agreed to store the material under specified conditions and to hire Burden's archivist/librarian, Miss Gwendolyn Lloyd, to manage the collection at a salary to be mutually determined. Burden agreed that the IAS could assign the librarian additional related duties; Gardner agreed that Burden would have approval of any new librarian.[43]

"By this generous action," the secretary remarked, "the Institute became a center for aeronautical reference and research equal to any to be found n the United States."[44] More than that, the Institute was launching an entirely new venture. Meeting on September 29, 1939, the IAS Council established a new organization, the Aeronautical Archive, to house, care for, and service their growing collection of historical treasures. "The Aeronautical Archive will be directly under the financial supervision and ownership of the Institute," Gardner explained, "thus enabling patrons or estates to deduct the value of gifts from taxes and also to permit aeronautical corporations to charge a large part of their contributions to manufacturing costs under government contracts."[45]

The Archive would be "conveniently located" in offices adjoining the IAS headquarters area at 1505 Rockefeller Center. It should have come as no surprise to find that W.A.M. Burden would chair a 23-man board of directors that included some of the most distinguished names in American aviation and high finance.[46] Gardner was president of the new organization, with Elmer Sperry, Jr., serving as secretary. Initially, the Aeronautical Archives contained 18 divisions: the Burden library, the IAS library, magazines, reports, pamphlets, photographs, etchings and engravings, biographical files, bibliographical indexes, chronologies, air mail stamps and covers, index of indexes, airplane models, medallions and trophies, historical documents, aviation relics, moving picture films, and lantern slides.

The Archive would offer a variety of services to IAS members or to researchers introduced by members. A circulating library of several thousand books was available to members who were unable to visit New York in person. Through the Technical Information Service (TIS), "experienced personnel" would undertake "investigations requiring lengthy searching" at "the customary fee" of $2 per hour. Institute employees would provide photostatic copies of any item in the collection. The official photographer of the Archive would provide copies of any photographs desired by members at a standard fee. Microfilms, drawings, and tracings could be prepared. Translations, "carefully edited by trained engineers," would be undertaken at the rate of one cent per word.[47]

Newspapers noted that the move of Burden's collection from 33 Rector Street to its new home in Rockefeller Center began on March 1, 1940. Asked if armored cars would be required to transport the valuable materials, Burden responded that the job was in the hands of the company that had moved Columbia University's library, "without guns or anything."[48]

The already large collection soon grew larger. Even before he had shipped the bulk of his collection, Burden purchased the Phillips collection of 50,000 newspaper clippings on the history of flight. On April 1, 1940, the National Aeronautic Association (NAA) shipped its entire library on loan to the IAS. The move was necessitated by the NAA's move from Washington, DC's Dupont Circle to new and smaller quarters in the Willard Hotel. The library included very rare volumes originally collected by the Aero Club of America.

A steady flow of historic aeronautical treasure continued into Gardner's collection during the summer of 1940. In August, Gardner announced that IAS founding member Harry Frank Guggenheim was loaning his collection of aeronautical prints to the IAS. The number of Guggenheim prints in the IAS collection would eventually rise to 233. It was, quite simply, the world's finest collection of its kind in private hands.

The oldest material included prints produced in the 17th century illustrating early concepts of flying machines. The bulk of the collection consisted of prints and etchings produced from the late 18th through the 19th century. The strengths of the collection were the several score of images chronicling the decade from the invention of the balloon in 1783 to the first flight of a human being aboard a free balloon in the United States in January 1793. Guggenheim had acquired copies of some of the rarest of all balloon prints, including an image of the launch of the first Montgolfier balloon from the town square of Annonay, France, on June 4, 1783. The philanthropist had also acquired important manuscript materials, including original tickets to the launch of the world's first balloons and a note from President George Washington declining an invitation to attend a lecture on ballooning.

Gardner reeled in the third of his great donors (or "loaners") in the spring of 1942. Bella Clara Landauer (1885–1960), the widow of textile manufacturer I. Nathan Landauer, would stand close to the top of any informed list of great American collectors of the 20th century. The fever to acquire came to her fairly late in life. At the age of 38, on the advice of a physician that she find a hobby, Bella Landauer purchased a scrapbook filled with bookplates as a favor to an impoverished young man. Before she was finished, Mrs. Landauer would amass 50,000 book plates Over the next 30 years she collected trade cards, U.S. lottery tickets, cigar bands, wine labels, railroad passes, sheet music, and a host of other paper ephemera.[49]

Fascinated by the amount of social and cultural information contained in advertising items, she specialized in the preservation of materials that were produced to be discarded. She was best known for a collection of almost half a million items of advertising ephemera preserved at the New York Historical Society in 1926. She would serve as honorary curator of that institution for the next 30 years.

That gift was only the beginning of her largesse. "I have gathered together what must be hundreds of thousands of items purely for the love of this hobby," she once told a reporter, "never to possess them for my own use. When I find some place where they naturally belong, I give them away."[49] In addition to her gift to the New York Historical Society, she presented

24 volumes of printer's marks to the New York Public Library. Her collection of engravings illustrating various occupations went to the Baker Library at Dartmouth, her son James's alma mater, as did a series of scrapbooks, programs, posters, tickets, letters, and books documenting the career of her favorite playwright, Eugene O'Neill.[50–53]

Bella Landauer originally intended to build an aeronautical collection as a gift for her son William, a U.S. Marine who was fascinated by aviation. Already a sophisticated collector of paper objects, however, she was immediately captivated by the richness of the topical material on aviation. Her collection grew so rapidly that it could support a large and important loan exhibition of aeronautical books and prints at the New York Public Library in the fall of 1934.[54]

Mrs. Landauer collected the entire range of paper aeronautica, from 18th- and 19th-century engravings and prints to aviation trade cards, calendars, Christmas cards, and ephemera of all sorts. In some areas of specialization, she simply extended her existing collecting interests into the field of aeronautics. A long-time collector of song sheets about New York City, for example, she gathered the first of what would eventually become some 1,300 pieces of sheet music on aviation. Having begun her collecting activity with book plates, she developed a small collection of aeronautical "ex libris" and published a privately printed volume on the subject.[55,56]

Perhaps Bella Landauer's most important contribution to the aeronautical collecting tradition, however, is to be found in her pioneering focus on the theme of flight in American children's literature. Her enthusiasm was based on recognition of the impact that printed accounts and images of flight had exercised on the imagination of generations of youngsters as well as an appreciation for what the material said about attitudes toward aviation long before the invention of the airplane. Later collectors owed a considerable debt to Bella Landauer for establishing juvenile aeronautica as an important collecting area.

Gardner's Aeronautical Archive was like a snowball rolling downhill, attracting more and more material along the way. The leading donors continued to add new material to the collection. Burden, for example, purchased the vast archive amassed by the Tissandier family. Bella Landauer added 572 "air transport labels," to her existing donation.

Aviation pioneers were anxious to donate their own collections. Albert Francis Zahm, who had accompanied Jerry Hunsacker to Europe in 1913, presented the letters that he had exchanged with Octave Chanute during the planning of an aeronautical conference at the World's Columbian Exposition in 1893. Aeronaut A. Leo Stevens contributed a priceless collection of papers, clippings, and photos documenting a long and colorful aeronautical career that had begun before the turn of the century.

The descendants of Thaddeus Constantine Lowe, who had headed the Union observation balloon corps during the Civil War, turned over a treasure trove of papers documenting the first use of aeronautics by the U.S. Army. S. Paul Johnston, editor of *Aviation*, the journal that Gardner had founded so many years before and the man who would eventually replace him as executive secretary of the IAS, negotiated the gift of the papers of the Anglo-American flying machine pioneer William Samuel Henson to the Aeronautical Archive.[57]

Not content to wait for new acquisitions to arrive, Gardner combed other libraries for aeronautical rarities, which he photocopied and added to the IAS collection. He dispatched WPA workers to the New York Public Library with instructions to scour 18-th and 19th-century American newspapers in search of articles on balloon flights and other aeronautical doings, which could be copied to expand the archive.

The Aeronautical Archive of the IAS collection would grow to 23,000 items including, Gardner claimed, a copy or a photocopy of every book and pamphlet on aeronautics published

Glenn Luther Martin, IAS President for 1936, presents Lester Gardner with the first installment of Martin Company stock that will establish the Minta Martin Fund. (Courtesy AIAA.)

in America prior to 1900. Just to be sure, he commissioned N. H. Randers-Pherson and A. G. Renstrom of the Library of Congress Division of Aeronautics to prepare a special bibliography identifying all books and pamphlets on the subject published in the United States before 1900. Gardner published the resulting list under the title *Aeronautic Americana* in 1943, using funds from the Sherman Fairchild Fund.[58]

By the end of 1941, Lester Durand Gardner had realized his dream. The IAS was not only the leading American organization representing the interests of aeronautical engineers and scientists, it was one of the world's premier sources of information on the history of flight.

Generous Friends

Financial gifts continued to flow into the IAS coffers during the early war years. In January 1942, Gardner announced that Paul Kollsman and the Square D Company had raised their total donation to a $65,000 endowment, the interest on which would support the continued operation of the lending libraries. At the same time, Institute officials accepted a donation of $25,000 from Sherman Fairchild to support the limited publication of papers, reports, and translations of interest to "select engineers during and after the emergency."[59]

The events of June 1942, however, were to put all previous benefactions to the IAS in the shade. On June 17, Executive Vice President Gardner announced that Florence Guggenheim, widow of the late philanthropist Daniel Guggenheim, had presented her 162-acre estate at Sands Point, on the North Shore of Long Island, to the IAS. Hempstead House, a 40-room mansion, was the central feature of the property, which stretched for two-thirds of a mile along Long Island Sound just north of Port Washington. Inspired by Ireland's Kilkenney Castle, it was designed and built for Howard Gould, the son of financier Jay Gould, between 1901 and 1912. Daniel Guggenheim had acquired the property in 1917 and renamed the mansion Hempstead House.

The estate had strong links to the history of aviation. It was here that Daniel Guggenheim and conceived his famous Fund for the Promotion of Aeronautics, which had exercised such a profound impact on aviation in the 1920s. Moreover, the land being given to the IAS was part of a much larger estate that included Falaise, the home of Daniel and Florence Guggenheim's son Harry, an IAS founding member and benefactor. Lindbergh had stayed at Falaise following his return from Paris in 1927. He had written his first book, *We*, while a guest here. He and his wife Ann Morrow Lindbergh spent considerable time with the Guggenheims over the next two decades. As a metal plaque still mounted on the fireplace at Falaise attests, it was here, in 1929, that Carol Guggenheim called an article on rocket pioneer Robert Goddard to Lindbergh's attention, an event that led directly to Guggenheim support for Goddard's research.

Although Mrs. Guggenheim admitted that she was "no specialist in the art or science of aeronautics," that subject had been a familiar topic of discussion at her dinner table for many years. As a result of conversations with her husband and son, she expressed "great interest and faith" in the work of the IAS and "even larger hopes for its future . . . ." She expressed a hope that the Institute would make use of the site for its "scientific projects."[60]

As a result of discussions regarding Hempstead House, Glenn L. Martin offered to create a $450,000 endowment ". . . for use in conjunction with the Guggenheim gift." The fund, to be created through gifts of Martin Company stock over a five-year period, would be known as the Minta Martin Aeronautical Endowment Fund. "I hope that from the large laboratory which the council of the institute has named the Minta Martin Aeronautical Laboratory," the donor remarked, "developments will come which will make flying in the substratosphere more comfortable and safer for our combat pilots."[61]

Gardner agreed, explaining to reporters that the gift would allow the Institute to make "a direct contribution to the war effort by providing exceptional aeronautical facilities for experimental aeronautical investigations by specialists." After the war, he was quick to add, "the Institute plans to advance the art and science of aeronautics by using the estate to enlarge the scope of its aeronautical archives so that its large libraries, collections of prints, photographs and other aeronautical material will be available to the aeronautical industry and others for research . . . ."[60]

There was no desire to transform the IAS into a research outfit in competition with the NACA or universities. There was, however, a discussion of allowing Columbia University to establish a research site at Hempstead House. That was never a serious possibility, however. In March 1943 the *New York Times* reported that ". . . the Institute must postpone its plans for the establishment of an aeronautical laboratory until the specialists in this field are relieved of pressing duties with the Army Air Forces . . ." For the time being, the reporter noted, the IAS ". . . proposes to establish on this same property a museum and library which will be a branch of the endowed Paul Kollsman Library, now loaning aeronautical books by mail to thousands of persons in all parts of the nation."[62]

Indeed, the Paul Kollsman Library was moved to Sands Point during the war years, along with the bulk of the Aeronautical Archives. Rather than supporting cutting-edge aeronautical research, the $325,426. 90 of the Minta Martin Fund received to date was invested in U.S. Government stocks and bonds, the interest of which would help to build the restricted endowment funds of the Institute.[63] The Guggenheim and Martin gifts, Gardner reported with justifiable pride, established the Institute as ". . . one of the best endowed scientific societies in the country."[64]

The War Years

"To a large extent, Gardner explained to readers of the *Aeronautical Review* in February 1942, the aeronautical press was afflicted with 'information anemia.'" In the wake of Pearl Harbor, organizations like the IAS and publishers like Gardner were conscious of the need to pay special attention to security. "During the period required for clarification and codification of regulations on publishing information, the publishers have . . . endured restrictions and complied with orders with a minimum of grumbling." The inability to print the latest news from the far-flung world of aviation, Gardner noted, ". . . hurts the editors and publishers [more] than it hurts the readers."[65]

In fact, wartime restrictions on paper would have a far greater impact on the IAS than the need to be responsible in reporting news of the aviation industry. When the War Production Board imposed initial cuts on the availability of paper in March 1943, the Institute reduced both the weight of the paper and the trim size of the *Journal of the Aeronautical Sciences* and *Aeronautical Engineering Review*.

The IAS issued its *Aeronautical Engineering Catalog*, as planned, in 1944, but in a very limited printing. Members who were in need of a copy of the 500-page publication, which listed contact information for the manufacturers of 2,100 materials, parts, and accessories, had to request the volume in writing. When additional restrictions were imposed beginning in January 1944, the Institute was forced to discontinue offering subscriptions to individuals and institutions who were not IAS members.[66]

As might be expected during an era of expansion for aviation and the aeronautical industry, however, the IAS prospered during the war years. Nowhere was this more apparent than in the jump in corporate membership. In June, 1944, the *Aeronautical Engineering Review* acknowledged the support of 203 corporations and divisions. If the columns of IAS publication are any indication, the Student Branches remained very active in spite of the number of men who had put their college careers on hold for the duration.

The Institute maintained the normal schedule of six meetings. The period July 1944 to April 1945 was typical. The annual summer meeting was held, as usual, in Los Angeles. The air transport meeting, which was also emerging as an annual event, was staged in Washington, DC, also the traditional location of December's Wright Brothers Lecture. In October 1944, IAS members were invited to Washington for a tour of the Navy Special Devices Section and both the Anacostia and Pautuxent River Naval Air Stations. The following month, a fall meeting in Dayton was combined with a special tour of Wright Field, the U.S. Army Air Forces research and development center. The Annual Meeting in New York in January 1945 was followed by an April, "National Light Airplane Meeting," in Detroit.

The IAS also arranged a number of meetings in cooperation with other organizations in 1944, including the American Meteorological Society, the Engineering Society of Detroit, the Engineers Clubs of both Dayton and Cincinnati, and the Aeromedical Association of the United States. Local Sections and Branches had held joint meetings with their colleagues in the ASME, the American Institute of Electrical Engineers, and the SAE.

Each month brought several columns in the *Aeronautical Engineering Review* announcing still more, "Gifts to the Aeronautical Archives." Badges, insignia, buttons, and other items of war memorabilia joined the continued flow of model aircraft from proud designers and manufacturers. In addition to the displays of aeronautica overflowing the IAS/Aeronautical Archives areas of Rockefeller Center, Gardner filled available display space around New York with a series of loan exhibitions.

As early as the summer of 1939, he created an exhibition of some 50 model airplanes for the New York Museum of Science and Industry. The display ranged from a model of the original Wright Flyer of 1903 through the latest transport aircraft to warplanes like the Bristol Blenheim, Fairey Battle, Hawker Hurricane, Potez 63, and Savoia-Marchetti SM-79 that would soon be battling one another in the skies over Europe.

Gardner mounted an even larger exhibition in the lobby of New York's University Club in March 1942. He filled 16 cases with trophies, models, and other items and covered the walls with prints, paintings, photos, bookplates, medals, and song sheets. The proud curator shared his creation with four photos in the *Aeronautical Engineering Review*.

That spring, the Aeronautical Archive mounted a major exhibition of children's books on flying drawn from the Bella Landauer collection. "It may be asked," Gardner admitted, "why, in times like these, such an exhibition should be held." He argued that the "psychological well-being" of the industry required "some semblance of balance." The leaders of aviation, he suggested, had not been attracted to the field, "... by scientific works on the possibility of flight, for these were neither generally available nor intelligible, but by children's books and dime novels, which stimulated the imagination of the youth of two and three generations ago, and fanned their desire for the conquest of the air."[67]

Lester Gardner regarded these exhibitions of aeronautica as an important contribution to the war effort. "It is altogether fitting," he wrote less than a month after Pearl Harbor, "that during this period, when the industry surges forward at the greatest speed in its history, space be devoted somewhere to a link to the past to which we owe so much." In a time of trial, a little perspective was a healthy thing. It was perhaps the best explanation he could offer for his own dedication to preserving the history of flight.

References

[1] Edmund Preston, ed., *FAA Historical Chronology: Civil Aviation and the Federal Government, 1926–1996*, Federal Aviation Administration, Office of Public Affairs, Washington, DC, 1998, pp. 15–16.

[2] Charles Lawrance, "The President's Annual Report," *Journal of the Aeronautical Sciences*, March 1935, p. 78.

[3] "The Thurman H. Bane Award," *Journal of the Aeronautical Sciences*, Feb. 1944, p. 207.

[4] "Fourth Annual Meeting of the Institute of the Aeronautical Sciences," *Journal of the Aeronautical Sciences*, Vol. 3, No. 4, Feb. 1936, p. 111.

[5] "Corporate Members," *Journal of the Aeronautical Sciences*, Vol. 3, No. 9, June 1936, p. 296.

[6] "Institute of the Aeronautical Sciences," undated brochure, AIAA files, p. 8.

[7] "Proposed Code of Ethics," *Journal of the Aeronautical Sciences*, Vol. 6, No. 3, 1939, p. 118.

[8] "Fourth Annual Meeting of the Institute of the Aeronautical Sciences," p. 111.

[9] "Functions of the Institute," typed manuscript in the files of the AIAA.

[10] "Annual Summer Meeting," *Journal of the Aeronautical Sciences*, Vol. 7, No. 6, 1940, p. 259.

[11] "Air Transport Meeting," *Journal of the Aeronautical Sciences*, Vol. 6, No. 2, 1939, p. 72.

[12] Information on "Conquest of the Air" is drawn from http://www.timeout.com/film/69630.html, cited June 9, 2005.

[13] "The Institute Library," *Journal of the Aeronautical Sciences*, Vol. 3, No. 6, April 1936, p. 221.

[14] Gaston Tissandier, *Bibliographie aéronautique: catalogue de livres d'histoire, de science, de voyage et de fantasie, traitant de la navigation aérienne ou des aerostats*, H. Launette et cie, Paris, 1887.

[15] *Journal of the Aeronautical Sciences*, Vol. 2, No. 2, March 1935, p. 79.

[16] "The Institute's Aeronautical Index," *Journal of the Aeronautical Sciences*, May 1935, p. 126.

[17] "Addition to the Institute's Aeronautical Index," *Journal of the Aeronautical Sciences*, Vol. 2, No. 4, July 1935, p. 177.

[18] Western Union telegram, Lowell Thomas to Lester Gardner, Jan. 29, 1936, transcript of program, both in Gardner Scrapbooks, Library of Congress.

[19] Reginald Cleveland, "Air Index in the Making," *New York Times*, April 28, 1935.

[20] "The Aeronautical Index," *Journal of the Aeronautical Sciences*, Vol. 4, No. 12, 1937, p. 514.

[21] "Aeronautical Index and Bibliographic Files," *Journal of the Aeronautical Sciences*, Vol. 6, Feb. 1939, p. 170.

[22] "Square D Gives Library to Aid Aero Progress," *Detroit Free Press*, June 27, 1940.

[23] "Air Institute to Set Up Lending Library; 100,000 Volumes Made Available by Gift," *New York Times*, July 19, 1940.

[24] Lee Shippey "Lee Side o' LA," *Los Angeles Times*, May 14, 1942.

[25] "The Aeronautical Reader's Guide," *Journal of the Aeronautical Sciences*, Nov. 1940, p. 49.

[26] "Presses With Wings," *Aeronautical Review*, Dec. 1940, p. 3.

[27] Gutzon Borglum to Ned, Sept. 10, 1908, Gutzon Borglum Papers, Manuscript Division, Library of Congress, Box 52.

[28] *A Descriptive and Historical Catalogue of a Collection of Engravings, Drawings, Portraits, Autographs and Other Materials Illustrative of the History of Ballooning and Flying* . . . . , Messrs. Hodgson & Co., London, 1917, National Air and Space Museum Archive, hanging files. National Air and Space Museum. See also, identifying note bound in the Upcott/Cuthbert Scrapbook, Library, Ramsey Rare Book Room.

[29] "A Great Balloonist—Story of Gaston Tisssandier and His Aerial Adventures," Unidentified Newspaper Article, Box 5, Tissandier Papers, Manuscript Division, Library of Congress. Biographical details in this account are drawn from an undated manuscript autobiography in Tissandier Paper.

[30] Marvin W. McFarland, *The Papers of Wilbur and Orville Wright*, McGraw-Hill, New York, Vol. 2, 1953, p. 1022.

[31] *Musee retrospectif de la classe 34 Aerostation a l'exposition universelle internationale de 1900, a Paris, Rapport du comite d'installation*, National Air and Space Museum Archive.

[32] *Fuhrer durch die historische abteilung der internationalen luftschiffahrt-ausstellung, Frankfurt A.M., 1909*, Druck Der Kunstanstalt Wustrn & Co., Frankfurt A.M., 1909.

[33]Herbert George Wells, "Preface," in, *Kathleen Burn Moore, The Countess of Drogheda, Catalogue of Paintings and Prints of the Earliest and Latest Types of Aircraft* . . . . , The Grosvernor Gallery, London, 1917.

[34] C. L. Marsh, "In Search of Treasure: Thrills and Humours of Twenty Years Spent in Collecting an Aeronautical Library," *Airways*, June 1928, pp. 367–368.

[35] J. E. Hodgson, *A History of Aeronautics in Great Britain*, Oxford Univ. Press, Oxford, 1924.

[36] "The Society's Exhibit at the International Aero Exhibition, Olympia, July 16th–27th, 1929: A Summary by the Editor," *The Journal of the Royal Aeronautical Society*, Oct. 1929, pp. 980–1015.

[37] Lester Gardner to Orville Wright, Jan. 3, 1938, National Air and Space Museum Registrars File 5676.

[38] Lester Gardner, "Gifts to the Institute," *Journal of the Aeronautical Sciences*, July 1938, p. 286.

[39] "The Hart O. Berg Collection," *The Journal of the Aeronautical Sciences*, Aug. 1939, p. 427.

[40] W.A.M. Burden, *Peggy and I: A Life Too Busy For a Dull Moment*, New York, 1982, p. 140.

[41] Lester D. Gardner, *The Aeronautical Archive*, Institute of the Aeronautical Sciences, New York, nd, p. 1.

[42] William A. M. Burden to Institute of Aeronautical Sciences, Agreement Re: Loan of Aeronautical Library, Burden Collection files, AIAA headquarters, Jan. 29, 1940.

[43] On March 12, 1942, Burden approved the appointment of Maurice H. Smith as librarian, Burden to Gardner, Burden Collection files, AIAA headquarters, March 12, 1942.

[44] Lester D. Gardner, *The Aeronautical Archive*, Institute of the Aeronautical Sciences, New York, 1939 [?], p. 2.

[45] *The Aeronautical Archives*, Institute of the Aeronautical Sciences, a pamphlet in the collection of the AIAA, New York, 1939 [?].

[46] Charles Colvin served as the first treasurer of the Aeronautical Archive. The Board of Directors included Thomas H. Beck, Donald L. Brown, James H. Doolittle, Donald W. Douglas, Sherman M. Fairchild, R. H. Fleet, Jack Frye, Robert Gross, Harry F. Guggenheim, Howard R. Hughes, J. C. Hunsaker, P. G. Johnson, Grover Loening, Glenn L. Martin, Thomas Morgan, W. A. Patterson, E. V. Rickenbacker, Lawrence S. Rockefeller, Howard Scholle, C. R. Smith, Frank Tichenor, Guy W. Vaughan, and Edward P. Warner.

[47] *The Aeronautical Archive*; "The Technical Information Service of the Institute of the Aeronautical Sciences," *Journal of the Aeronautical Sciences*, April 1942, p. 194.

[48] "Finest Air Library Donated to Aero Science Institute, *Flying Time*, March 9, 1940.

[49] "Old Scraps of Paper Produce A Rich Harvest for Collector," *New York Times*, Nov. 12, 1939.

[50] Bella C. Landauer, "Collecting and Recollecting," *New York Historical Society Quarterly*, July, 1959, pp. 335–349.

[51] Bella C. Landauer, "Literary Allusions in American Advertising as Sources of Social History," *New York Historical Society Quarterly*, July, 1947, pp. 148–149.

[52] James J. Heslin, "Bella C. Landauer," in Clifford Lord, *Keepers of the Past*, Univ. of North Carolina Press, Chapel Hill, NC, 1965, pp. 180–189.

[53] Undated clippings, Bela C. Landauer Biographical file, National Air and Space Museum Archive.

[54] William T. Sniffen, *Aeronautics: An Exhibition of Books and Prints on View to October 15, 1934*, a bound manuscript in the Ramsey Rare Book Room, National Air and Space Museum, TL 506 A1S67.

[55] E. Jay Doherty, "Of Planes They Sang!" *Flying and Popular Aviation*, Feb. 1942.

[56] Bella C. Landauer, *Bookplates: From the Aeronautica Collection of Bela Landauer*, privately printed at the Harbor Press, New York, 1930.

[57] "Gifts to the Aeronautical Archives," *Journal of the Aeronautical Sciences*, Jan. 1944, p. 175.

[58] N. H. Randers-Pherson and A. G. Renstrom, *Aeronautic Americana: A Bibliography of Books and Pamphlets on Aeronautics Published in America Before 1900*, The Sherman Fairchild Fund, Institute of the Aeronautical Sciences, New York, 1943.

[59] "Library Expansion," *Aeronautical Review*, Jan. 1942, p. 5.

[60] "Sands Point Estate is Guggenheim Gift," *New York Times*, June 17, 1942.

[61] "$500,000 Given for Air Studies," *New York Times*, June 21, 1942.

[62] "More Air Research," *New York Times*, March 20, 1943.

[63] "Report of the Administrative Committee," *Aeronautical Engineering Review*, March, 1945, p. 71.

[64] *History of the Institute of the Aeronautical Sciences, Tenth Anniversary, 1932–1942*, Institute of the Aeronautical Sciences, New York, 1942, p. 15.

[65] "Restriction Hit Press," *Aeronautical Review*, Feb. 1942, p. 5.

[66] "Additional Paper Cuts," *Journal of the Aeronautical Sciences*, Jan. 1944, p. 173.

[67] "Keeping Balance," *Aeronautical Review*, Feb. 1942, p. 5.

Chapter 5—
Turning Professional: The American Rocket Society, 1941–1952

Reorganizing for the Future

By the fall of 1941, the leaders of the American Rocket Society (ARS) were floundering. The rocket experiments that had been their focus for a decade were at an end. Their work had finally produced a practical liquid-propellant motor, James Wyld's regeneratively cooled rocket. With that quart-can sized power plant in hand, Wyld and a handful of ARS colleagues—Lovell Lawrence, John Shesta, and Franklin Pierce—had established Reaction Motors, Inc. (RMI), in the New Jersey suburbs. There were tough times. In 1944, philanthropist and venture capitalist Laurance Rockefeller had to rescue the floundering company from financial collapse. Still, only six years after the firm was established, an RMI rocket motor boosted the Bell X-1 through the "sound barrier."

RMI was only a few months old when Theodore Von Kármán and his team of young rocketeers from the Guggenheim Aeronautical Laboratory of the California Institute of Technology (GALCIT) sent a light plane roaring aloft, powered by one of their early jet assisted take-off (JATO) rockets. They quickly followed the RMI pattern, spinning off their own firm, the Aerojet Engineering Corporation. By the fall of 1942, Robert Goddard, Robert Truax, and others were working for the U.S. Navy at Annapolis, Maryland, boosting seaplanes off the water with their own rockets. After the war, Curtiss–Wright would use those wartime designs as the starting point for work on a rocket motor that would propel the X-2 to record speeds and altitudes. An embryonic industry was taking shape.

In 1940, looking back over the 10-year history of ARS, Secretary Max Krauss remarked:

> ... we now see this field getting serious consideration by even Mr. John Q. Public, while technicians, governments, institutions of learning are devoting a good deal of energy and money to further the advancement of this science. Perhaps, as was the case with the airplane, the war will give the great impetus that will lead to realization of the dream of those early pioneers—the rocket as a practical medium of transportation.[1]

Indeed, rocket weapons would play a major role in every theater of the war that was already underway. New names—Bazooka, Katyusha, Nebelwerfer—entered the vocabulary of Allied and Axis nations alike. In Germany, a rocket-propelled interceptor, the Me-163, went zipping through the Allied bomber formations. By 1944, A-4 (V-2) rockets were falling on London, Brussels, and Amsterdam. The first large ballistic missile, the V-2, could deliver one metric ton of high explosive over a distance of more than 200 miles. At the peak of its trajectory, the rocket was coasting along at the edge of space, 60 miles up. Dropping straight down on its target at hypersonic speed, the weapon announced a new era in warfare and heralded the coming of the space age.

That was only the beginning. As early as November 1945, General Henry H. "Hap" Arnold, wartime commander of the U.S. Army Air Forces, not only predicted the advent of nuclear-tipped ballistic missiles but forecast a day in the not-too-distant future when the nations of the world would build "spaceships capable of operating outside the atmosphere."[2] General Electric won the contract to assemble and fly a series of captured V2 rockets in order to provide America with experience in handling and launching large ballistic missiles. Wernher von Braun, the brilliant engineer who had spearheaded the development of the V2, was assisting that effort, along with more than 100 other key German rocket engineers who had been hired by the U.S. Government. Behind the Iron Curtain that had descended across Central Europe, Soviet technologists were working with "their" German engineers to prepare and launch their own supply of captured German rockets. No longer the business of amateurs, rocketry was an infant industry on which the fate of nations would depend.

Ironically, the influence and reputation of the ARS was expanding at the very moment when the leaders of the organization were struggling to define a new role for themselves. The war years were a time of uncertainty for the ARS. Treasurer Samuel Lichtenstein reported a deficit of $106.23 when he closed the books on 1940, the Society's 10th year.[3] Secretary Max Krauss could point to a pitifully small increase in membership. The organization had recruited one new Active Member that year, increasing the total to 33. Sixteen of the 89 Associate Members were newcomers. Six new Junior Members joined the single youngster enrolled in 1939. The ARS recognized two affiliate societies, each of which had six members. On the occasion of its 10th anniversary, the membership roster had fewer than 130 members, many of whom were no longer paying dues.[4]

By the spring of 1941, President Lovell Lawrence attempted to revitalize the Society, suggesting that the group establish an office, reinstitute its experimental program, prepare a handbook of basic technical information on rocketry, and take other steps to attract new members. The organization did pursue some of those goals. The ARS Board of Directors convened on August 29, 1941, in their first real office in Room 606 at 1 East 42nd Street. They appropriated $25 for the purchase of a reconditioned typewriter (guaranteed for one year) and $40 for an addressograph machine. Discussions of resuming rocket development and testing continued during the early war years.

The notion of developing a much-needed engineering handbook on liquid-propellant rocketry proved more difficult to achieve. Constantine Paul Lent, who served as vice-president of the ARS in 1944, set to work on such a volume under the auspices of the organization, but the two resulting books, which he published himself, were poorly reviewed and ultimately rejected by the ARS.[5,6]

Pearl Harbor dashed any hope of an immediate expansion of the ARS. When the Board met for the second and last time in their new office on Thursday, December 17, 1941, Franklin Pierce suggested suspending activities "for the duration of the present crisis in the country." Following his lead, the members of the Board immediately voted not to schedule any further public meetings and to suspend publication of *Astronautics*. Members in good standing as of January 1, 1942, would be credited with the balance of their dues upon the resumption of normal activities.[7]

The concern on the part of ARS leaders that the continued operation of their organization might provide valuable information to the enemy proved to be an overreaction. In response to a query from Acting Secretary Cedric Giles, Commander W. B. Howe of the Third Naval District assured the group that the government had no objection to the activities of the ARS or the publication of *Astronautics*. Furthermore, "insomuch as future issues of this publication will not contain technical reports," the organization would not even be required to clear their publication with intelligence officials. After a three-month hiatus, the regular business of the society resumed with the annual meeting on April 9, 1942.[8]

After only a few months, the luxury of a dedicated office proved too expensive for the ARS budget. Members were asked to direct all correspondence to a Board member's office, Room 382, 50 Church Street, New York. Roughly four regular meetings were held annually during the early war years, usually at the Engineering Societies Building at 29 West 39th Street.

During 1942, most Board meetings were held in the offices of one or another of the members, often at Dr. Samuel Lichtenstein's dental surgery at 156 West 86th Street. On February 11, 1943, perhaps for old-times sake, the Board met at Nino and Nella's restaurant, now relocated to 49 Charles Street.[9] At some point that year, general meetings were suspended "... because many members had been engaged in confidential work for the armed services, and were unable to disclose the nature of their work."[6]

By the spring of 1943, with requests for information on rocketry increasing, the ARS rented yet another office, this time at 130 West 42nd Street. Still, the Society remained the loosest of organizations. In addition to paying Mrs. D. Lawrence, a part-time secretary, Pendray attempted to persuade Miss Murray, his own secretary, to handle some of the office work. While Lent, Wyld, Giles, and others tried to put in time responding to the increasing flood of letters and telephone calls inspired by news reports of rocket and other reaction-propelled weapons, most correspondence went unanswered. When James Wyld served as treasurer, he kept the "books" in a pocket notebook and seldom collected dues unless he ran into a member on the street.

The primary goal of the wartime ARS leadership was to assure the continued appearance of quarterly issues of *Astronautics*. By 1940, as the war in Europe closed down the informal international contacts that had linked the rocket groups of the 1920s and 1930s, the ARS journal became an important source of unclassified information on the latest advances in reaction propulsion. There were stories about Frank Whittle, the Royal Air Force (RAF) officer who had pioneered the turbojet engine, and photos of Axis and Allied jet- and rocket-propelled missiles and aircraft. Readers of *Astronautics* could study cut-away drawings of the A4 ballistic missile; photos of the Gloster E28/39, Bell P-59, DeHavilland Vampire, and Me-262 and 163; and schematic drawings of the V-1, Hs-293, and Japanese Ohka suicide aircraft.

In the months leading up to American entry into the war, subscribers included key government agencies in France, England, Russia, Japan, China, and Germany. Once hostilities began, *Astronautics* went to the National Advisory Committee for Aeronautics (NACA), the Institute for the Aeronautical Sciences (IAS), the Office of the Army Chief of Staff, Wright Field, California Institute of Technology (Cal Tech), West Point, Annapolis, and other major research libraries across the nation.

Interest in rocketry was on the rise, and the ARS was increasingly seen as a principal source of information on the subject. By the middle of the war, the makeshift office was receiving an average of 30 letters a week. The group prepared an exhibit on rockets for the Hayden Planetarium, displayed the refurbished test stand at the New York Museum of Science and Industry, and loaned the motor for ARS rocket #3 to Philadelphia's Franklin Institute. When Walt Disney began work on an animated film inspired by Alexander de Seversky's best-selling *Victory Through Air Power*, he consulted the ARS for information on the future of flight.

In spite of the fact that general meetings had been suspended, the organization began to grow. Membership rose from 177 in 1942, to 251 in 1943, to 334 in 1944. While an "ever increasing number of members" were serving with the armed forces, it was clear that "a greater proportion of members are technicians than formerly, and many are occupied at present with rocket research."[10]

Although the ARS remained a relatively small and ill-managed organization, it was the only game in town. The continued appearance of *Astronautics* and the absence of many other trustworthy sources of information were attracting the growing number of professionals employed in the infant industry into membership. Moreover, it was apparent that rocketry was a growth business. When "... considering the great technological advances in rocket development now being made in all parts of the world," Secretary Cedric Giles commented in the spring of 1943, "it appears that this war will be instrumental in providing the needed impetus to rocketry that the last war gave to aviation. In the present prelude to space flight, the Society should foresee and prepare now its postwar course, so as to be in an advantageous position as the "era of flight" is superseded by an "age of rockets."[10]

How would the ARS contribute to the creation of such a future? There were those who looked forward to a resumption of the prewar rocket experiments. Richard Hartley Wills offered his thoughts to the editor of *Astronautics*. "Every sincere rocket enthusiast is looking forward to the new era of peacetime rocket experimentation that the conclusion of the present war will herald." He suggested that the Society prepare for work on a "meteorological rocket," an effort that would be chronicled by "motion pictures, radio, and articles in reputable magazines." Success with such a rocket capable of reaching altitudes of up to 100,000 feet could be achieved "at moderate cost" and would attract the funds with which to build "... a large testing field with a modern machine shop [that] could be maintained with full time experts on rocketry collaborating for the eventual realization of interplanetary travel."[11]

Most members, however, realized that the era of the amateur rocketeer had passed. The realization of the dream of space flight would depend on the expenditure of vast sums of government money on rocket-propelled weaponry and on the rise of a new generation of professional scientists and technicians who could develop the required technologies and manage complex programs. The real business of the ARS, they believed, was to support the work of those professionals.

The drive to redefine the Society began as early as August 16, 1943, when Secretary James Wyld suggested the creation of a new class of membership, "... for experts only, with its own publication, etc., but within the ARS machinery." Perhaps, he noted, the time had come for a new generation of leaders, representing, "... a more technical group (with scientific standing ....)" When Pendray asked Wyld if he thought the current leadership should step down in favor of men with advanced technical training, the young engineer admitted that "... the present rocket technical group" was too small to run the organization. Lawrence Manning persisted, suggesting the creation of a new class of Fellows of the ARS who would have voting control of the organization. Although no decisions were made, Cedric Giles was instructed to "... go over the existing constitution and draw up suitable changes to be considered later."[12]

A fundamental transformation was underway. On May 17, 1944, the Board asked Dr. Lichtenstein to study the constitutions of other "scientific societies" and to draft a new ARS constitution that would reflect "new objectives ..., and provide for a national organization." Having helped to transform astronautics from the stuff of dreams to a budding industry, the leaders of the ARS decided to reorganize into a technical society that would meet the professional needs of the growing number of scientists and engineers working in this exciting new field. "The general consensus," Pendray explained to the members of the Board of Directors on November 28, 1944, was that the ARS should seize "... the opportunity to organize on a really national basis the rocket and jet propulsion engineers of the country, and thus fulfill its destiny."[13]

An essential first step was to establish an efficient office following sound business procedures. As had so often been the case in the past, Ed Pendray came to the rescue. In March 1945, he convinced Mrs. Agnes C.D. "Billie" Slade, one of his former secretaries, to spend two days a week in "the small, windowless cubicle in a midtown New York building" that served as the ARS office. After a six-week probationary period at $20 per week, the organization's first employee was hired at $50 a week. By the time "Billy" Slade retired 16 years later, she had become a legend in the organization. "She deserves recognition," James Harford, the organization's first executive secretary, notes. "She held the ARS together."[14]

When Mrs. Slade arrived on the scene, the ARS had only 237 dues-paying members. Many others were enjoying at least some of the privileges of membership free of charge. The "assistant to the secretary" slashed costs by immediately cutting all nondues payers. At her suggestion, a new dues structure was established. Active members would pay $10 per annum. Associate and Junior members would pay, respectively, $5 and $3. A new class of corporate members would be accessed $250 per year.

Mrs. Slade answered questions from the members, solved their subscription problems, instituted a rational bookkeeping system, inaugurated promotional activities, created the packages of introductory materials that went to all new members, kept the leaders in line, and generally transformed a loose and undisciplined organization into a taut ship.

She launched new projects as well. At a Board meeting on March 27, 1945, Pendray agreed to arrange for the publication of a new ARS book containing both of Robert Goddard's published Smithsonian reports.[15,16] In the end, Mrs. Slade would do the lion's share of the work. *Rockets,* with a new foreword prepared by Robert Goddard before his death in August 1945, appeared in 1946.[17]

"Billie" Slade soon proved to be indispensable. Even with the addition of a file clerk, Mrs. Mildred Dame, on May 6, 1946, the workload remained heavy. On June 1, 1947, Mrs. Slade tendered her resignation, "in order to take a summer vacation ... as a rest from the strenuous [sic] work of the past year." The Board immediately voted her a three-month vacation, increased her salary, and placed her in complete charge of the society's business affairs.[18]

She had a profound impact on at least one young visitor to her tiny office. In the late 1940s, a high school student from Rahway, New York, fascinated by astronomy and the prospect of space travel, called at the ARS office to purchase back issues of *Astronautics* and British Interplanetary Society publications. As Billie Slade added up the total, it became apparent that the youngster had selected more than he could afford. When he began rummaging through the pile, trying to decide which issues to return, she smiled and sent him on his way with all of the magazines. It was an act of kindness that Carl Sagan would never forget.[19]

With the office in good hands, at least for the moment, the next step was to orchestrate the expansion of the ARS into a genuinely national organization. The prewar Society claimed members from "the majority of states in the U.S."[20] Of course, most of those states had supplied only two or three members apiece. Although the leadership had made half-hearted efforts to encourage affiliate groups in other parts of the nation, metropolitan New York remained the focal point of ARS activity.

Now "hot spots" of professional interest in rocketry were popping up in universities, military bases, and industrial centers around the country. If anything, the center of gravity of American rocketry was shifting to the West Coast. In the spring of 1935, Theodore von Kármán, the emigrant Hungarian aerodynamicist who was heading the GALCIT, accepted graduate student Frank Malina's proposal for doctoral work that would involve a theoretical study of rocketry as well as serious experiments with both solid and liquid propellants.

Within five years, that project had grown into a full-fledged program attracting the first U.S. military contracts for rocket research. Like the ARS researchers, the GALCIT team established a company, the Aerojet Engineering Corporation, that would produce

Theodore von Kármán (1881–1963). (NASM Archive, Smithsonian Institution, 76-19208.)

rocket-assisted takeoff units for aircraft and missiles like WAC and Corporal. General Tire and Rubber Company acquired the operation in 1944 and renamed it the Aerojet General Corporation. Now there was talk of establishing a rival society for West Coast rocket engineers.

In a successful effort to prevent such a step and to draw California engineers into the fold, Lovell Lawrence and Pendray planned "a small dinner" for "a few key men among the nation's rocket engineers," including some of the Aerojet crowd.[21] The dinner, arranged for December 1944, was a low-key affair that achieved all that the ARS Board of Directors had hoped. Although there was no formal agreement, the "leaders in rocket development from all parts of the country, particularly the West Coast, began to join [the ARS]."[22]

Pendray was convinced that affiliation with an established engineering society would represent a major step forward for the ARS. He met three times with representatives of the American Society of Mechanical Engineers (ASME) before reporting to the ARS Board of Directors on August 16, 1945. The proposed arrangement offered advantages for both parties. The mechanical engineers would be investing in the future and associating themselves with a new field of great promise that would be of interest to their members. The ARS would enjoy the prestige of formal association with a leading professional society; access to a much larger pool of interested engineers and potential members; and the prospect of cost savings in office space, publications, and other areas where economies of scale would be at work.

President James Wyld appointed Pendray, Lawrence, and Shesta as a committee of three to negotiate an arrangement with the ASME. After meeting with their ASME counterparts (R. F. Gagg, Assistant to the General Manager of the Wright Aeronautical Corporation; R. Thomas Sawyer, an engineer/manager with the American Locomotive Company; and Louis N. Rowley, the associate editor of the journal *Power*) the group recommended affiliation to the Board of Directors on November 2. Following an affirmative vote of the active members, the ARS and ASME began their affiliation on December 1, 1945.

The leaders of the ARS recognized that this was a pivotal moment in the history of the organization. Alfred Africano, who chaired the new Committee on National Organization, noted that, "in the next year or two," they might well decide simply to become "a rocket section of the ASME." If they did pursue the notion of transforming the ARS into a genuine engineering society, however, the association with the established organization could only help them improve their own organization. The first step would be to revise the classes of membership to emphasize professional standing and achievement. Trained engineers, scientists, and administrators ("... degree or equivalent required ...") would be admitted as full Members. Technicians "... and others sufficiently interested to further the aims of the Society" would be accepted as Associate Members, and "technical college students" would qualify as Student Members.

"While most of us who did confidential research during the war are still unable to write freely or even at all," he continued, ARS publications should meet the highest professional standards and become the primary outlet for the presentation of "actual scientific data." The Society should establish prizes and awards to match those offered by other professional organizations and establish committees of technical specialists. Finally, Africano noted, the ARS should take full advantage of the opportunity to "gain some experience with the ASME method of arranging annual conventions and running local Section affairs."[23]

Turning Pro

Thanks to the new arrangement, Mrs. Slade and her new assistant, Mrs. Brickley, could move out of the cramped ARS offices at 130 West 42nd Street and into more prestigious, if not much larger, quarters in Room 617 of the Engineering Societies Building at 29 West 39th Street. The Society had its own phone number for the first time: Pennsylvania 6-9220. Other professional links were also beginning to fall into place. Late in 1945, ASME leaders requested that the ARS appoint a representative to their Gas Turbine Coordinating Committee. The organization invited all 600 members to attend their first 'Dutch Treat' dinner at New York's Hotel Lombardy on June 21, 1946. Nineteen enthusiasts showed up. Even more important, the rocketeers held their first annual convention, complete with nine papers in three technical sessions, in conjunction with the ASME annual meeting at New York's Hotel Pennsylvania held December 5–6, 1946 (Ref. 24).

The leaders of the ARS were not content to simply ride the coattails of an older and larger organization. While considering an abortive proposal by the American Institute of Electrical Engineers to create an all-embracing American Association of Engineers, the rocketeers came to grips with what would be required to achieve their goal of becoming the national society for engineers in their field. Their program of expansion would rest on the foundation of a new set of bylaws. The document began with a restatement of the goals of the organization in a run-on sentence even longer than the 1930 original:

> The purpose of the Society shall be to aid and encourage by all suitable means the development and application of the principle of jet propulsion to rockets, aircraft, water and underwater craft and to all other appropriate and practical devices; to aid and encourage the development of the sciences and engineering techniques pertaining thereto, and to create increasingly wide interest in the field of jet propulsion and rocketry among both technicians and laymen, to the end that jet propulsion in all its various forms shall readily and permanently be developed for the ultimate good of man; the preparation, collection, correlation and dissemination, by publication or otherwise, of facts, information articles, books, pamphlets and other literature pertaining to jet propulsion, rocketry and subjects relating thereto; the establishment of a library containing such literature for the information of members, scientists and others to whom the privileges of such library may be granted by the Society; the collaboration or affiliation with other organizations, whether technical or otherwise, in any manner, and to any extent which, in the judgment of the Society or of the Board of Directors, will best aid in accomplishing its objectives; the raising of funds for research and experimentation, and such other activities as the Society or the Board of Directors may from time to time deem necessary or desirable in connection with the foregoing.[25]

The new bylaws were approved at the second Annual ARS Convention, held in Atlantic City, New Jersey, on December 19, 1947, once again in conjunction with the ASME. They established Africano's suggested classes of membership and expanded the Board of Directors to nine members to strengthen the central organization. The addition of four regional sections underscored their determination to create a national society: New York; Washington, DC, (including Virginia, West Virginia, Maryland, Delaware, and North Carolina); Ohio; and California.

Africano was also appointed chair of the Technical Section Committee. The major task of the committee, he explained, was to establish subcommittees covering the various technical specialties: reaction motor development, fuels and combustion, instrumentations and communications, and aerodynamics and space problems.

The chairs of the subcommittees, Africano continued, would address other technical issues of interest and importance to engineers in the field, including the development of standards.[26]

In view of the fact that one purpose of a professional society was to recognize and honor the achievements of members, the ARS would offer three national awards:

1) The Robert H. Goddard Memorial Lecture: The honoree would be nominated by the members from among the leaders of the profession. The winner was to receive a medal, certificate, and cash prize before the annual lecture.

2) The C. N. Hickman Award: Named for a one-time associate of Robert Goddard who had also served as research director of Bell Laboratories. A medal, certificate, and $100 cash award would go to the member who had made an outstanding contribution to the field during the preceding year.

3) The American Rocket Society Junior Award: The member who had submitted the best paper for publication during the past year would receive a medal, certificate, and cash prize.

Membership continued to climb, increasing 61% from the 318 members of 1945 to 759 in the spring of 1947. Just a year later, ARS publications were distributed to 1,157 members and subscribers. The leaders of the Society renewed their arrangement with ASME in the spring of 1951.[27]

There were opportunities to cooperate with other groups, as well. In December 1950, S. Paul Johnston, who followed Lester Gardner as Director of the IAS, suggested to Ed Pendray that they discuss an affiliation between their two organizations. Pendray raised the subject with the ARS Board, presenting as a model the existing contract between the IAS and the American Helicopter Society. Provided that the legal independence of the ARS was guaranteed, the members of the Board agreed that the additional affiliation might be useful in opening another potential pool of members and easing the administrative load carried by the small Society. The Board instructed a newly appointed Finance Committee to confer with IAS leaders and offer recommendations.[28]

The ARS committee suggested an arrangement in which the current staff of the organization, its office space, and its relationship to the ASME would remain unchanged. Johnston, as director of the IAS, would also serve as director of the ARS. Other members of the IAS professional staff would also help manage ARS affairs. Finally, a Joint Policy Committee with representatives from the ARS, IAS, and ASME would provide oversight. Johnston rejected the proposal as too cumbersome. "If the ARS cannot see its way clear to establish its own directorate," he commented, "an outright integration with the management of one or the other of the larger societies would appear to be the most satisfactory arrangement." He was quick to add that joint meetings and other sorts of cooperation remained a possibility.[29]

In June 1953, Martin Summerfield suggested that the ARS take advantage of that possibility and explore the potential of joint national meetings with the IAS. Discussions between ARS President Fred Durant and IAS Executive Director S. Paul Johnston led to the first joint technical sessions scheduled for the IAS annual meeting at New York's Hotel McAlpin on January 29, 1954. In addition to arranging a single session of papers, the ARS was asked to provide a luncheon speaker for that day and to fill an exhibition area on the 24th floor, where the meeting would be held. It was the initial step in the long and sometimes complex association between the two organizations.[30]

Improvements to ARS publications were critical to the long-term success of the Society. Beginning with the March 1945 issue, *Astronautics* became *The Journal of the American Rocket Society*. The title seemed more appropriate for the flagship publication of "... a national engineering organization in the field of rocket and jet propulsion similar in form to the present great engineering societies."[31] The change reflected the desire to attract serious professionals rather than enthusiastic amateurs dreaming of space flight. Nevertheless, the editor offered an off-hand apology for the change and promised that "... the name *Astronautics* is to be held in abeyance for possible future use."[31]

Considerable thought went into upgrading the journal. Ed Pendray asked Frank A. Parker, a Princeton professor involved in Project Squid, a cooperative research program in jet propulsion, for his thoughts on the subject. He responded in November 1948 with a 5-page, "Prospectus for a New Scientific Journal to be Published by the American Rocket Society." Parker suggested that the *Journal of the Aeronautical Sciences* was a model worthy of emulation. He underscored the need for technical rigor, including the use of an editorial board and referees, and recommended against publishing "news" or information on the general business of the ARS in the technical journal. The new journal should be the work of in-house professionals who were paid for their efforts.

ARS leaders, struggling to match their ambitions to their pocketbook, accepted some, but not all, of Parker's advice. Taking advantage of ASME experience in the field, they negotiated an editorial service arrangement and an advertising agreement with the senior organization. Beginning in March 1948, the journal was published by the firm handling ASME printing. In addition, the editorial staff producing the monthly issues of *Mechanical Engineering* would do most of the day-to-day editorial work on the journal. ASME staff also solicited the first advertising to grace the pages of the ARS publication.[32]

Everyone seemed pleased with the editorial arrangement and with the new and more professional appearance of the journal. In addition to advertising, the publication continued to keep readers up to date on news in the field, ARS doings, and employment opportunities. The editors did follow Frank Parker's admonition to apply professional rigor in selecting papers for publication, however. Through the late 1940s and into the early 1950s, breezy historical articles and general descriptions of wartime developments in reaction propulsion gradually gave way to technical papers that had passed muster with an editorial panel that included some of the biggest names in the field.

Early on, the editorial duties were shared by several members. Louis Bruchiss was editing the journal on February 10, 1947, when 34 New York residents met at Miller's Restaurant on Fulton Street for the second Dutch Treat Dinner of the American Rocket Society.[33] Major James Rudolph succeeded him on December 4, 1947. Among their other duties, the members of the Technical Sections Committee, established in January 1948, also supervised work on the journal and "acted as the technical nucleus of the Society." Long-time member and

Massachusetts Institute of Technology (MIT) graduate Robertson Youngquist chaired the committee and served as editor until 1951.[34]

On January 15, 1951, Frank Parker, now the chair of the Ad Hoc Reviewing Committee on the *Journal of the American Rocket Society*, submitted yet another report recommending that the journal be "reorganized" into a publication "devoted entirely to the interests of the jet propulsion field." The present scope of the journal was too narrow. A new tagline should be added to the cover identifying the publication as, "A Journal Devoted to the Jet Propulsion Sciences." Every care should be taken to elevate the level of the articles, improve the format, and expand it to a monthly publication. Above all, the committee pointed to the need for "the immediate appointment of a highly qualified technical editor well known in the field of jet propulsion and willing to earnestly devote himself to the task of successfully revising the Journal on the highest technical level."[35]

The notion of selecting a real leader in the field to edit the journal had been discussed as early as 1950, when President William L. Gore offered the job to Dr. Hsue Shen Tsien. A native of Shanghai, born in 1909, Tsien had earned his doctorate at Cal Tech, where Theodore von Kármán had remarked that the young Chinese scientist was the brightest of his graduate students.[36]

During the 1930s, Tsien had joined a group of fellow graduate students led by Frank Malina who were conducting theoretical and experimental rocket studies. A combination of mishaps with corrosive chemical propellants and the occasional roar of rocket motors arising from a desert arroyo near the Pasadena campus led to their identification as the "suicide squad."

One of the theoreticians of the group, Tsien continued to work with Malina and others on a variety of wartime projects before accepting a full professorship at MIT in 1946. He was lured back to Pasadena three years later, when he was named Goddard Professor of Jet Propulsion. That summer Tsien, his wife Yin, and their two children took the first steps toward becoming American citizens. Their world came crashing down around them in June 1950, when federal officials accused Tsien and other members of the suicide squad of attending meetings of a local Communist Party cell during the 1930s.[37,38]

When he attempted to return to China for a visit with his ailing father two months later, customs agents seized eight cases of personal notes. Tsien explained that he had planned to carry on his work while he was overseas. The scientist was arrested and jailed for two weeks. When he was released, his security clearance was lifted and he was barred from leaving the county.

Facing an uncertain future, Tsien declined the invitation to edit the *Journal*. Indeed, things went from bad to worse in the weeks and months to follow. In spite of the ardent support of friends and colleagues, some of the most distinguished scientists and engineers in the United States, the government refused to reinstate Tsien's clearance, effectively barring him from active participation in cutting-edge technology. In 1955, he was allowed to return with his family to China. He was quickly appointed to positions of high authority in a number of leading scientific societies and played a major role in shaping science, technology policy, and progress in the People's Republic for several decades to come.[39]

A year after Gore had discussed the editorship with Tsien, Frank Parker approached another potential editor on behalf of the Ad Hoc Committee struggling to improve the journal. A native New Yorker, Martin Summerfield (1916–1996) had earned a B.Sc. in physics from Brooklyn College in 1936 and then traveled west to pursue graduate studies at Cal Tech, where he received an Sc.M. in 1937 and a magna cum laude Ph.D. in physics in 1941. In June 1940, he joined Malina, Tsien, and other members of the suicide squad. As the group specialist on liquid-propellant motors, Summerfield played a key role in the project. He was one of the founders of the Aerojet Engineering Corporation, where he served as chief of the Special Engines Project, 1943–1945. During this period, he was a leader in the development of liquid-propellant JATO engines for the U.S. Army Air Forces.

Summerfield returned to Cal Tech in 1945 to head the Rocket Research Division of the Jet Propulsion Laboratory. His duties, Frank Malina recalled, included "planning and research analysis of possible applications of rocket propulsion to extraterrestrial space flight." It was Summerfield who suggested mounting a WAC Corporal rocket on the nose of a V-2 to create a two-stage rocket capable of arching up into space. On February 24, 1949, one of the resulting Bumper-WAC vehicles reached a record altitude of 244 miles.[40]

The young engineer left Cal Tech to join the faculty of Princeton University in 1949. He was just settling into his academic duties in the spring of 1951 when Parker offered him the opportunity to serve as editor of the *ARS Journal*. After checking to ensure that Tsien still regarded himself as "unavailable," Summerfield accepted. It was an interesting situation, indicating the extent to which technical capacity and intellectual talent mattered more to the leaders of the ARS than security concerns. Like Malina and Tsien, Summerfield was accused of links to communism in Depression-era Pasadena and stripped of his security clearance. With the assistance of his university, he waged a long and ultimately successful fight for vindication.[41]

The decision of the ARS leadership to put Martin Summerfield in command of the publication program was wise as well as courageous. Under his supervision, the journal finally emerged as *the* publication in the field, an essential prerequisite for the development of the ARS into the leading professional society for scientists and engineers in an exciting new set of disciplines. Summerfield took over as editor-in-chief beginning with the September 1951 issue. Six months before he officially took the job, he suggested the employment of Mr. Bohenek as managing editor at an initial salary of $200 for each 32-page issue, with extra fees for additional work.[27] Summerfield remained on the job until the merger with the IAS in 1962, after which he served as editor-in chief of the *AIAA Journal* through 1963.

In an editorial celebrating Summerfield's 10th anniversary as editor, Theodore von Kármán remarked that a "... journal of the highest quality with an agreeable style of presentation and sustained vigor does not just happen." Summerfield, he noted, had "... a deep understanding of the field, ... the persistence to find good manuscripts and to stimulate authors to prepare them, ... [and] the strength to refuse poor manuscripts and to accept manuscripts with new ideas." Nothing was more important to the growth of the ARS than the quality of its flagship publication.[42]

In January 1949, Treasurer Ed Pendray submitted a plan for a substantial expansion of the organization. He opened the report with an analysis of the current budget.

Salaries	$8,000
Secretary (part-time)	
Clerk-typist	
Rent, heat, light	$240
Office expenses, including postage	$600
Telephone	$150
Printing and stationary (except *Journal*)	$800
Journal (editing, printing, distribution)	$3,600
Returns to Sections for local expenses	$300
Total Annual Expense	$13,690

Income did not completely cover those expenses.

800 members @ $10.00	$8,000
50 Junior members @ $5.00	$250
3 Institutional members @ $250.00	$750
Books sales and other income	$800
Total annual income	$9,800

Pendray then presented a budget for the additional expenses that would be required if the ARS was to grow. The real key would be the addition of a full-time professional, an executive secretary to guide the organization.

Executive secretary	$8,000
Additional clerical help	$3,000
Travel and organizational expenses	$1,000
Improvements to *Journal*	$8,000
Additional publications	$2,500
Membership promotion and service	$2,000
Total additional expenses	$24,500

Pendray then projected the total expenses that a healthy ARS would face within three years.

Executive secretary	$8,000
Secretary	$3,000
Clerical and stenographic help	$4,000
Travel and organizational expenses	$1,500
Rent and office expenses	$1,500
Printing and stationary	$1,000
Journal (six issues)	$10,000
Other publications	$2,500
Membership promotion and services	$2,000
Support for local Sections	$2,250
Contingencies and accumulations	$4,250
Total expenses	$40,000

How would the ARS pay the cost of expansion? Pendray had worked that out, as well.

2,000 members @$10	$20,000
200 Junior members @ $5.00	$1,000
50 Institutional members	$12,500
Sale of books and publications (net)	$2,500
Advertising ($500 per issue)	$3,000
Total income	$39,000

The treasurer suggested that the ARS launch a concerted fundraising campaign, the goal of which would be to raise $75,000 to fund a period of healthy growth. In addition to launching a membership campaign, he suggested a four-pronged effort:

1) Approach the roughly 3,000 firms thought to be involved in one phase or another of reaction propulsion. A $750 pledge from just 100 of those companies would meet the need.
2) Offer ARS Life memberships at $1,000 each

3) Solicit large gifts of $2,000 to $10,000 from individuals, corporations, and foundations

4) Create a new class of JATO Members by inviting current members to double their dues for three years.

Five years after the end of the war, the leaders of the ARS were developing a new vision. They were charting a course into the future and were poised for liftoff.

A New Generation

By the early 1950s, the ARS was emerging as a growing and vibrant organization representing the interests of engineers, scientists, and industrialists who commanded a set of technologies that might govern the rise and fall of nations. Ed Pendray remained an important figure in the organization during the immediate postwar years. Other veterans of the prewar experimental era—including Alfred Africano (president 1938, 1939), James Wyld (president 1944, 1945), Lovell Lawrence (president 1946), and Roy Healy (president 1947)—gradually faded from the scene as leadership passed to a new generation of professionals. Some well-known prewar pioneers, including Robert Truax, who had begun his own rocket experiments while an Annapolis Midshipman, signed up for the first time. Scientists with an interest in the possibility of space flight, like the Harvard astronomer Fred Whipple, also joined.[43]

The effort to fold the Californians into the ARS had brought some of the nation's leading rocket engineers and scientists into the organization. With the possible exception of Martin Summerfield, none of the West Coast's technical professionals would be more influential than a lawyer, Andrew G. Haley. Born in Tacoma, Washington in 1904, Haley was educated at George Washington University and the Georgetown University Law School. Following graduation, he worked as an attorney for the Federal Radio Commission and the Federal Communications Commission until 1939, when he entered private practice.

Haley also had a considerable reputation as an authority on immigration law, an area of expertise that Theodore von Kármán required in 1938. von Kármán had recently become a U.S. citizen, but his sister Josephine was facing deportation because she was unable to convert her temporary visa to a document that would allow permanent residency. At the suggestion of a friend in the Navy, von Kármán consulted a bright young legal miracle worker named Andy Haley, who succeeded where the other attorneys had failed.

Impressed, the professor invited Haley to visit Pasadena in January 1942, where he walked von Kármán, Frank Malina, Martin Summerfield, Edward S. Forman, and John W. Parsons through the legal intricacies required to incorporate the Aerojet Engineering Corporation, a process that was complete by March. Although all of the founders contributed to the capitalization, Haley, who was named secretary of the corporation, was by far the biggest investor and controlled the majority of stock.

With no experience in engineering or rocketry, Haley immediately returned to Washington and enlisted in the U.S. Army Air Forces, serving as a Major in the Office of the Judge Advocate General. The other founders, including von Kármán, proved to be far better engineers and scientists than business executives. Within a few months, the Air Force was charging Aerojet with poor management and threatening to cancel the firm's contracts for the development of JATO rockets. In desperation, Kármán approached his friend General Henry H. "Hap" Arnold, commander of the USAAF, and secured Haley's release from the service and his appointment as president and general manager of Aerojet.

As Kármán later noted, Haley proved to be "an incredible administrator" who "held the company together by guts and audacity."[44] He remained at the helm through the war years.

Alfred Africano, ARS President, 1937–1939, and an architect of the ARS post-war transformation. (Courtesy AIAA.)

Having established Aerojet as a viable company, he sold the firm to General Tire and Rubber in 1944, making millions in the process, and then returned to his Washington, DC, law firm, Haley, McKenna & Wilkinson.

Andy Haley had caught the space bug, however, and recognized rocketry as a key defense industry that had no place to go but up. Participation in the ARS offered him an opportunity to keep his hand in. He immediately rose to a leadership position, serving on the Board (1951–1952), as vice president (1953), and as president (1954) of the ARS and president of the International Astronautical Federation (IAF) (1958–1959). A pioneer of space law, Haley is remembered for his space age reformulation of the Golden Rule. Doing to an intelligent being from a different world as you would have him do unto you, Haley suggested, might harm him. Rather, the rule should be, "Do unto others as they would have you do unto them."[45]

A successful lawyer and businessman, Haley also struggled with demons. Jerry Grey, a young Navy veteran who joined the ARS in 1946, remembered him as "probably the heaviest drinker I have ever met" and an occasionally incomprehensible public speaker. "I have listened to long, detailed Haley speeches in which I understood all the words," Grey notes, "but had no idea of what he had said." On one memorable occasion, Haley stumbled to the podium to introduce a distinguished speaker and promptly fell backwards off the platform. "Undaunted, he clambered back up, grabbed the table to haul himself to his feet, and continued right on with the introduction."[46]

If Jerry Grey and others smiled and shook their heads at Haley's failings, they also respected the shrewd political instincts and managerial skills that had led to his extraordinary business success and rapid rise to the inner circle of ARS leadership. Perhaps most important, as one ARS colleague recalled, they appreciated Halley's "big Irish heart."[47]

Immediately after joining the ARS board, Haley accepted a position as chairman of the finance committee, the man most responsible for pursuing Pendray's funding suggestions. He launched a campaign that began with the solicitation of advertising and contributions from corporations that might find potential business in rocketry. He offered book sellers special discounts on the sale of the ARS Goddard publication. Most important, he worked to introduce professional business practices to the publication of the *Journal*, "the most important function of the national organization and our main hope for continued and outstanding success." With the content of the *Journal* entrusted to Martin Summerfield and the business side of the operation in good hands, the Board of Directors could focus on improving their financial prospects and pursue their plans for the future.[48]

Others migrated into the new field as a result of their engineering duties. At a single board meeting on June 2, 1947, new applicants for membership included employees of both Republic and Grumman aircraft, Sperry Gyroscope, Curtiss–Wright, and RMI. Richard William Porter (president, 1955) was typical of this group. One reporter described him as an unruffled, pipe-smoking, family man who raised orchids in a home greenhouse and played clarinet in the Schenectady Symphony Orchestra.[49] Porter had worked on the General Electric team developing the complex fire control system for the B-29 bomber and joined the ARS when he was placed in charge of the company team supervising the launch of captured German V-2 rockets under an Army contract.[50]

Martin Summerfield. (Courtesy AIAA.)

William Gore (president 1949–1950) had been a Marine pilot in 1938, when he wrote a paper advocating the use of rockets to boost aircraft off carrier decks. As a wartime test pilot, he flew the first tests of solid-propellant JATO units mounted on an F4F fighter. Upon leaving the service in 1946, he became Aerojet-General's director of government operations with offices in Dayton, Ohio; and Washington, DC.

After four years of graduate training at Cal Tech, Charles W. "Chuck" Chillson had gone to work at Curtiss–Wright in 1936 as a propeller engineer. When the Propeller Division spun off a Rocket Department in 1947, Chillson was named chief engineer. He joined the ARS in 1949 and was immediately named to the Board of Directors. He was elected vice president the next year and became president the year after that in 1952.

Frederick C. Durant, III, one of the most visible and influential of the newcomers, was born in 1916 to an old Philadelphia family and graduated from Lehigh in 1939 with a degree in chemical engineering. After a short stint with the E. I. DuPont Company, he earned his wings as a naval aviator and spent the war years as a flight instructor and test pilot. Discharged in 1946, he took a job with Bell Aircraft, where he discovered that engineers were ignoring the government reports on wartime German rocket research, studies that fascinated him. He was especially intrigued by Austrian Eugen Sänger's proposal for an "antipodal bomber" that would circumnavigate the globe by bouncing its way along on the top of the atmosphere. Durant left Bell, reactivated his reserve status as a lieutenant commander, and became director of engineering at the U.S. Naval Air Rocket Test Station in Dover, New Jersey.

Thin as a rail, with short-cropped sandy hair, freckles, a ready grin, and an ever-present pipe, Durant was young, outgoing, ambitious, and fully convinced of the future for human beings in space. He was accepted as a member of the ARS on October 4, 1948, not long after moving to Dover, and was sitting on the ARS Board of Directors before the year was out. He was elected vice president in 1952 and moved into the presidency the following year.

In March 1951, Durant changed jobs one more time, establishing himself as a consultant with an office at 1424 K Street in Washington, DC. In fact, he was working as a Central Intelligence Agency (CIA) analyst, specializing in tracking foreign rocket activity and personnel. His involvement as a leader of the ARS and, ultimately, the IAF, placed him

C. W. Chillson, ARS President for 1952, and H. R. J. Grosch, ARS President for 1951, greet Undersecretary of the Army A. S. Alexander. (Courtesy AIAA.)

in the ideal position to pursue and promote space ventures and keep an eye on international competitors at one and the same time.[51]

Keeping Secrets

Rocketry, the business of wild-eyed dreamers only a decade before, was now a matter of vital national interest and cloaked in secrecy. As an Iron Curtain descended across eastern Europe, hostilities broke out in Korea, and the atomic spy scare dominated the headlines, security issues became a matter of real concern in ARS circles. When Fred Durant returned from a visit to the West Coast in May 1951, he noted that most of the meetings being planned by the California Section would be closed sessions, open only to members with appropriate security clearances. When the program committee proposed scheduling closed sessions at a national meeting in 1953, there was some opposition from engineering professors who would be barred from attending because they were not eligible for security clearances. That notion was overruled by the members of the ARS Board, who recognized that a great many engineers working at the cutting edge of the field were employed on secret government projects and could only share new information in closed sessions.[52]

In June 1952, Frank Parker returned from a visit to Washington with bad news for his fellow directors. "Apparently," he remarked, "a number of people in the Armed Forces feel that the Rocket Society may be an easy source for security leaks." Worse, he had "... heard rumors that the Society is under general security scrutiny by one or more of the Defense Department Intelligence Agencies." Although some officials suspected that the ARS might have published "borderline classified material," the real concern was that secrets were being revealed "... through meetings where classified material is discussed in the presence of persons not having necessary clearance."[53]

Dr. Noah Davis, 1953 chair of the program committee, admitted that security had become "... the biggest headache to the writer and to the chairman." In an effort to meet a short deadline for the Los Angeles Semi-Annual Meeting that year, he had sent out for review and preprinting papers that had not yet been cleared. As a result, one paper contained some classified material that had to be reported to the authorities. Only papers that had been completely cleared would be sent out for preprints.[54]

Frederick C. Durant, III, ARS President for 1953. (Courtesy AIAA.)

The directors had always argued that, in the future, the author of an article of paper was responsible for clearing his work with security officials. As security concerns grew, however, the Board recognized that the ARS had a moral responsibility in the matter. In the spring of 1951, Andrew Haley, newly appointed the first general counsel of the ARS, drafted a form indicating that appropriate security officials had approved material for use by the Society. Martin Summerfield argued that the security notifications published in each issue of an ARS publication should cover the matter. Fred Durant further suggested that the editor of the "Jet Propulsion News" section of the journal be asked to document every item that he published in an effort to avoid publishing classified material.[55]

Summerfield, who was waging his own battle to recover his security clearance, assured the Board that articles submitted by H. S. Tsien did not represent a problem because they were entirely theoretical and had passed muster with security officials. Although the directors were reluctant to encourage Tsien to take an active role in the governance of the Society, neither did they ostracize the man who just a few months before had been regarded as the best brain in the field.[56] At the 1953 honors banquet, for example, they sat the embattled professor at the head table along with Charles Lindbergh, James Doolittle, Harry Guggenheim, Mrs. Robert Goddard, and Laurance Rockefeller.[57]

Oddly, in view of the temper of the times, it was not until the spring of 1954 that the ARS Board discussed the fact that their technical journal was recommended reading behind the Iron Curtain. J. B. Cowan, of the membership committee, pointed out that two members of the Soviet Embassy in Washington were members in good standing. The IAS screened its membership and did not accept applications from communist nations. Neither did the ASME. Cowan discussed the matter with several "outstanding members," who agreed "that we should not accept memberships from anyone connected with a country with whom we are discussing war."[58]

General Electric's Richard Porter asked if all applications from Iron Curtain nations should be rejected, pointing out that the qualifications for membership established by the constitution and bylaws referred only to educational and occupational qualifications and did not mention security status. He called for an amendment to the bylaws clarifying the situation. In the meantime, Porter suggested that the membership committee screen the current roster and forward the names of any members who seemed undesirable to the Board along with a paragraph explaining the perceived problem. Finally, Martin Summerfield moved that the ARS secretary inform Colonel Leonid Pivnev, the Soviet Embassy figure whose membership had precipitated the discussion, that he was no longer eligible for ARS membership.[58]

The inimitable Andy Haley, ARS President for 1954. (Courtesy AIAA.)

The problem arose again in January 1955, when the ARS received a membership application from Yugoslav citizen. The executive secretary explained that, although the State Department had recently approved the application of two Yugoslavian engineering students who had applied for admission to Princeton, university officials had decided not to accept them. The ARS board decided, oddly enough, to reject the application but to allow the individual to subscribe to the *Journal*. When the University of Zagreb requested a copy of the ARS publication for "permanent exhibit," the Board voted to send a copy to the U.S. Embassy, which would present the journal to the university if State Department officials thought it appropriate.[59]

For all of their care and concern over security issues, The ARS remained suspect until the end of the decade. In a 1957 article headlined with photos of Klaus Fuchs, Julius Rosenberg, and Rudolph Abel, a *New York Times* reporter pointed out "... that the Soviet Union has gathered more useful information from ordinary intelligence work in the broad open daylight than it ever has from cloak and dagger operations." Recently, it seems, one Soviet agency had spent $316,277 "for the purchase of American books and scientific and engineering treatises and journals." Precisely what did the Soviets buy? "As a sample, one representative on June 28, 1955 ordered from the American Rocket Society a number of papers, including 'Toxicity and Health Hazards of Rocket Propellants,' 'Gas Torch Igniters for Rocket Motors,' 'Ballistics of an Evaporating Droplet,' 'Propellants and Special Fluids Versus Valve Seal Design,' and 'The Role of Plastics in Liquid Propellant Rocket Field.'" Although each of those reprints had been cleared by security officials, the titles just sounded like something that should be kept from the Russians.[60]

Growing Pains

For the first two decades of its history, the ARS seems to have had few internal political problems. Even the shift from the science fiction period to the era of experimental rocket development had occurred with remarkably little internal dissent. The founding members who were not interested in building rockets simply faded away. The organization was tiny and, for much of its history, the activity of the ARS was of very little interest to any but the mere handful of space and rocket enthusiasts scattered across Europe and America who knew or knew of one another.

By 1950, the ARS was taking its place among the older engineering and scientific societies. Although relatively rapid growth, increased visibility, and involvement in vitally important technologies were the elements of success, they also brought with them new management problems. When coupled with the advent of a new generation of ambitious and sometimes volatile technical professionals, squabbling was probably inevitable.

Dissent was certainly the hallmark of the 1952 election. Some of the problems were related to the need for better geographic representation on the Board of Directors. Martin Summerfield, although now residing on the East Coast, noted that, with an upcoming resignation, there would be only one Californian on the Board. Responding to the argument of the Northeast New York Section that members of the Board should be able to attend most meetings, which were always held in New York, Summerfield noted that attendance at meetings should not be the only criteria for membership. The makeup of the Board should also reflect the point of view of West Coast members.

Summerfield also reported that some members thought that academics were underrepresented on the Board. He pointed to the critical importance of schools such as Cal Tech, MIT, Princeton, and Johns Hopkins to the field and noted that better representation of university departments "might . . . broaden the base of the Society."[61]

Finally, Summerfield reported that he had heard objections to the tradition of automatically promoting the vice president to president each year. Specifically, Fred Durant, who had served as vice president in 1952, was about to take over the reins of the Society. General Counsel Andrew Haley admitted that even H.R.J. Grosch and C. W. Chillson, who had served as president in 1951 and 1952, regarded Durant as ". . . a little too positive."

The comment probably referred to a growing fear on the part of some members that vocal enthusiasts like Durant, who spoke up for some attention to the more distant goals of space flight, might somehow tarnish the image of the ARS as a "serious" technical organization. It was an ongoing, usually low key, debate that would continue through the middle of the decade. In addition, Durant, one of the first ARS representatives to the IAF, had ruffled President Chillson's feathers when he complained that the New York office was not giving sufficient attention to correspondence from foreign nations, particularly when the correspondence involved space travel.[62]

Haley disagreed with those who sought to alter the course of the election, arguing that the tradition of stepping up to the higher office was valuable as a means of rewarding those who had labored on behalf of the group and helped to prevent an administrative coup by "haphazard politicians." Some day, when the organization had a membership in excess of 3,000 and "a permanent Secretariat with an efficient executive secretary," he continued, they might consider changing some current policies. Until then, they would do well to maintain the tradition. Haley asked Summerfield to join his campaign by urging Bernie Dorman, a member of the nominating committee, to support Durant's promotion to president. Finally, Haley suggested that he had no intention of accepting another term on the Board. He would, however, be willing to stand for election as vice president, in which position he promised to "act as a pretty effective 'balance wheel' for Fred." In view of his argument in favor of tradition, of course, this meant that he would follow Durant as president in 1954.[63]

The election went as Andy Haley had hoped.[64] That was a good thing because 1953 was a critically important year in the history of the ARS. Having transformed the *Journal* into a full-fledged professional publication, created a stable administration that could be passed from hand to hand under the oversight of the Board of Directors, and increased membership and otherwise established the ARS as the major organization representing scientists and engineers involved in critically important new technologies, the group finally decided to hire a full-time director. On the evening of Monday, September 14, 1953, President Durant asked the members of the Board to consider his proposal to hire James Joseph Harford as executive secretary.[65]

References

[1] American Rocket Society, Secretary's Report for year Ending March 31, 1940, AIAA Headquarters Files.

[2] Arnold quoted in R. Cargill Hal, "Origins of U.S. Space Policy: Eisenhower, Open Skies and the Freedom of Space," *Exploring the Unknown: Selected Documents in the History of the U.S. Space Program*, edited by John Logsdon, NASA, Washington, DC, 1995, pp. 1:213–215.

[3] American Rocket Society Financial Report, April 1, 1939 to March 31, 1940, ARS Board Minutes, AIAA Headquarters Files.

[4] Secretary's Report, Year Ending March 31, 1939, ARS Board Minutes, AIAA Headquarters Files.

[5] Constantine Paul Lent, *Rocket Research*, Pen and Ink Press, New York, 1944 [?]; new edition, 1945 and *Rocketry*, Pen and Ink Press, New York, 1947. *The Journal of the British Interplanetary Society*, July 1948, pp. 173–174, advised readers to take *Rocket Research* "... with a few molecules of sodium chloride," and regretted that the author "... did not lavish the same loving care on the compilation of the text of this work as evidently he expended in the preparation of the illustrations." The same prestigious journal (May 1948, p. 133) judged *Rocketry* to be "... completely useless as a serious work on rockets." The quality of the two volumes was a matter of concern in the inner circles of the ARS. At the suggestion of Commander Robert Truax, the Board removed them from the list of books for sale by the organization on June 7, 1948. (Minutes of the Meeting of the ARS Board of Directors, June 7, 1948, AIAA Headquarters).

[6] Untitled historical notes on the American Rocket Society, G. E. Pendray Papers, Department of Rare Books and Special Collections, Princeton Univ. Library.

[7] Special Board of Directors Meeting, Dec.17, 1941, ARS Meeting Minutes, AIAA Headquarter Files.

[8] W. B. Howe, Commander U.S. Navy (Ret.) to Cedric Giles, March 11, 1942, in ARS Meeting Minutes, AIAA Headquarters Files.

[9] Minutes of Meetings of the ARS Board of Directors, 1940-1944, AIAA Headquarters Files.

[10] Report of the Secretary, Minutes of the ARS Annual Meeting, May 24, 1943, AIAA Headquarters Files.

[11] Letter quoted in Untitled historical notes on the American Rocket Society, G. E. Pendray Papers, Department of Rare Books and Special Collections, Princeton Univ. Library.

[12] Minutes of the Meeting of the ARS Board of Directors, Aug. 16, 1943, AIAA Headquarters Files.

[13] Minutes of the ARS Board of Directors, Nov. 28, 1944, AIAA Archive.

[14] Tom Crouch, interview with James Harford, Aug. 4, 2004. Tape in the collection of the AIAA.

[15] Robert H. Goddard *A Method of Reaching Extreme Altitudes*, Smithsonian Instituion, Washington, DC, 1919.

[16] Robert H. Goddard *Liquid Propellant Rocket Development*, Smithsonian Instituion, Washington, DC, 1936.

[17] Robert Goddard, *Rockets* American Rocket Society, New York, 1946.

[18] Minutes of the Meeting of the ARS Board of Directors, April 14, 1947, p. 2, AIAA Archive.

[19] Sagan told the story to Frederick C. Durant, III, who related it to the author in an unrecorded interview on March 16, 2005.

[20] Report of the Secretary, Annual Meeting, April 9, 1942, Minutes of the ARS Board of Directors, AIAA Headquarters Files.

[21] Minutes of the ARS Board of Directors, Nov. 28, 1944, AIAA Archive.

[22] G. Edward Pendray, "The First Quarter Century of the American Rocket Society," *Jet Propulsion*, Nov. 1955, p. 589.

[23] Alfred Africano, "Report to the Board of Directors on Subject of National Organization," Aug. 23, 1946, ARS Board of Directors Minutes, AIAA Headquarters Files.

[24] The G. Edward Pendray Papers, Princeton Univ. Libraries, Box 12.

[25] Minutes of the Annual Meeting of the ARS, April 24, 1947, AIAA Archives.

[26] Minutes of the Second ARS Dutch Treat Dinner, Feb. 10, 1947, in Minutes of the Meetings of the ARS Board of Directors, AIAA Headquarters Files.

[27] Minutes of the Meeting of the ARS Board of Directors, April 21, 1951, AIAA Headquarters Files.

[28] Minutes of the Meeting of the ARS Board of Directors, Dec. 11, 1950, AIAA Headquarters Files.

[29] S. Paul Johnston to R. T. Sawyer (chair of the ARS Finance Committee), Jan. 9, 1951, in Minutes of the Meeting of the ARS Board of Directors, Jan. 16, 1951, AIAA Headquarters files.

[30] Minutes of the Meeting of the ARS Board of Directors, Oct. 5, 1953, AIAA Headquarters Files.

[31] "Name of the Official Organ is Changed," *Journal of the American Rocket Society*, March 1945, p. 15.

[32] An ASME-ARS Co-Coordinating Committee was established to handle normal relations between the two organizations. At a meeting of this group on Oct. 19, 1948, Roy Healey, representing the ARS, complained of ASME handling and costs of advertising in the *Journal*. "Memo of Decisions Made at Meeting of ASME-ARS Co-Coordinating Committee," ARS Board Minutes, AIAA Headquarters Files.

[33] Untitled report, ARS Board Minutes, Fall 1948, AIAA Headquarters Files.

[34] Minutes of the ARS Board Meeting, Jan. 12, 1948, ARS Board Minutes, AIAA Headquarters Files.

[35] Frank Parker, "Report of the Ad Hoc Reviewing Committee of the *Journal of the American Rocket Society*," in Minutes of the ARS Board Meeting, Jan. 12, 1948, ARS Board Minutes, AIAA Headquarters Files.

[36] Minutes of the Meeting of the ARS Board of Directors, Jan. 3, 1950, AIAA Headquarters Files.

[37] James Harford, "Martin Summerfield, 1916–1996," *AIAA Bulletin*, Sept. 1996, p. B8.

[38] Assorted references, Summerfield biographical file, National Air and Space Museum Archive.

[39] For more on Tsien, see Iris Chang, *Thread of the Silkworm*, Basic Books, New York, 1995.

[40] Frank J. Malina, "America's Long-Range Missile and Space Exploration Program: The ORDCIT Project of the Jet Propulsion Laboratory, 1943–1946: A Memoir," *Essays on the History of Rocketry and Astronautics: Proceedings of the Third Through the Sixth History Symposia of the International Academy of Astronautics, Volume II*, edited by R. Cargill Hall, NASA Washington, DC, Conference Publication 2014, 1977, p. 349.

[41] Jerry Grey, *Enterprise*, William Morrow, New York, 1979.

[42] Theodore von Kármán, "Martin Summerfield: Ten Years as Editor," *ARS Journal*, 1961.

[43] Whipple was accepted by an ARS Board Meeting on Oct. 25, 1946. Truax was accepted on Dec. 4.

[44] Theodore von Kármán with Lee Edson, *The Wind and Beyond*, Little, Brown, Boston, 1967, p. 259.

[45] Andrew Haley biography, New Mexico Museum of Space History, Alamogordo, New Mexico.

[46] Grey, *Enterprise*, p. 118.

[47] Telephone interview with F. C. Durant, III, Feb. 6, 2005, notes in the author's collection.

[48] Andrew G. Haley Papers, Box 18, National Air and Space Museum Archive contains Haley's early correspondence regarding the ARS.

[49] "Architect of the Space Age: Richard William Porter," *New York Times*, March 20, 1958, p. 12.

[50] Minutes of the Meeting of the ARS Board of Directors, June 2, 1947, AIAA Headquarters Files.

[51] The details of Frederick C. Durant, III's, CIA career were the subject of a telephone interview with the author, February, 2005. For documentary evidence of the connection, see F. C, Durant, III, "Report of Meetings of Scientific Advisory Panel on Unidentified Flying Objects Covered by Office of Scientific Intelligence, CIA, Jan. 14–18, 1953," Document I-18, *Exploring the Unknown: Selected Documents in the History of the U.S. Civil*

Space Program, Volume 1: Organizing for Exploration, edited by John M. Logsdon, NASA, Washington, DC, 1995, p. 201.

[52] Minutes of the Meeting of the ARS Board of Directors, May 7, 1951 and Jan. 5, 1953. AIAA Headquarters Files.

[53] Minutes of the Meeting of the ARS Board of Directors, June 2, 1952, AIAA Headquarters Files.

[54] Minutes of the Meeting of the ARS Board of Directors, Dec. 1, 1953, AIAA Headquarters Files.

[55] Minutes of the Meeting of the ARS Board of Directors, May 7, 1951, AIAA Headquarters Files.

[56] Minutes of the Meeting of the ARS Board of Directors, Feb. 19, 1951, AIAA Headquarters Files.

[57] Minutes of the Meeting of the ARS Board of Directors, June 2, 1952, AIAA Headquarters Files; on the 1953 honors night banquet see James Harford,: *The American Rocket Society, 1953–1963: A Memoir,*" IAF-88-605, author's collection.

[58] Minutes of the Meeting of the ARS Board of Directors, April 5, 1954, AIAA Headquarters Files.

[59] Minutes of the Meeting of the ARS Board of Directors, Feb. 7, 1955, AIAA Headquarters Files.

[60] E. W. Kenworthy, "Espionage in Real Life Can Be Duller Than Fiction," *New York Times,* Oct. 20, 1957, p. E6.

[61] Martin Summerfield to C. W. Chillson, Nov. 28, 1952, Andrew Haley Papers, NASM Archive, Box 18.

[62] F. S. Durant, III, to C. W. Chillson, Jan. 18, 1952; Chillson to Durant, Jan. 25, 1952; Durant to Chillson, Jan. 29, 1952; all letters in the Charles W. Chillson Papers, Collection XXXX-0008, Box 1, National Air and Space Museum Archive.

[63] Andrew Haley to Martin Summerfield, July 10, 1952, Haley Papers, Box 17.

[64] The election went exactly as Haley had hoped. When the ballots were counted, he received 441 votes for vice president. Durant had received only 439 votes for president. ARS Annual report for 1952, Chillson Papers, Box 1.

[65] Minutes of the Meeting of the ARS Board of Directors, Sept. 14, 1953, AIAA Headquarters Files.

Chapter 6

Engineering the Space Age: The American Rocket Society, 1953–1958

Dawn of the Harford Era

Born in Jersey City in 1924, Jim Harford graduated from Yale University with a degree in mechanical engineering in June 1945 and immediately entered the U.S. Navy Reserve as an ensign. Discharged in 1946, he worked as a sales engineer with the Worthington Corporation from 1946 to 1949, marketing pumps and other machinery. "I was a lousy engineer," he is quick to admit. He had a way with people, however, and a talent with words. A job as associate editor of *Modern Industry* (1950–1952), later *Dun's Review*, was much more to his taste.

When Harford proposed to his future wife in 1952, she accepted on the condition that he take her to Paris. Fortunately, *Modern Industry* was sending a team to Europe to report on the progress of the Marshall Plan. The couple settled in Paris, where Harford supplemented the money that he received from reports to *Modern Industry* by signing on as a contract writer for the U.S. Mutual Security Agency (MSA), which administered the Marshall Plan. Supplied with a government car, he and his wife toured the French countryside, visiting various plants and preparing reports for the MSA. In addition, he brought in extra money by writing the occasional article for a French industrial journal under an assumed name, "so it wouldn't look like an American preaching to the French."[1]

After 15 months, with their first child on the way, the Harfords were ready to come home. They were sharing a salami sandwich and a bottle of wine in the Jardin de Tuileries when Harford opened a letter from his brother Tom, a partner in the firm of Emery and Harford, who were selling advertising space for the *Journal of the American Rocket Society*. The American Rocket Society (ARS), he noted, was looking for an executive director, someone with experience in both engineering and publishing. Would Jim be interested? "I remember laughing ... The American Rocket Society! Can you imagine that!" Several weeks later, his brother followed up with a second package containing the March 21, 1952, issue of *Collier's Magazine*, with artist Chesley Bonestell's cover painting of a rocket streaking into orbit and the headline, "Man Will Conquer Space Soon." "And Wow!" Harford remembers, "That just about turned me on."[2]

Harford was hooked. The couple returned to the United States, where he was interviewed by Ed Pendray, President Richard Porter, and Fred Durant. With the approval of the Board, he began work at a salary of $800 per month on Monday, October 3, 1953. Harford was enormously pleased. He would now be bringing home more than three times the $3,000 that he had been earning in Europe. What's more, it was a job he relished. "These guys were full of zeal," he recalled half a century later, "and I picked up that zeal." He quickly discovered, however, that he would have to exercise all of his considerable people skills to prosper in his new environment.

Mrs. Slade and her current assistant, Mrs. Catherine Beck, made room for him in their small office, but they did not extend the warmest of welcomes. "Billie [Slade] was not easy to work with," Harford recalls. She saw him as a highly paid, inexperienced, "whipper-snapper with a Yale degree." He "had to learn diplomacy to keep Billie happy," and she had to understand "that that was the deal." Over time, they became an effective team.

Then there was Martin Summerfield. Harford knew that he was "crucial to the development of the ARS as a professional society." He had put the *Journal* on its feet; further strengthened the publication program by hiring an effective associate editor, Irvin Glassman; and was a persuasive voice in ARS internal affairs. For all of that, Summerfield could be "a difficult character" who argued for a larger publication than the ARS could afford. "We had," Harford recalls, "a lot of shouting matches." In the end, it was a healthy tension that resulted in a stronger organization.[1]

Harford had no illusions as to his primary task. His salary of $9,600 per annum represented more than 20% of the total annual budget of $47,000. The Board expected him to focus on signing up new corporate members with deep pockets. The new executive secretary would have to earn his keep. The ARS had 10 corporate members when Harford arrived.[3] He began by following up on East Coast contacts supplied by ARS members. One month after taking the job, he met with 20 business leaders, soliciting both corporate memberships and advertising. Next, he launched a national campaign in which ARS leaders directed letters of solicitation to friends and acquaintances in industry.

Thirty-five years later, at the time of his retirement, Harford remained grateful to the Avco Corporation and IBM, "because I remember vividly persuading those two companies to be the first Harford-generated Corporate members."[4] Laurance Rockefeller was also critically important. The first individual to sign up as a Corporate member in order to provide maximum support to the ARS, he was an angel to whom Harford would return time and again. When Rockefeller died in 2004, Harford wrote a letter to the editor of the *New York Times* chiding the newspaper for neglecting to mention the philanthropist's many contributions to rocketry in their obituary. Following World War II, he saved Reaction Motors, Inc. (RMI), from bankruptcy. He was a heavy investor in other aerospace firms, from Eastern Airlines to Piasecki Helicopter, McDonnell Aircraft, and the Marquardt Corporation. His support for the ARS was rooted in his deep fascination with the possibility of space travel, which also led him, late in life, to fund several "flying saucer" studies.[4,5]

On February 1, 1954, the members of the ARS Board of Directors paused to consider their financial position and expressed genuine concern over a drop in cash on hand and accounts receivable over previous months. Fiscal concerns ranged from delinquent membership accounts to a slow growth in corporate memberships and a slump in advertising revenues. Dr. Kurt Berman, chair of the Finance Committee, was particularly worried.

Harford, who had been on the job for only five months, asked the Board to be patient. His programs, notably the effort to increase corporate memberships, were just beginning to bear fruit. Rather than retrenching, he argued that the success of the organization depended on their willingness to expand operations. He pointed out that the Institute of the Aeronautical Sciences (IAS) maintained a permanent staff of almost 100 people to service an organization with 12,000 members. At the ARS, 4 people attempted to meet the needs of 2,500 members.[6]

The staff was overworked. "Society accounting alone," he noted, "requires one qualified person. Additional tasks included arranging meetings, assisting with Section affairs, handling publications issues, processing memberships, taking care of the roughly 200 changes of address received every month, dealing with the mail (including the envelopes for 3,500 issues of the Journal every two months), and responding to a wide range of requests for information." Mrs. Slade, he noted, had "to work several evenings a week and parts of virtually every weekend," to keep abreast of her duties. Moreover, all four overworked people shared a one and one-half room office crowded with stacks of publications, correspondence, and files. "We are too large an organization," he concluded, "to be run in [this] hand-to-mouth, penny-pinching fashion."[7]

Harford pointed to *Journal* advertising as a source of untapped revenue to finance an expansion that would enable the ARS to better serve its members. He indicated that 60% of IAS operating revenues came from advertising, but the ARS drew only 33% of its operating funds from that source. A chart brought home the rapid rise in the number of ARS members and subscribers in recent years. He summed up by outlining a six-part plan:

1) Expand the *Journal* to a monthly that would carry industry news and general information articles in addition to papers aimed at technical professionals
2) Stimulate continued membership growth by encouraging the creation of new Sections and Student Sections
3) Create a book and film library to serve the members
4) Continue the effort to boost corporate membership
5) Take advantage of public interest in space flight
6) Expand the office and the staff to improve morale and efficiency

At the next meeting of the Board, Finance Chairman Berman proposed that they allow Harford to proceed and to solicit $30,000 from foundations to support the first steps toward expanded operations. The Board agreed to expand the *Journal* to a monthly and approved a new organization chart recognizing the executive secretary as the office supervisor and the individual responsible for business and financial operations under the supervision of the Board. Harford also announced that he and Mrs. Slade had located a new and larger office space in Room 840 at 500 Fifth Avenue. The move was complete by the time of the next Board meeting on August 27, 1954. Although the subject was not discussed, it was clear to Harford that the move away from the Engineering Societies Building was the beginning of the end of their dependence on the American Society of Mechanical Engineers (ASME). The ARS would continue to cooperate with that organization and other engineering groups, but it would do so as an equal.[8]

Takeoff

As Harford promised, things began to look up in the weeks and months following the perceived crisis and the decision to expand. Within six months of taking the job, the executive secretary had signed up five new corporate members, a 33-percent increase, with two additional companies reporting that the matter was still under consideration.[9] By September 1955 the 10 corporate members of 1953 had grown to 61, 25 of which had come aboard since January 1.[10]

Harford was able to combine the effort to increase corporate support with considerable attention to the needs of ARS members. It was an essential task in an era of rapid growth. On April 1, 1945, the organization had 318 members (51 Active, 221 Associate, 46 Student). Five years later, the total had climbed to 1,002 (572 Active, 286 Associate, 144 Student), a 315% increase![11] The number of institutions subscribing to the *Journal* had jumped from 60 in 1945 to 394, for a 650% increase. At the beginning of 1953, the ARS counted 2,546 members, a 39% increase in less than three years.[12] By March 1, 1954, just four months after Harford had taken the reins, the total number of members and subscribers had climbed to 3,328. The grand total of members and subscribers reached 4,282 (3,125 members, 1,157 subscribers) by January 1, 1955. The total would reach 21,000 by the end of the decade.[4,13,14]

As membership grew, the Sections became ever more important. Between 1948 and 1950, the ARS established five Sections, so few that many members living outside New York City stood alone as at-large members.

ARS Section, 1950 (Ref. 15)

	Member Type			
Section	Active	Associate	Student	Total
New York	265	90	45	400
Washington, DC	25	25	7	57
South California	114	19	12	145
New Mexico/West Texas	58	16	2	76
Indiana (Purdue University)	4	3	23	30
At Large	108	138	55	301

By the fall of 1955, the ARS had expanded to include 26 regional sections.

ARS Section, September 1, 1955 (Ref. 16)

	Member Type			
Section	Active	Associate	Student	Total
Alabama	73	1	7	81
Arizona	8	3	0	11
Central Texas	20	2	0	22
Chicago	54	11	11	76
Cleveland/Akron	78	12	8	98
Detroit	66	13	16	95
Florida	40	11	4	55
Fort Wayne	5	2	38	45
Indiana	16	5	25	46
Maryland	79	4	6	89
National Capital	81	16	8	105
New England	31	7	8	46
New Mexico/West Texas	123	17	20	160
New York	371	77	100	548
Niagara Frontier	111	7	8	126
Northeastern New York	47	0	4	51
Northern California	35	6	12	53
Pacific Northwest	51	13	8	72
Philadelphia	37	11	11	59
St. Louis	30	1	29	60
Southern California	572	40	33	645
Southern Ohio	52	13	10	75
Twin Cities	3	2	29	34
Others	90	54	78	222
Canadian	12	2	1	15
Foreign	37	9	3	49
Total	2,122	339	477	2,938

Mrs. A. C. "Billie" Slade and Jim Harford (front row center) and the ARS staff, 1957: Dean Roberts (left front); Bill Chenoweth (right front); (back row left to right) unknown, unknown, Eugenia Ott, Peggy Brookfield, Catherine Beck, Wilma Blake, Ruth Locke, Walter Brunke. (Courtesy AIAA.)

Harford recognized that the Sections were the heart of the organization, the essential units through which individual members connected to the national program. He made it his business to attend as many Section meetings as he could. He got to know local leaders, worked hard to establish programs that would meet the needs of the Sections, and made good communications his highest priority.

A strong program was the key to success. Walter Brunke joined the staff in 1954 with a portfolio that included everything from meeting arrangements to running the mail room. His real specialty, Harford recalled, involved "bootlegging services from ASME and SAE and any other society in the building that he could charm." He would remain on the staff of the ARS/AIAA until his retirement in 1988. Some notion of his immediate impact can be seen in the expanded list of meetings that the ARS planned or participated in during 1955: (Ref. 4)

- A full session of four papers at the IAS annual meeting in New York on January 27
- Spring meeting with three sessions of eleven papers in Baltimore on April 18–21
- Boston meeting with two sessions plus two additional sessions cosponsored by ASME
- International Astronautical Federation (IAF) in Copenhagen on August 1–6
- Aerothermochemistry Symposium in Evanston, Illinois, on August 22–24
- Los Angeles meeting with six to ten sessions on September 19–21
- Annual meeting in Chicago on November 13–18
- Nuclear Congress in Cleveland on December 12–17

With other changes underway, the time had come to resolve some of the organizational problems that had created internal difficulties in the early 1950s. One step aimed at alleviating some of the concerns voiced during the 1953 election was to double the size of the 1954

Nominating Committee from 7 to 14 members. That small effort toward democratization was not enough to prevent a minor revolt.

As the immediate past president, Fred Durant chaired the 1954 committee, which included all of the 1953 Section presidents. He outlined his plans in a letter to members of the group in the summer of 1954. Durant took a traditional line, urging that Richard Porter, the current vice president, be moved up to the presidency. Always before, the nominating committee had suggested a single name chosen from the members of the Board of Directors to serve as vice president. Durant favored Noah W. Davis of the Buffalo Electrochemical Company. He did suggest adding a blank space to the ballot so that members could write in a candidate for either office.

Durant emphasized that nominees for the Board of Directors should be "of the working variety." Anyone elected to the Board, he believed, should commit to attending at least 6 or 7 of the 10 or 11 meetings held each year. A member living within 300 miles of New York, he argued, should try to attend almost every meeting. Durant, who lived in Washington, held himself up as an example, noting that he had missed only one meeting in three years, and that a result of family illness. He provided a list of men who, in his opinion, were qualified to serve as directors but invited other members of the committee to suggest names of their own.[17]

Upset by the letter, Edward F. Francisco, past-president of the New Mexico–West Texas Section, asked for a special meeting of the Section Board of Directors. A spirited discussion produced a long letter from Section officers to national president Andy Haley. The group fundamentally disagreed with the practice of automatically moving the vice-president up to the highest office. Nor did they buy the argument that, "in these days of modern communications," a member of the Board of Directors should have to attend all meetings in person. Moreover, they felt "constrained" by Durant's suggestion of a single name for vice president and by his suggested candidates for the national Board. They expressed their fear that the election was "loaded," and pointed to the need for "new blood" and an open election.[18]

Dick Porter suggested that all of the material be sent to Durant, as chair of the Nominating Committee, without recommendation. Pendray preferred calling Durant's "attention to the seriousness of the problem," and asking that he make recommendations to the national Board. Still others preferred empowering Haley to write to the Section, defending Durant's approach and pointing out that Mr. Francisco, as president of the Section, had voted to continue all standing practices. Noah Davis voiced fears of a serious split in the Society and counseled that all Sections be informed that the national Board was considering new bylaws that would address precisely the issues raised by the New Mexico–West Texas Section. Martin Summerfield's recommendation carried the day:

1) The letter would be referred to Durant with instructions that the Nominating Committee should not consider itself bound by tradition or any consideration other than good judgment.
2) The Nominating Committee was empowered to spend any funds required for travel necessary to obtain a consensus on the proposed slate of candidates.
3) President Haley was asked to communicate the action of the Board to the Section and to inform them that a revision of the bylaws reflecting their general recommendations was under consideration.[19]

The Board supported Durant's position, with only the most modest gesture toward compromise. The final ballot included Porter and Davis as the sole candidates for president and vice president. The candidates for membership on the Board of Directors were drawn entirely from Durant's list, with the addition of Edward Francisco, Jr., the chair of the New Mexico–West Texas Section.

As promised, new bylaws were drafted, discussed, and approved. The older policies that had created divisions were often traditional practices, not rules codified in print. There was, for example, nothing on paper that said that, having served a year in office, the vice president would succeed to the presidency. Nor were there rules insisting that members of the Board be East Coast men to ensure that they could attend a maximum number of meetings. That being the case, the revision of the bylaws did not have to deal directly with those issues.

Rather, the ARS members created a new governing unit, the Executive Committee. For the first year of operation, beginning in January 1955, the group would include President Richard Porter; Vice President Noah Davis; Treasurer Robert Lawrence; Andy Haley and Milton Rosen, representing the Board of Directors; Jim Harford; Billie Slade; and Martin Summerfield. The Board would remain the primary governing body, but it would now meet quarterly, whenever possible as part of a national meeting. The concentration of leadership in a smaller body combined with simple alterations to the traditional ways of doing things that had generated problems would work wonders to guarantee a smoother course into the future.[20]

Debating Space

On December 29, 1952, William L. Gore, a one-time Navy test pilot now responsible for marketing Aerojet products to the government, dropped a line to Andy Haley. Gore, who had been president of ARS in 1950, wanted to share his concerns with his old friend, a member of the ARS Board of Directors who had recently been appointed chair of the organization's new Ad Hoc Committee on Space Flight. "There appears to be a trend of thinking pulsating throughout a certain segment of the membership," he began, "which, if continued, could revert the Society to its former role of the past, instead of the continuous progress now being made toward the future." The problem, he continued, "is the von Braun inspired vogue of the Satellite Vehicle and Space Travel."[21]

The tension between those who regarded any discussion of space flight initiatives as potentially dangerous to the reputation and image of the ARS and those members who believed that serious technical studies of space flight issues were entirely appropriate and critically important to the future of the organization was a major thread running through the history of the organization during the years 1950–1955.

The situation was easy enough to understand. The ARS had begun 20 years before as the American Interplanetary Society (AIS), an organization founded to nourish and promote the dream of space travel. For a decade, the members had dedicated themselves to developing a successful liquid-propellant rocket motor, the progenitor of powerful rockets that would one day propel human beings away from the earth. During World War II, however, rocketry passed from the hands of the amateurs to those of the industrialists.

The ARS had transformed itself into an organization representing the needs and interests of a new generation of scientific and technical professionals who were flooding into the field. For a time, the dream of space flight was overwhelmed by the difficult engineering problems that had to be overcome if rocket weapons suited for the Cold War were to be perfected. In such an atmosphere, it seemed to some that any discussion of travel beyond the atmosphere was Buck Rogers stuff, an unworthy distraction from more important matters at hand.

Young engineers with an interest in space flight were frequently advised to cool their enthusiasm. That was a problem for Frederick I. Ordway, III, who had discovered science fiction as a youngster and joined the ARS as a student member in 1941, while still in his early teens. In 1951, following military service and graduation from Harvard, Ordway went to work for the Engineering Division of RMI, where a superior reacted to his talk of interplanetary voyaging with the remark that he personally did not care whether he was designing bath tubs or thrust chambers. Moving on to Republic Aviation in 1954, Ordway found that any discussion of space travel made the leaders of the Guided Missile Division very nervous indeed.[22]

Kurt Stehling, an irrepressible "space cadet" since his high school years in Canada, faced the same problem. His engineering colleagues at Bell Aircraft kidded him when local newspapers interviewed him on space issues. His supervisor, Walter Dornberger, who had been the military commander of the German rocket effort during World War II, "grumbled and carped about my satellite ideas" and advised against becoming involved in ARS space discussions. When Stehling did join the ARS in 1948, the organization was not, he emphasized, "a rallying point for space protagonists." Propulsion, he notes, "was the main concern."[23]

Ironically, just as the professional rocket engineers were focusing their attention on near-term technical problem solving, the general public was catching the space bug. Willy Ley, who had squired the Pendrays through the Rakentenflugplatz so many years before, published the first edition of his best-known and most influential book, *Rockets*, in 1944. Based on the author's 20 years of research, his own experience in Germany, and his correspondence with virtually all of the pioneering figures in the field, the book traced the history of rocketry from the black powder era through the 1930s and explained the basic physical principles that would govern space flight. Ley would update the book through edition after edition over the next 25 years. He succeeded in capturing the attention and the imagination of readers young and old.

Artist Chesley Bonestell introduced the reader's of *Mechanix Illustrated* to the idea of a "Rocket Moon" in September 1945. It was the first in a series of popular magazine pieces illustrated with his stunning paintings of space travel and lunar and planetary surfaces.[24] It was impossible for anyone with half an imagination and an adventurous spirit to look at Bonestell's art and not dream of standing on another world. In 1949, he collaborated with Willy Ley to produce the beautifully illustrated book *Conquest of Space*. The next year, Bonestell teamed with producer George Pall and science fiction writer Robert Heinlein to create the classic space flight film *Destination Moon* (1950).

The film was loosely based on *Rocket Ship Galileo* (1947), the first in a long string of Heinlein's juvenile novels that would infect American youngsters with interplanetary wanderlust. If that was not enough, Fletcher Pratt, an ARS founder, and artist Jack Coggins published the first edition of *Rockets, Jets, Guided Missiles and Spaceships*, their own glossy and richly illustrated book on the subject in 1951. For those lucky young Americans whose parents were among the first on their block to invest in a television set, *Tom Corbett, Space Cadet* took to the airwaves for the first time in October 1950. This was just the beginning.

Bill Gore was certainly correct in identifying Wernher von Braun (1912–1977) as a man who was doing all that he could to inspire enthusiasm for space travel. An active member of the German Rocket Society while still a teenager, he had earned a Ph.D; played the key role in selling the German military establishment on the idea of rocket weaponry; and led the team that developed the A4 (V2), the first large ballistic missile, and the progenitor of the first space launch vehicles. As gifted a manager as he was an engineer, von Braun had no doubt that government funding for guided missile development would open the way to space. So it had.

As the Third Reich collapsed around them, von Braun and his team surrendered to the American Army and negotiated their transfer to the United States, where he knew they would have the best opportunity to continue their work. They spent several years in West Texas and

New Mexico, supporting Dick Porter's team of General Electric (GE) technicians who were reassembling and launching captured German A4s. Only when the Army shifted them to the Redstone Arsenal in Huntsville, Alabama, were they finally able to pick up where they had left off in 1945.

Richard Porter, ARS President for 1955. (Courtesy AIAA.)

In 1948 and 1949, while living in exile in the Southwest, von Braun kept himself busy developing a detailed plan for the first expedition to Mars. What preliminary steps would lead up to an interplanetary voyage? What sort of ships would be required? How would they be constructed? How long would the journey take? How could the crew be protected from the manifold dangers that they would encounter? The result was complicated treatise crowded with differential equations. Convinced that popular support would speed the advent of the space age, von Braun then wrote a long science fiction novel about such a trip. His engineering treatise would be published as an appendix to *The Mars Project*.

He sent the manuscript to a New York agent in February 1950. It was passed from one editor to another until as many as 18 publishers had rejected it. Clearly von Braun was a far better engineer than novelist. One of the editors apparently asked Bill Gore for his opinion. Then serving his second term as president of the ARS (1949, 1950), Gore told Fred Durant that the book was so fascinating that he could not put it down.[25] Chuck Chillson read it as well. He found it "extremely interesting ... fascinating and informative ... ." Enthusiasm was so high that the Society actually considered publishing the book. After "much discussion," on December 11, 1950, the members of the Board instructed Gore to advise von Braun that they could not afford to undertake the project. In a follow-up letter in the spring of 1952, Gore again explained that he "... was personally extremely disappointed that it was not financially possible to arrange for the English publication of your fascinating work."[26]

For more than a decade, the ARS had ignored the problems of space flight, focusing attention on technical issues of more direct interest and benefit to their engineer members. Kurt Stehling recalls that the discussion of space flight was much more common at the Section level, which "often sponsored talks and lectures on astronautics that would have been banned at national meetings." There was, he recalls, "a feeling that other societies and groups, such as the High Altitude Panel, were gradually relegating the ARS into an "artificers, plumbers, fire sprayers, and injection head drillers society ... ."[27]

That began to change on December 11, 1950, when Mrs. Slade presented a letter from the British Interplanetary Society (BIS) inviting the ARS to participate in the Second International Congress of Astronautical Societies, the predecessor of the IAF, to be held in London in September 1951. The Board instructed her to respond with an apology for the failure of the ARS to participate in the first Congress, which had been held in Paris, and to promise to supply representation to the London meeting. Perhaps one of the British members of the ARS could be persuaded to represent the United States.

William L. Gore, ARS President for 1950. (Courtesy AIAA.)

In fact, Fred Ordway, an enthusiastic, if very junior, member of the ARS had attended the organizing meeting the year before. Honeymooning in Paris in August 1950, he saw a notice of the meeting and arranged to attend. As a "walk-in" delegate to this first international space meeting, he met engineers from across Europe as well as the leaders of the BIS, including both Arthur C. Clarke and Eric Burgess. Within a few years, Ordway would establish himself as a member of the inner circle of American space enthusiasts.[22]

Billie Slade received a second letter from the BIS early in 1951, urging the Americans to attend the London meeting and asking them to submit papers to be read at the gathering. Karman and Frank Malina, both of whom were in Europe, would attend. Haley and Durant announced that they were also thinking about attending. In fact, Durant had suggested to his Central Intelligence Agency (CIA) managers that the London conclave would be an ideal opportunity to get to know foreign rocket experts and to learn something of their programs. The CIA agreed to fund the trip.

Soon thereafter, von Braun informed the group that he wanted to present a paper at the gathering, "The Importance of a Satellite Vehicle as a Step Toward Interplanetary Flight." It was an early section of his *Mars Project* describing the need for a large space station in earth orbit from which to launch voyages into deep space. Security officials would not allow him to leave the country, however. Although he could have asked a member of the BIS, of which he was a member, to read the paper for him, Von Braun through it more politic to ask for the assistance of the ARS. Durant offered to discuss the matter with von Braun, and if they both agreed, to read the paper for him. The pair met in New York and struck up an immediate friendship that would last until von Braun's death in 1977.

Founded in Paris in 1950, the IAF was the first association established to advocate space flight through cooperative discussions between nations. The organization planned to meet periodically in a host nation. Over the years, the biannual meetings offered the perfect occasions to announce plans and achievements to the world. Other organizations, notably the Space Science Symposium sponsored annually by the Committee on Space Research (COSPAR) of the International Council for Science, founded in 1958, attracted the heavyweight scientific papers. The IAF planned sessions that were broader and designed to appeal to those who were not narrow specialists. The meetings became especially important as the one place where space scientists and engineers from around the world and both sides of the Iron Curtain could meet one another and discuss issues of mutual interest. It was an opportunity to make and cement personal contacts, buttonhole foreign colleagues, hold private conversations, and fish for information.

Sixty-three delegates attended the 1951 IAF Congress in London. The theme was the "Earth-Satellite Vehicle." The real purpose of the meeting, which attracted considerable international publicity, was to underscore the fact that space flight would be a reality in the next few decades. As Chairman Arthur C. Clarke explained, "Space Flight is likely to be the next technical achievement of our species."[28]

Fred Durant made a considerable splash in London. He presented von Braun's paper, one of the highlights of the meeting, while dressed in his uniform as a Lt. Commander in the U.S. Navy Reserve. He was invited to appear on the BBC program, *In Town Tonight*. True to form, he made new friends and contacts with his ready smile and constant good humor. This was precisely what his CIA employers expected of him, and it suited Durant right down to the ground. He made such a good impression that the members of the IAF elected Fred Durant the second president of the organization (1953–1956), following President Eugen Sänger, whose plans for a world-circling spaceship had inspired his own interest in the subject. Andy Haley must also have made an impression. He was elected vice president of the IAF for American Affairs at the London meeting and would become fourth president of the organization (1957–1959).[29]

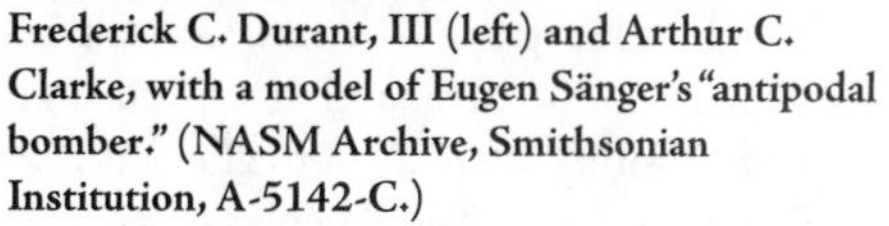
Frederick C. Durant, III (left) and Arthur C. Clarke, with a model of Eugen Sänger's "antipodal bomber." (NASM Archive, Smithsonian Institution, A-5142-C.)

Durant was especially pleased to meet Arthur Clarke. The two took to one another immediately. For decades to come, Durant's Washington home would become one of Clarke's principal American bases of operation. At the time, Clarke was just beginning to emerge as a promising science fiction writer. His nonfiction book *Interplanetary Flight* (1950) had done well in Britain and was about to be published in the United States as *The Exploration of Space*. Durant knew of him as a result of reading the *Journal of the British Interplanetary Society*.

The American delegate was an admirer of the BIS. Founded in Liverpool in 1933 by Phillip Cleator, the organization had been in part inspired by the AIS. Like the early AIS, it was a refuge for science fiction fans and dreamers. Clarke and other early members were avid readers of American science fiction and of ARS President David Lasser's *The Conquest of Space*, the first widely available English language book on the subject. Although the members of the AIS followed the German pattern, transforming themselves into a rocket research group, the BIS remained true to its original principles. Rather than building and flying rockets, the members preferred to conduct paper studies of future spacecraft and missions, which they published in their *Journal*. Some notion of the romantic vision that inspired them is apparent in a 1939 BIS promotional brochure written by Arthur Clarke:

> Go out beneath the stars on a clear winter night, and look up at the Milky Way spanning the heavens like a bridge of glowing mist . . . . Looking out across that immensity to the great suns and circling planets, to worlds of infinite mystery and promise, can you believe that man is to spend all his days cooped up and crawling on

the surface of this tiny Earth—this moist pebble with its clinging film of air? Or do you, on the other hand, believe that his destiny is indeed among the stars, and that one day our descendants will bridge the seas of space?[30]

Durant had joined the BIS in 1948, the same year he signed up with the ARS. "The AIS had lost some of the courage of its conviction," he noted, "and changed its name to the American Rocket Society. The word interplanetary was too far out for industrial support. The BIS stuck by their guns, and I have always lauded them for their courage."[28]

Durant and Haley returned from the London conference eager to share their enthusiasm at an ARS Board meeting in Atlantic City on November 28, 1951. Haley opened the discussion, remarking that "... every society must have some future concept which must not be overlooked." For the ARS, that concept was space travel. "This concept is not preposterous," Haley insisted, "it is essential; it is going to be." The *Journal*, he suggested, should devote one-quarter of its space to the subject.

Durant concurred, noting that although "the idea of satellite stations and space travel ... were treated as a thing far distant in this country," the Europeans took these notions much more seriously. He pointed out that the BIS, "... with a membership very close to our own," pays far more attention to "interplanetary projects." He urged the Board, "... not to take lightly the developments in this field which well may become a reality by the end of the century." Although he recognized "... that a certain balance must be maintained," the "... inclusion of a certain amount of astronautical material in the *Journal* would do much to encourage the interest of young students in the field as a whole."[31]

Richard Porter offered an immediate response, remarking that "interest in interplanetary travel is without doubt something which will eventually develop into specific problems." At the moment, however, it was "... the duty of the Society and the function of the *Journal* to devote all of its efforts to the engineering problems facing the rocket and jet propulsion personnel." The government, he concluded, needed, "... every bit of help it could get, and particularly in the engineering field, to meet the defense mobilization program," adding that "the American Rocket Society should post-pone its explorations into space travel until such time as rocket engineers are not so vitally needed by Uncle Sam."[31]

Fred Durant rose to his feet once more, arguing that "the satellite vehicle" would have very real military applications. In any case, he remarked, increasing security might restrict the publication of more current research in the *Journal*. Interesting papers on space flight could fill the gap. Not to be outdone, Haley added that there "was no irreconcilable difference" between the two points of view. At the same time, he argued, an organization like the ARS needed to look toward the future, and that future lay beyond the atmosphere.[31]

Some elements of the ARS membership were certainly willing to jump into the space age with both feet. RMI engineers proposed a "minimum satellite launch vehicle" that they described as a sort of space age "Model T."[27] On November 17, 1951, Rollin W. Gillispie, who had been employed by Bell Labs, GE, and Aerojet, wrote the ARS Board, suggesting that they sponsor his Project Arc, "a non-governmental undertaking to prepare complete engineering specifications for a circumlunar rocket and related facilities." That sort of enthusiasm can only have increased the caution of Dick Porter and others like him.

Twelve days later, on November 29, Robert Truax, a long-time member who had participated in the prewar ARS rocket tests while still a Midshipman at Annapolis, sent another proposal to the Board. He pointed out that the original purpose of the AIS had been to promote space flight and suggested that "the obvious first step is the development of satellite vehicles." That was nothing new, he reminded the Board, alluding to military satellite studies, all of which were classified.

Indeed, at least some members of the ARS Board were familiar with the story. As early as 1944, Commander Harvey Hall, Lloyd Berkner, and several other U.S. Navy scientists studied the possibility that the Germans might launch a satellite with a V2 rocket. Their calculations indicated that no existing rocket could perform that task. The following year, the Navy Bureau of Aeronautics appointed Hall to head a Committee for Evaluating the Feasibility of Space Rocketry, which studied the requirements for launching a satellite that would collect data on conditions in space and relay it back to earth.

When Homer Stewart of Cal Tech's Jet Propulsion Laboratory confirmed the Bureau of Aeronautics study, the Navy contracted with North American Aviation, Inc., and the Glenn L. Martin Company for preliminary studies of the earth satellite vehicle. Although the Navy decided that the project would be too expensive for a purely scientific return, the U.S. Army Air Forces commissioned the Rand Corporation, a spin-off of Douglas Aircraft, to conduct its own study. The Rand analysts failed to suggest a practical use for such a vehicle, but they underscored the psychological advantages that would accrue to the first nation to accomplish the feat.

Unwilling to completely abandon the effort, Navy researchers developed plans for a high-altitude test vehicle (HATV), a rocket that could boost an instrument package to altitudes of up to 400 miles, gathering data on the upper atmosphere and near-earth space and demonstrating technology that might convince military planners of the feasibility of an orbiting satellite. That project, along with surviving Air Force and Army paper projects, died when a public outcry against fiscal irresponsibility greeted a paragraph in the 1948 annual report of the Secretary of Defense admitting that all three services were considering earth satellite vehicles programs.[32]

By 1951, Bob Truax was growing impatient with the half-hearted military efforts that remained cloaked in secrecy. "The time has come," he announced to the ARS Board, "to stop talking about space travel and start doing something about it." Within a year or two, he predicted, "the military rocket program will have carried rocket development about as close to the space vehicle as it is likely to come for some time." He had no doubt that "a unified program, representing both civilian and military interest, would be successful in obtaining from the government the required funding." Here he suggested, was a job for the ARS.

Truax could "... think of no other agency that has the breadth of scope and talent within its membership to do it." The cost of the program, he estimated, would be $100 million, "Andy Haley knows the ropes around Capital Hill," while "Laurance Rockefeller has wide contacts in business and finance." Surely Colonel Holger Toftoy of the Army missile program and his counterparts in the Navy and Air Force would cooperate. In order to get things started, he was sure that "... the membership at large ... will be much more willing to chip in when there is a concrete objective at hand."[33]

It was nothing less than a call for the ARS to abandon its role as a traditional professional society in favor of managing a national space effort. Truax had first raised this issue late in 1950 when he accepted the Goddard Award, the most prestigious prize that the ARS could bestow. His proposal, radical as it must have seemed to those members who were nervous about the very mention of space flight, could not be written off as easily as Gillispie's notions.

The debate continued at a Board meeting on February 4, 1952. Durant argued that the ARS should "hold leadership in the field of space travel." If they did not, he warned, another group would. Haley added that a space travel initiative would make a good "peg" on which to hang the new ARS public relations program that Ed Pendray was developing. Professor Maurice Zucrow of Purdue University urged restraint, pointing out that "developments in space travel are still not of a scientific nature, and it would be most unsuitable for the Society to wander too far from technical problems." C. C. Ross argued that the ARS would have to work through another organization, like the National Science Foundation (NSF). Perhaps their most useful function was to reignite military and Congressional interest in the project.

President Chillson closed the discussion by appointing an Ad Hoc Space Flight Committee, which was to "... investigate and recommend to the Board the way in which the American Rocket Society can best serve its members and fulfill the aims outlined in the by-laws for activity in the field of space flight, satellite space stations, astronautics, etc., and to define the extent to which the American Rocket Society should become active in these fields." Andy Haley would chair the group, which would include Fred Durant and Richard Porter as representatives of the national Board and four interested members of the Society: Kurt Stehling of Bell Aircraft, Milton Rosen of the Naval Research Laboratory, Aerojet's William Zisch, and John R. Youngquist of the Glenn L. Martin Company.[34]

Truax responded with another letter in February 1952, offering advice to the members of the Ad Hoc Committee and including a suggestion that they simply ignore the Project Arc proposal. The group consisted of Aerojet engineers who might be useful members of a technical team, he explained, but who were not capable of commanding the support required for such a program. He also expressed his surprise at the negative reaction of some ARS members.

"It did not occur to me," he wrote, "that the members of the Society would require any great convincing to warmly support my idea." Although he recognized that "the old moon rocket enthusiasts" of the 1930s had been "well diluted with people who merely regard rocketry as an interesting way to make a living," he had been certain that a good businessman "would welcome any program which might result in more money in the industry and a broader field of application." He pointed to officials who were by no means sold on liquid-propellant rocket weapons. Space flight would provide a second market for their products.[35]

Tomorrowland

While such talk was bubbling along within the ARS, the public discussion of space flight was gathering momentum. On Columbus Day, 1951, the Hayden Planetarium of the American Museum of Natural History hosted a Space Travel Symposium. Arranged by Willy Ley, it attracted an audience of 200 to hear talks on space science, space medicine, and space law. When the Planetarium offered to take reservations for the first trip to the Moon, 25,000 people signed up.[36]

Fascinated by the reaction to the symposium, Gordon Manning, the editor of *Collier's* magazine, dispatched an associate, Cornelius Ryan, to attend a space medicine conference to be held in Houston on November 6–9. Wernher von Braun, one of the presenters, captivated him, as he had so many others. Ryan returned to New York planning a spectacular series of articles on space flight.

Published in eight issues of the magazine from March 22, 1952, to April 30, 1954, the articles covered the future of space travel from the launch of the first orbiting earth satellite to the first human voyage to Mars. Wernher von Braun, Willy Ley, Heinz Haber, and other authorities in the field worked closely with *Collier's* writers and a team of illustrators that included Chesley Bonestell, Fred Freeman, and Rolf Klep. Seth Moseley, the magazine's publicist, planned an unprecedented media blitz that included television appearances for von Braun with John Cameron Swayze on the *Camel News Caravan*, Dave Garroway on *Today*, and on the *Gary Moore Show*. There were store window displays in New York and Philadelphia and press releases blanketing the nation's newspapers and broadcasting outlets. Cornelius Ryan spun two books out of the series, *Across the Space Frontier* and *Conquest of the Moon*. Ley and von Braun turned out a book, *The Exploration of Mars*.[37–39]

The series caught the imagination of Walt Disney, who produced three Disneyland television programs based on the *Collier's* articles. Nearly 100 million viewers watched the premiere of *Man in Space* on March 9, 1955. *Man in the Moon* followed later that year, and *Mars and Beyond* aired on December 4, 1957.[40]

All of that lay in the future when, on December 8, 1952, *Time* magazine published a cover article on the "Journey Into Space." The magazine presented a more cautious view, focusing on the division among rocketeers who regarded von Braun as a "major prophet and hero," and those who saw him as a "wild propagandist." Obviously, the German émigré had become a symbol of the division that was apparent in the ARS.

As Jonathan Leonard, the author of the *Time* piece, noted: "The practical rocket men fear that their gradual march toward space may disappoint an oversold public," who, "happily mixing fact & fiction, apparently believes that space travel is just around the corner."[41] Although the von Braun plan as outlined in *The Mars Project* and covered in the *Collier's* series, books, and television programs began slowly with the launch of an unmanned earth satellite, the focal point was on the creation of a large orbiting space wheel where astronauts would live and work. The occupants of such a station, he argued, "could dominate the world."

> Every two hours it would circle the earth .... A 100-inch telescope parked in space and manipulated by remote controls could distinguish objects on the earth only 16 inches apart. This would permit ... U.S. observers to report ... every change of the Kremlin guard. Large objects, such as Russian air bases would show up plain as day. The station would also be useful for launching atom-armed guided missiles. They would spiral downward red hot, and their descent would be timed to keep them in view of the station. Their targets on earth would be visible too. As the missile approaches its target, its course could be corrected by radio from the station, making a square hit inevitable. Once a supply of such missiles had been stockpiled in orbit, potential aggressors below would be forced to keep the global peace.[41]

As Leonard noted, "most rocket engineers, even the hard-headed practical ones, are deeply moved by the idea of space flight." When they looked at von Braun's proposal, however, "each man sees the worst difficulty in the specialty he knows best." Dr. Milton Rosen, who was heading the Navy's Viking sounding rocket program and who was a member of the ARS Ad Hoc Space Committee, was characterized as "the most articulate critic of the Von Braun plan." Described as "a careful, meticulous man," Rosen was said to be "frankly aghast at the difficulties that Von Braun lightly brushes aside."

A far more bitter critic, identified as "an important missile expert," remarked that Von Braun had "lost the war for Hitler."

> His V-2 was a great engineering achievement, but it had almost no military effect and it drained German brains and material from more practical weapons ... he was thinking of space flight, not weapons, when he sold the V-2 to Hitler .... He is still thinking of space flight, not weapons, and he is trying to sell the U.S. a space flight project disguised as a means of dominating the world.

In answering the critics, von Braun admitted that "his writings have been deliberate attempts to arouse public enthusiasm and warm the cold feet of timid military planners." He reminded the members of the ARS that he, too, was "a hard-boiled development man, who had his ample share in setbacks and disappointments." He was deeply involved in missile development and regarded space flight as something that would be achieved on the foundation of work presently underway.[42] He agreed with Rosen and others that the first step toward space would be to create a special commission, above political and interservice rivalries and the "self-seeking pressure of missile manufacturers," to study the matter, make decisions, and chart a course for the future. In the end, von Braun, Rosen, and Truax would have found themselves very much in agreement on the key issues.

Inside the ARS, however, the publicity only increased the concerns of those who regarded space flight advocacy as a dangerous tactic. Having struggled to obtain ARS Board approval of a $300 annual payment in support of the IAF, Andy Haley complained to a friend that the members of the Board were "very cool" toward the international organization with its "Buck Rogers connotations."[43]

Fred Durant admitted that "things did get worse for the development of space when I was President of the ARS." Corporate support was crucial to the ARS. Not only did advertising dollars and corporate memberships keep the Society afloat, industry support enabled young engineers to participate in ARS activities. In 1953, as Durant knew, "the guided missile industry was where the money was . . . ."[44]

Remarking on the Truax proposal in a letter to Andy Haley in January 1952, one-time ARS President Chuck Chillson complained that he found "it difficult to wax intense over sponsorship of immediate efforts to establish satellite space stations and interplanetary voyages when we still have many years, if not decades, of work to do before we can reach all parts of the earth with rocket powered vehicles." With the world "in its present state," he concluded, "we can ill spare . . . the time, money and effort, for extra curricular, extra-terrestrial projects."[45]

Other industry leaders were even more adamant. Bill Gore told Fred Durant that Dan Kimball, head of Aerojet-General and a future Secretary of the Navy, had warned him that "if the ARS doesn't stop this foolishness, Aerojet would withdraw its support."[46] Gore, who had praised von Braun's vision just two years before, now sounded a different tune in a letter to Andy Haley. He admitted that "the sound thinking rocket technician and his administrative counterpart, which in general compose the majority of the membership," looked forward to the day when space travel would be as commonplace and television and jet airliners had become. He doubted, however, "whether such trends should be advocated and supported officially by a Society when its membership is working full time endeavoring to solve the problems of making the cotton gin work."[47]

Gore argued that the ARS could not afford to "jump the gun of propriety, bear the brunt of publicity and weather the storm of criticism" that would result from "revolutionary" rather then "conservative" thinking. If they were not careful, he warned, ARS leaders faced "a risk of membership withdrawal . . . if there is an official advocation of the Society towards the 'Moon Rocket' trend, when the majority of the field is engaged in solving the problems of a 50,000 ft. rocket." Gore warned Haley that the "corporate, scientific and technical membership" did not approve "the von Braun trend" and urged that "extreme caution and sound thinking must prevail if continued advancement of the field and of the Society is to be assured."

The ARS leadership dealt a symbolic blow to space flight initiatives when Martin Summerfield persuaded the Board to change the name of the *Journal of the American Rocket*

Society to *Jet Propulsion* beginning with the issue of January–February 1954. The notion was to build industry support and attempt to stretch the ARS to attract members from across the reaction-propulsion industry, including the booming turbojet engine firms. President Fred Durant was quick to assure readers that the ARS had not lost interest in space flight. The change simply recognized that "... many of the scientific problems relating to space flight are being studied in all realms of jet propulsion." The effort was a failure. Most propulsion engineers remained quite content to make their professional homes in the IAS, Society of Automotive Engineers (SAE), and the ASME.[48]

If the change of title of the journal pleased the conservatives in the ARS, it sent an uncomfortable message to members who believed that the organization should pay more attention to space flight. Just before Christmas 1953, Walter Hurd, vice president of Philippine Airlines, informed Mrs. Slade that he did not intend to renew his membership. He was primarily interested in the future of space flight, "and I feel this side of rocketry has not been given sufficient attention ... ."[49] Fred Durant responded, pointing out that the ARS was officially involved in the work of the IAF, describing the work of the Ad Hoc Space Flight Committee, and calling attention to a 3-hour space flight session at the last annual meeting, including the participation of Wernher von Braun and other leaders in the field.

The American Astronautical Society

Whatever progress Durant, Haley, and the members of the Space Committee had made, it was not enough for some members. At 8 p.m. on November 20, 1953, 35 people gathered in Room 129 of the American Museum of Natural History for the organizational meeting of the American Astronautical Society (AAS), "an American counterpart of the British Interplanetary Society."[50] The founder, Hans Behm, a space enthusiast and member of the Museum's Department of Micropaleontology, had organized a Staten Island Interplanetary Society in April 1952, while he was an employee of the Staten Island Museum. With that experience under his belt, he was establishing a new organization devoted to promoting space flight on a national level.

A worried Jim Harford, who had been on the job only five weeks, brought Andy Haley up to speed on the new group on November 10, after dining with Behm; James Rosenquist; Joseph Chamberlain, Chief Astronomer of the Hayden Planetarium; and Frank Forrester, the Museum's chief of public relations. "It was," he said, "easy to sense an anti-ARS feeling, a feeling that we had perhaps 'sold out' to commercial expedience."[51] The new group would be linked to the BIS, he reported. They had invited all American members of the BIS but very few members of the ARS to the meeting. "They've obviously been encouraged by Arthur Clarke to tap BIS membership in the U.S., which according to them numbers almost 500." On the basis of some "heavy late-afternoon coaching by Mrs. Slade," Harford made a case for sticking with the ARS, pointing to the work of the Ad Hoc Space Committee and noted the leading role the Durant and Haley were taking in the work of the IAF.

Harford attended the first meeting as well, attempting without success to persuade the attendees to work on behalf of the ARS rather than establishing a new organization. He "was received politely but negatively by a group of young people who were determined to form a society, expressing the conviction that the ARS was too industry-oriented and too large."[4]

Indeed, the AAS was founded by space enthusiasts who regarded the ARS as entirely too cautious. James Rosenquist, an RCA employee, believed that "... the American Rocket Society was not doing for space thought in the United States what the BIS was doing in England, or

Robert Truax, ARS President for 1957. (Courtesy AIAA.)

other groups in their respective countries."[52] Aerospace writer Martin Caidin remarked that he became involved with the new organization because, "... the ARS was getting so stuffy you could choke on the stuff." Like Rosenquist, he was anxious to create "... an outfit to match the free-wheeling and exciting spirit of the BIS."[53] Dr. S. Fred Singer, a University of Maryland physicist anxious to loft scientific instruments into orbit, recalled that he joined because the new group was serious about exploring space. "Clearly, at that time, the American Rocket Society was not."[54]

Haley was quick to follow up on Harford's comment that the founders of the AAS had the blessing of the BIS. Arthur C. Clarke was just as quick to reassure him that it was all news to the BIS leadership. "I should be sorry, after all this time," he remarked, "if you could have felt that that the BIS was up to anything behind the back of the ARS."[55] Eric Burgess, another BIS friend, confirmed Clarke's reassurance, commenting that he would "disagree with any attempt to set up a branch of the BIS in New York, exactly the same way that I would condemn any attempt by the ARS to establish a branch in London."[56]

The AAS survived and prospered. Some ill feeling remained.[57] When the new group titled their journal *Astronautics*, lawyer Haley protested. The AAS promptly shifted to the *Journal of the Astronautics*. It soon became clear that the AAS would not replace the ARS as the principal technical society in the field, nor would it replace the older organization as the primary U.S. representative to the IAF. Still, the AAS did appeal to members of the space contingent of the ARS. Fred Durant, Andy Haley, and Frederick I. Ordway were among the first 200 members of the AAS, as were leading rocket builders like von Braun and Krafft Ehricke.

"I did not spread the word about that I was an AAS member," Durant recalls. "But I also thought the AAS represented an important minority view .... Some space ideas and opportunities were pushed which could not get expressed elsewhere."[57] Like the ARS, the AAS suffered from internal political problems. A 1959 conflict over management issues led to the resignations of von Braun, Ehricke, and Singer, who apparently felt that their advice was being ignored.[58]

The AAS soldiered on, gradually focusing on both technical issues related to space flight and broader considerations of the social, cultural, and legal implications of astronautics. During the period 1958–1960, members debated the possibility of a merger with the ARS. The members of the boards of the two organizations were scheduled to meet at the AAS western national meeting on the morning of August 4, 1959, to agree on the final terms of a merger. The evening before, AAS negotiators delivered a note to the ARS contingent:

> The basic assumption in considering a merger of the ARS and AAS has been to establish a National Society that would better serve the Astronautical Sciences in the U.S.A. and the world. We feel that this aim is best served by a society dedicated

> solely to the advancement of the astronautical sciences. We therefore conclude that further discussions of an AAS-ARS merger are not warranted.[59]

Looking back on the episode many years later, George Arthur, an AAS leader, remarked: "I do hope most of us remain glad that we did not merge." The AAS, he believed, continued to play an important role, functioning as "... very much a tempering influence ... upon the giant societies." The organizations continue today, running their parallel courses.[60]

On the Utility of an Artificial Unmanned Earth Satellite

Unable to thwart the establishment of the AAS, Jim Harford urged a counterattack. "I certainly think that we [the ARS] should capitalize on the space flight angle before a group like this [the AAS] swipes it from us." The ARS, he argued, should tap the growing lay audience of space enthusiasts by offering solid and interesting information, not science fiction. The Hayden Planetarium, he had learned, received 500 letters a month inquiring about space travel. The ARS should play a role in meeting that need.[51]

The members of what was now called the Space Flight Committee aimed to do a good deal more than simply publicize space flight. At the very least, they were determined to bring the ARS into the space age and to influence national policy. The group met over a two-year period, both at the ARS offices at 500 Fifth Avenue and at Andy Haley's Washington offices. The ARS Board had finally given the committee firm marching orders:

> The American Rocket Society should act as a "catalyst" and should promote interest and sound public and professional thinking on the subject of space flight. It should not attempt to evaluate the merits of individual proposals or undertake work on the subject on its own accord. It should, however, encourage such activity on the part of other organizations.[61]

What sort of activity would they encourage? They would ignore all of the von Braun-style large-scale proposals and select an initial goal that was within the reach of current technology. "Although many proposals ... have been put forward," Rosen noted, "the small, unmanned, earth satellite is the only one for which feasibility can be shown." Everyone agreed that the successful launch of an earth satellite would be the first important step toward space flight and would have important implications for science and for national prestige. Clearly, the ARS was not in a position to build and launch a satellite, to select the group that would, or even to conduct an official assessment of the value of such a project. What they could and would do was to determine who was in a position to set the official wheels in motion.[61]

In 1953, Richard Porter had invited Dr. Alan Waterman, director of the National Sciecne Foundation (NSF), to attend a meeting of the Space Flight Committee. President Truman had created the NSF in 1950, "to promote the progress of science; to advance the national health, prosperity, and welfare; [and] to secure the national defense ... ." As an independent government organization charged with broad responsibility to encourage basic research and education in the sciences, the NSF seemed the ideal organization to conduct an unbiased study of the utility of a satellite and advise the White House on the matter. Would the Foundation welcome a proposal from the ARS justifying an NSF study of the potential utility of a satellite? Waterman responded with an enthusiastic "Yes."

Milton Rosen was appointed to head the Space Flight Committee on January 1, 1954, when Andy Haley took office as president. The members of the new committee included

Three ARS members savor a moment of triumph: (left to right) William Pickering (ARS President for 1962, AIAA President for 1963), James Van Allen, and Wernher von Braun with a model of the first American satellite. (NASM Archive, Smithsonian Institution, 78-340.)

Harry J. Archer; William J. Barr; B. L. Dorman; Andy Haley; Kenneth Jacobs; Chester M. McCloskey; Keith K. McDaniel; William P. Munger; James R. Patton, Jr. Richard W. Porter; Darrell C. Romick; Micahel J. Samek; Howard S. Seifert; Willis Spratling, Jr.; Kurt R. Stehling; and Ivan Tuey.[61]

The task at hand was to produce the proposal. The form and content were hammered out in a meeting of the full committee. The details were ironed out by letter and telephone in the weeks that followed. Rosen first sent Waterman a confidential draft calling for the NSF to study "the utility of an unmanned satellite vehicle to science, commerce and industry, and national defense." When the NSF approved, Rosen set to work to craft a formal document.[62]

The final proposal was a short, crisp, and persuasive justification for an earth satellite program. It consisted of a very short introduction and background statement followed by a series of short essays in which a group of leading researchers each outlined the potential value of such a satellite to their field. The contributors included

- Communications — John R. Pierce, Bell Labs
- Astronomy — Ira Bowen, Mount Wilson and Palomar Observatories
- Geodesy — John O'Keefe, U.S. Army Map Service
- Upper Atmosphere — Homer Newell, Naval Research Laboratory Research
- Space Biology — Herman Schaefer, U.S. Navy School of Aviation Medicine
- Metoerology — Eugene Bollay, North America Weather Consultants

Although Rosen aimed for a conservative document, outlining only what was immediately possible, the individual presenters proved to be both farsighted and persuasive, none more so than John Pierce, who would pioneer the communications satellite over the next decade. In 11 short paragraphs, he predicted everything from transoceanic broadband links to satellite relays for international radio and television broadcasts.

Titled *On the Utility of an Artificial Unmanned Earth Satellite*, the document was transmitted to Waterman on November 24, 1954, and released to the press at the ARS annual meeting. Rosen remained in close touch with Waterman, who was entirely encouraging. *The New York Times* approved, as well. Reporting that the first satellite might weigh "as little as a hundred pounds" and "have a life expectancy of only a few weeks." Nevertheless, "... even such momentary freedom from the earth would mark the beginning of a new freedom for man—freedom from gravity, the first step toward the stars."[63]

Following the annual meeting, Waterman asked Rosen how the committee proposal had been received by the membership as a whole. He seemed pleased with the positive response and especially with the information that high-ranking military officers were now aware of the proposal. The ARS, he noted, "could look forward to definite action on his part."[64]

Obviously, the ARS was not the only group promoting the benefits of an earth satellite. In 1952, the International Council of Scientific Unions (ICSU), noted that 1957–1958 would be a period of maximum solar activity and established a committee to plan for an International Geophysical Year (IGY), during which scientists around the globe would undertake special geophysical studies. Lloyd Berkner, vice president of that committee, created two subcommittees under Fred Singer and Homer Newell to consider the scientific utility of a satellite. On October 4, 1954, the ICSU Committee endorsed the launch of small scientific satellites as a suitable contribution the IGY activities.[65]

From MOUSE to Orbiter

The news came suddenly on July 28, 1955, when Eisenhower administration officials, including Alan Waterman, announced that the United States would launch "small earth-circling satellites" as part of the national contribution to the IGY. Waterman and others informed the ARS that Rosen's report "was ... one of the persuasive documents leading to President Eisenhower's decision ... ." Richard Porter, who had been closely involved in developing and circulating the ARS proposal, would become Chairman of the Earth Satellite Panel of the U.S. National Committee for the IGY.[4]

Even more gratifying than the notion that the ARS had helped to shape national policy was the fact that the news media recognized the organization as the place to go for information about space flight in general and earth satellites in particular. Jim Harford was in Atlantic City in discussions with ASME officials when the news broke on July 28. It seemed to Billie Slade that every newspaper, magazine, radio, and TV station in greater New York besieged her office that afternoon in search of information. She passed out copies of the Rosen committee report as press releases and briefed reporters on satellite information that had appeared in ARS publications. Back in the office the next day, a Saturday, Harford provided additional material upon request and appeared on an NBC radio program.

So many inaccurate stories appeared that Harford prepared a comprehensive press release on the subject of earth satellites. The response prompted him to schedule a press conference at the Hayden Planetarium, complete with visuals drawn from ARS papers. Following a briefing on the subject, ARS members and representatives of the NSF would answer questions. Such an event, he suggested to the Board, would not only provide reporters with solid information and background material, it would also, "identify the Society with the satellite properly in the public mind."[4]

Outside the bounds of the organization, individual ARS members were hard at work developing specific proposals for an American satellite. Fred Singer had captured much attention when he unveiled his concept for a Minimum Orbital Unmanned Satellite of Earth (MOUSE) at the fourth IAF conference in Zurich in the summer of 1954. Although it was only a paper concept and a cardboard model, it was the first serious attempt to demonstrate how the instruments to conduct several key studies could be packed into a relatively small, light-weight satellite design.

Some of the most vocal ARS space partisans played key roles in Project Orbiter, which came very close to being selected to put the first U.S. satellite in orbit. The story begins with Lt. Commander George Hoover and Alexander Satin, farsighted planners with the Air Branch of the Office of Naval Research (ONR). By the spring of 1954, the pair had become convinced that the technology required to orbit an earth satellite was in place and that such a vehicle would bring international prestige to the United States and produce scientific and, ultimately, even military benefits.[66]

Hoover, who had a reputation for supporting imaginative and sometimes risky projects, approached another Lieutenant Commander, Fred Durant. Recognizing an opportunity when he saw one, Durant put together a meeting at the Washington offices of the ONR on June 25, 1954, at which he introduced Hoover to von Braun, perhaps the leading authority on large launch vehicles; Fred Singer, of MOUSE fame; and Harvard/Smithsonian astronomer Fred Whipple, an authority on satellite tracking.

Von Braun traveled from Huntsville with a ready-made proposal suggesting that one of his Redstone rockets, topped with upper stages made up of 31 clustered Loki antiaircraft rockets, could launch a five-pound object into an orbit with a minimum altitude of 200 miles. Whipple argued that even such a small satellite could be visually tracked from the Earth, so that no onboard telemetry would be required. Because von Braun planned for just such a minimum vehicle, without even a tracking beacon, the effort was originally known as Project SLUG.[67]

Intrigued by the notion of an interservice satellite project, Hoover visited Huntsville to meet with von Braun and his commander, General Holger Toftoy. The result was an informal agreement to cooperate on a satellite program. The Army would supply the booster; the Navy would provide the tracking facilities and data analysis.

Durant, who would continue to coordinate the effort, pulled even more ARS veterans, including Milt Rosen and Bob Truax, into subsequent meetings. With the Naval Research Laboratory and Cal Tech's Jet Propulsion Laboratory aware of the project, things began to move forward. By the end of 1954, the ONR had let three contracts totaling $60,000 for feasibility studies and actual component design.

As historian Michael Neufeld has noted and as Fred Durant's role suggests, the intelligence community was also intrigued by what was now known as Project Orbiter. In January 1955, a joint Army–Navy proposal to the Secretary of Defense pointed out that the "... Central Intelligence Agency has shown an intense interest in Project ORBITER. The Agency apparently thinks considerable psychological warfare value and scientific prestige will accrue to the United States if we launch the first artificial satellite."[68]

Although the Orbiter team seemed far ahead of any possible competing efforts, there were problems. Attempts to attract Air Force support that would make the project even more appealing to the Pentagon failed. The interservice rivalry for control of missile programs predisposed U.S. Air Force planners to reject cooperation with the Army and Navy. Initial Air Force support for an internal proposal originating with Holloman Air Force Base (AFB) researchers who called for the use of an Atlas/Aerobee combination was lukewarm, at best.

Orbiter was facing technical challenges, as well. Whipple suggested that optical tracking would be easier if the vehicle was in a stable orbit circling the earth's equator. Von Braun and his team admitted that taking advantage of the extra speed contributed by an equatorial launch in the direction of the earth's rotation would also increase their chances of success. It seemed that Orbiter would be launched from a Pacific Island or from a specially prepared ship, like the U.S.S. *Norton Sound*, from which Rosen had launched a Viking rocket, demonstrating the possibility of ship-launched ballistic missiles.

The more he thought about it, the more Milton Rosen was convinced that he could do better than Orbiter. He suggested that a next-generation multistage launcher built on the foundation of his own Viking sounding rocket, with Aerobee-derived upper stages, could orbit a heavier satellite than Huntsville's minimum Redstone-Loki combination. Rosen's satellite would carry some scientific instruments, have the ability to telemeter data to ground stations, and carry a radio tracking beacon. The satellite would be designed and built by Rosen's Naval Research Laboratory (NRL), which would not be participating in Project Orbiter after all.

The Orbiter proposal, which went to Secretary of Defense Donald Quarles on March 23, 1955, promised a launch in the fall of 1957 for a cost of $8.5 million. Milton Rosen's NRL proposal, which carried a $7.5 million price tag, went forward on April 15, followed by a vague U.S. Air Force offer for a launch at a later date costing $50 to $100 million. Quarles created a panel, chaired by Homer Joe Stewart of the Jet Propulsion Laboratory (JPL), to advise on the selection of one of the three satellite programs. The Stewart Committee had eight members, two nominated by each of the services and two by Quarles's Office. Richard Porter, who had played such a key role in assisting Rosen to develop the ARS proposal to the NSF, was a key member of the group, which met several times during the summer and took testimony from each of the teams. The Deputy Secretary of Defense announced the final decision on September 9, 1955. Milton Rosen and the Vanguard team would be charged with launching the first American satellite.

Why Vanguard instead of Orbiter? The obvious answer is that Rosen was offering a better product, a satellite that would return some useful scientific data, orbited by a launcher better suited to the task than the Redstone-Loki vehicle. The Orbiter launch of an inert, five-pound payload would only provide the information based on earth-based observations of the orbit. Moreover, although the NRL, a military organization, was sponsoring Vanguard, the launch vehicle would be a civilian product based on the Viking sounding rocket. The Orbiter launcher was composed of two modified guided missiles and managed by the Army's chief ballistic missile agency. Vanguard was far easier to describe as "civilian" and "scientific" than Orbiter and better met the political and diplomatic goals of the Eisenhower administration.

The Space Flight Program

In the wake of the Eisenhower decision, Andy Haley congratulated Milt Rosen and the members of the Space Flight Committee. Rosen noted that the work of the committee had only just begun. The Board instructed the group to help plan for the future of space flight discussions within the ARS and to "... maintain a continuing interest in the existing satellite program and explore the steps toward space flight that should logically follow the small scientific earth-satellite."[69]

They would go well beyond that limited charter in an attempt to shape public policy. Dr. Krafft Ehricke, one of the most imaginative of the Pëenemude Germans and one of the few to establish his own career away from the Huntsville circle, chaired a committee made up of men who had shaped the course of the ARS for the last decade, including Fred Durant, Andy Haley, Dick Porter, Milt Rosen, Fred Singer, Alexander Satin, Kurt Stehling, and Wernher von Braun. A number of other important figures rounded out the group. Karel Bossart, of Convair, was a key figure in developing the Atlas missile. Hubertus Strughold was an authority on aerospace medicine. Darrell Romick of Goodyear, George Clement of the Rand Corporation, and Air Force leaders Major General George D. Colchagoff and Colonel William David joined out the reorganized Space Flight Technical Committee.[70]

Rather than focusing on future space projects, the ARS group recognized that the outstanding need was for an organization to manage a sustained national space effort. A temporary program like Vanguard could not plan for the long-range future of a space program beyond the launch of the first earth satellites. Such planning was required to take full advantage of related military technology and economies of scale. Instead, Ehricke and his committee argued for a new Federal agency, "... not a lone association of operating activities and advisory committees such as is carrying on the present Vanguard program, but a permanent executive organization responsible to the Congress and equipped with full authority to carry out its decisions."

In the late summer of 1957, after two years of work, the Space Flight Technical Committee sent their proposal for a National Space Flight Program to be managed by an Astronautics Research and Development Agency to the ARS Board. The report called for an entirely civilian organization, independent of the military, operating under a strong board of directors. The agency would have four divisions. Astronautic Field Operations would be responsible for all ground-based tracking, laboratory, and operational facilities, including, eventually, a "space-borne station." Astronautic Sciences would manage all research on the ground and in space. Astronautic Systems would handle all engineering matters as well as problem areas like navigation and human factors questions, and the Astronautic Program Division would manage the specific projects.[71]

Funded at the level of $100 million the agency could be expected to achieve a steady string of successes:

1) Place payloads weighing thousands of pounds in earth orbit that would perform scientific missions, meet military needs, and revolutionize communications and meteorology within five years.
2) Place payloads of several hundred pounds on the moon or in lunar orbit within 5 to 10 years
3) Send scientific payloads to the orbits of Mars or Venus within 5 to 10 years
4) Send human beings into space within 10 years
5) Send human beings around the moon in 15 years
6) Land human beings on the moon and return them safely to earth within 20 years

During the discussion of the plan at an ARS Board meeting early that fall, von Braun remarked that the government might welcome such a plan as a means of getting more value from the funds already being spent on missile programs. Martin Summerfield suggested that the government would pay more attention to such a scheme if the ARS could enlist industry leaders to promote the plan. Milt Rosen thought that an existing government agency, perhaps the National Advisory Committee for Aeronautics (NACA) or the NSF, could manage such a program. Bob Truax, now ARS president, agreed, noting that the ARS report might "... provide the slight push necessary to start some ... agency of government on this project."[72]

A New Age Now Begins

Historic events intervened before ARS leaders could share the report with anyone outside the organization. The Soviets stunned the world by orbiting *Sputnik I*, a 184-pound satellite, on October 4, 1957. They sent the dog Laika into orbit aboard a second and even heavier spacecraft in November. As if the Communist triumphs were not enough, the Vanguard TV-3 vehicle was lost in a catastrophic explosion on the pad on December 1, 1957. The von Braun team rode to the rescue on January 31, 1958, with the launch of the first U.S. satellite, *Explorer I*. Ironically, it was an outgrowth of Orbiter. Rosen's team succeeded in launching their first satellite on March 17.

Robert Truax sent a copy of the ARS proposal for a national space program and an agency to manage it to the White House on October 14, 1957. The plan was released to the public at the ARS annual meeting on December 4, 1957.[73] In his letter to the President, Truax pointed out that the document was not a reaction to the *Sputnik* launch but the result of two years of thought by experts in the field. "We do feel," he explained, "... that the recommendations represent a course of action, which, if carried out, will insure the eventual superiority of the United States in this new field."[74]

President Eisenhower signed the *National Air and Space Act* into law on July 29, 1958. None of the many studies of the political decisions leading to the creation of the National Aeronautics and Space Administration (NASA) specifically mentions the ARS proposal. Eileen Galloway, an analyst with the Legislative Reference Service who was very close to the process recalls, however, that ARS member James Van Allen, head of the Rocket and Satellite Panel of the National Academy of Sciences (NAS), forwarded a report to one of the key Congressional Committees involved in shaping the legislation. "This report had gone to Eisenhower on October 14. It was all about setting up a civilian space agency ... pointing out how many benefits there were to space, especially communications, meteorology, and navigation." Surely this is a slightly confused memory of the ARS proposal.[75]

In January, Ms. Galloway added, Van Allen and George P. Sutton, now president of the ARS, had presented a second report, "which contains many words that are similar to the words that are in the Act, like 'leadership,' 'pre-eminence,' and 'separating from the military.'" James Killian, President Eisenhower's science advisor, also recalls that, "... even before *Sputnik I*, the American Rocket Society and the Rocket and Space Panel of the National Academy of Sciences, had begun the formulation of a joint report which they submitted to the administration in January 1958."[76]

Perhaps Sutton and Van Allen resubmitted the ARS report in January with a cover letter on NAS letterhead. Perhaps the two organizations did draft a joint report, which surely would have been based on the ARS original. In any case, it is clear that the ARS report was receiving attention in both the White House and the Congress, even if the origins of the document were less than perfectly clear to decision makers.

The IAS was looking to the future, as well. On January 21, 1958, S. Paul Johnston, now the executive director of the IAS, and a member of an advisory panel created by Nobel Laureate Edward Purcell of the President's Scientific Advisory Committee, prepared a critically important memo summing up the options for organizing a space agency. Creating a new organization would simply take too long. Handing the space program to the Atomic Energy Commission, as some Congressional leaders proposed, would distract an organization with no experience in the new field from its primary responsibilities. If the Advanced Research Projects Agency of the Department of Defense were placed in command, military requirements would surely outweigh the civil and scientific aspects of the space program. The NACA, which had been conducting flight research for more than four decades, was the natural choice to take on the job. So it would be.[77]

The creation of the NASA was the result of decisions and recommendations offered by a complex and confusing series of study panels, Congressional committee hearings, and executive office discussions. Whatever the impact of the ARS report, NASA certainly embodied many of the features that the Space Flight Technical Committee had proposed.

The tension between the missile men and the space enthusiasts within the ARS ended with the launch of *Sputnik I* and the drive to orbit the first American satellites. It was now apparent to everyone that space would be the arena in which the United States and the Soviet Union would demonstrate their technological prowess and contend for international power and prestige.

The members of the ARS had helped to shape early U.S. space policy by initiating the U.S. satellite effort, planning the two major competing satellite proposals, and offering their vision of a government agency that would orchestrate human journeys to the moon. For the moment, all of that lay in the future. Even at the dawn of this new era, however, there were triumphs to savor. In one of the best-remembered photos of the period, the leaders of the Explorer team—Wernher von Bruan, physicist James Van Allen, and William Pickering of JPL—three very happy members of the ARS, hold a full-scale model of *Explorer I* aloft at the press conference announcing the first successful launch of a U.S. satellite. The United States was a space-faring nation at last, and the ARS had helped to make it so.

References

[1]Author interview with James J. Harford, Princeton, NJ, tape in the collection of AIAA, Aug. 4, 2004.

[2]It was the first in a series of *Collier's* articles that appeared between March 22, 1952, and April 30, 1954. Supervised by editor Cornelius Ryan; written by Wernher von Braun, Willy Ley, and others; and illustrated by Chesley Bonestell, Fred Freeman, and Rolf Klepp. The articles had an enormous public impact.

[3]When Harford arrived, the ARS had 10 corporate members: Aerojet-General Corporation; Reaction Motors, Inc.; Curtiss Wright Corp., Propeller Division; Douglas Aircraft Company; Genisco, Inc.; Harvey Machine Tool Company; Haynes Stellite Co.; Oerlikon Tool & Arms Corporation; Laurence Rockefeller; and United Aircraft Corporation, from James Harford, "The American Rocket Society, 1953–1963: A Memoir," IAF-88-605, 39th Congress of the International Astronautical Federation, copy in the author's collection, 1988.

[4]Harford, "The American Rocket Society, 1953–1963: A Memoir."

[5]On Rockefeller's flying saucer interests, see: "UFO Updates," http://www.virtually strangenet/ufo/updates/2004jul/m 18-001 shtml. Cited by the author on Feb. 1, 2005.

[6]In February 1951, the ARS staff consisted of Harford ($800 per month); Mrs. A. C. Slade ($500 per month); her assistant, Miss Catherine Beck ($50 a week); and Ruth Locke, a file clerk ($40 per week).

[7]"Proposed American Rocket Society Expansion Plan," Mintues of the Meeting of the ARS Board of Directors, April 15, 1954, AIAA Headquarters Files.

[8]Minutes of the Meetings of the ARS Board of Directors, June 7, 1954; July 12, 1954; Aug. 27, 1954, AIAA Headquarters Files.

[9]Greer Hydraulics, Inc; Kearfott Corp.; Bell Aircraft Corp.; Linde Air Products; and the Aeronautical Digest signed up. General Electric and North American Aviation were still considering the matter. Minutes of the Meeting of the ARS Board of Directors, Feb. 1, 1954, AIAA Headquarters Files.

[10]Minutes of the Meeting of the ARS Board of Directors, Sept. 19, 1955, AIAA Headquarters Files.

[11]The American Rocket Society, 1945–1950, in the Charles W. Chillson Papers, RH XXXX-0008, Box 1, National Air and Archive.

[12]There were 1,289 Active members; 255 Associate members; 327 Student members; 9 Corporate members; 4 Affiliate Corporate members, and 659 individuals or institutions who only subscribed to the *Journal*. Minutes of the Meeting of the ARS Board of Directors, March 2, 1953, AIAA Headquarters Files.

[13]Minutes of the Meeting of the ARS Board of Directors, Aug. 27, 1954, p. 4, AIAA Headquarters Files.

[14]Minutes of the Meeting of the ARS Board of Directors, Feb. 7, 1955, AIAA Headquarters Files.

[15]"American Rocket Society, 1945–1950," Minutes of the Meeting of the ARS Board of Directors, Jan. 3, 1950, AIAA Headquarters Files.

[16]"ARS Membership. September 1, 1955," Minutes of the Meetings of the ARS Board of Directors, following the entry for Aug. 1, 1955, AIAA Headquarters Files.

[17]F. C. Durant, III to Members of the 1954 Nominating Committee, copy in Minutes of the Meeting of the ARS Board of Directors, Aug. 27, 1954, p. 10, AIAA Headquarters Files.

[18]Members of the Board of Directors of the New Mexico–West Texas Section of the ARS to Andrew Haley, July 22, 1954, copy in Minutes of the Meeting of the ARS Board of Directors, Aug. 27, 1954, p. 12, AIAA Headquarters Files.

[19]Minutes of the Meeting of the ARS Board of Directors, Aug. 27, 1954, p. 15, AIAA Headquarters Files.

[20]"Executive Committee Appointed," *Jet Propulsion*, Feb. 1955, p. 83.

[21]William Gore to A. Haley, Dec. 29, 1952, Andrew Haley Papers, Box 18, National Air and Space Museum Archive.

[22]Author interview with Frederick I. Ordway, III, June 1, 2005.

[23]Kurt Stehling, "Blazing a Trail to the First U.S. Satellites," *Astronautics & Aeronautics*, May 1982, p. 78.

[24]Ron Miller and F.C. Durant, III, *World's Beyond: The Art of Chesley Bonestell*, Donning, Norfolk, VA, 1983.

[25]Author interview with Frederick C. Durant, III, Feb. 9, 2005.

[26]Charles W. Chillson to Wernher von Braun, June 7, 1951. The technical portions of the Mars Project were published in German in 1952 and in English by the Univ. of Illinois in 1953.

[27]Stehling, "Blazing a Trail," p. 80.

[28]Neil McAleer, *Odyssey: The Authorised Biography of Arthur C. Clarke*, Victor Gollancz, Ltd, London, 1992, p. 78.

[29]"An apparent preponderance of members of the American Rocket Society here has caused some jealousy among other delegates, particularly as the executives of the American society are also the executives of the federation." John Hillaby, "Rocket Men Seek a Middle Course," *New York Times*, Aug. 3, 1954, p. 21.

[30]McAleer, *Odyssey*, p. 34.

[31]Minutes of the Meeting of the ARS Board of Directors, Nov. 28, 1951, AIAA Headquarters Files.

[32]Constance Green and Milton Lomast, *Vanguard: A History*, NASA, Washington, DC, 1970.

[33]Robert Truax to the ARS Board of Directors, Nov. 28, 1951, AIAA Headquarters Files.

[34]Minutes of the Meeting of the ARS Board of Directors, Feb. 4, 1952, AIAA Headquarters Files.

[35]Robert Truax to the ARS Board of Directors, Minutes of the Meeting of the ARS Board of Directors, March 3, 1952, AIAA Headquarters Files.

[36]For information on the Hayden Planetarium event, see Randy Liebermann, "The Collier's and Disney Series," *Blueprint for Space: Science Fiction to Science Fact*, edited by Frederick I. Ordway and Randy Liebermann, Smithsonian Inst. Press, Washington, DC, 1992, pp. 135–146.

[37]Cornelius Ryan (ed.), *Across the Space Frontier*, Viking, New York, 1952.

[38]Cornelius Ryan (ed.), *The Conquest of the Moon*, Viking, New York, 1953.

[39]Willy Ley and Wernher von Braun, *The Exploration of Mars*, Viking, New York, 1956.

[40]David R. Smith, "They're Following Our Script: Walt Disney's Trip to Tomorrowland," *Future*, May 1978.

[41]Jonathan Leonard, "Journey Into Space," *Time*, Dec. 8, 1952.

[42]Wernher von Bruan, Notes for an ARS Luncheon talk, May 6, 1953, Wernher von Braun Papers, Manuscript Division, Library of Congress, Box 46, Speeches and Writings, 1951–1955. My thanks to my colleague Michael Neufeld for providing this document.

[43]Andrew Haley to Gary Loeser, March 14, 1952, Andrew Haley Papers, Box 18, National Air and Space Museum Archives.

[44]F. C. Durant, III, "Perspectives on the American Astronautical Society," *Twenty-Five Years of the American Astronautical Society, 1954–1974*, edited by Eugene M. Emme, AAS Publications Office, San Diego, 1980, p. 158.

[45]C. W. Chillson to Andrew Haley, Jan. 22, 1952, Haley Papers, Box 18, National Air and Space Museum Archives.

[46]Telephone interview with Frederick C. Durant, III, Feb. 6, 2005. Notes in the author's possession. Durant tells the story in print, without using names, in F. C. Durant, III, "Perspectives on the American Astronautical Society," *Twenty-Five Years of the American Astronautical Society, 1954–1974*, edited by Eugene M. Emme, AAS Publications Office, San Diego, 1980, p. 158.

[47] William Gore to Andrew Haley, Dec. 29, 1952, Andrew Haley Papers, Box 18, National Air and Space Museum Archive.

[48]F. C. Durant, III, "ARS Journal to be re-named JET PROPULSION," *Journal of the American Rocket Society*, 1953.

[49]Walter Hurd to Secretary, ARS, Dec. 24, 1953, Andrew Haley Papers, National Air and Space Museum Archive.

[50]Hans J. Behm and J. G. Rosenquist to Invitees, Nov. 5, 1953, Andrew Haley Papers, Box 18, National Air and Space Museum Archive.

[51]James Harford to Andrew Haley, Nov. 11, 1953, Haley Papers, National Air and Space Museum Archive.

[52]James Rosenquist, "Founding of the American Astronautical Society, 1953–1954," *Twenty-Five Years of the American Astronautical Society, 1954–1974*, edited by Eugene Emme, AAS Publications Office, San Diego, 1980, p. 12.

[53] Martin Caidin, "More Ways Than One," *Twenty-Five Years of the American Astronautical Society, 1954–1974*, edited by Eugene M. Emme, AAS Publications Office, San Diego, 1980, p. 22.

[54]S. Fred Singer, "Enduring Challenges of Astronautics," *Twenty-Five Years of the American Astronautical Society, 1954–1974*, edited by Eugene M. Emme, AAS Publications Office, San Diego, 1980, p. 149.

[55]Arthur C. Clarke to Andrew Haley, Nov. 18, 1953, Haley Papers, National Air and Space Museum Archive.

[56]Eric Burgess to Andrew Haley, Nov. 22, 1953, Haley Papers, National Air and Space Museum Archive.

[57]F.C. Durant, III, "Perspectiveson the American Astronautical Society," *Twenty-Five Years of the American Astronautical Society, 1954–1974*, edited by Eugene M. Emme, AAS Publications Office, San Diego, 1980.

[58]Appendix 1, "Membership Roster and Renewals, 1953–1956," *Twenty-Five Years of the American Astronautical Society, 1954–1974*, edited by Eugene M. Emme, AAS Publications Office, San Diego, 1980, pp. 218–221.

[59]John Paul Stapp, "ARS-AAS Merger Negotiations Suspended," *Astronautics*, Sept. 1959, p. 19.

[60]George R. Arthur, "A Few Reflections, 1959–1960," *Twenty-Five Years of the American Astronautical Society, 1954–1974*, edited by Eugene M. Emme, AAS Publications Office, San Diego, 1980, p. 81.

[61]Milton Rosen, "On the Utility of an Artificial Unmanned Earth Satellite," *Jet Propulsion*, Feb. 1955, p. 71.

[62]Note to Document II-B, *Exploring the Unknown: Selected Documents in the History of the U.S. Civil Space Program, Volume 1: Organizing for Exploration*, edited by John M. Logsdon, NASA, Washington, DC, 1995, p. 281.

[63]Robert K. Plumb, "Science in Review," *New York Times*, Feb. 13, 1955, p. 177.

[64]Minutes of the Meeting of the ARS Board of Directors, Jan. 10, 1955, AIAA Headquarters Files.

[65]Green and Lomask, *Vanguard: A History*, p. 22.

[66]Few projects of the early space age have been so well researched and analyzed as the initial U.S. Satellite effort. My primary source for these paragraphs on Project Orbiter are based on the work of my colleague, Michael J. Neufeld, "Orbiter, Overflight, and the First Satellite: New Light on the Vanguard Decision," in Roger Launius, John Logsdon, and Robert Smith, *Reconsidering Sputnik: Forty Years Since the Soviet Satellite*, Harwood Academic, 2000, pp. 231–251.

[67]The Project was also known as "Orbit" for a short time. It officially became Project Orbiter in January 1955. Neufeld, "Orbiter, Overflight, and the First Satellite," p. 234.

[68]Neufeld, "Orbiter, Overflight and the First Satellite," Reconsidering Sputnik, p. 234.

[69]Minutes of the Meeting of the ARS Board of Directors, Nov. 14, 1955, AIAA Headquarters Files.

[70]"A National Space Flight Program," *Astronautics*, Jan. 1958.

[71]Krafft Ehricke, "Space Flight Program: Report by the Space Flight Technical Committee of the American Rocket Society," AIAA Headquarters Files.

[72]Minutes of the Meeting of the ARS Board of Directors, Sep. 1957, AIAA Headquarters Files.

[73]Richard Witkin, "Rocket Men Urge U.S. Space Agency," *New York Times*, Dec. 5, 1957, p. 5.

[74]Robert C. Truax to the President, Oct. 17, 1958, copy in *Astronautics*, Jan. 1958.

[75]Eilene Galloway, *Legislative Origins of the National Aeronautics and Space Act of 1958*, edited by John Logsdon, NASA, Washington, DC, 1998, p. 38.

[76]James R. Killian, *Sputnik, Scientists, and Eisenhower: A Memoir of the First Special Assistant to the President for Science*, MIT Press, Cambridge, MA, 1979, p. 125.

[77]http://www.hq.nasa.gov/office/pao/History/monograph10/nasabrth.html. Cited Feb. 19, 2005.

Chapter 7—
Years of Victory and Growth: The Institute of the Aeronautical Sciences, 1945–1957

Victory!

In January 1942, Frank Caldwell, president of the Institute of the Aeronautical Sciences (IAS), received a letter from Franklin Delano Roosevelt congratulating the organization on its 10th birthday and praising the members for their contribution to the war effort.

> In research laboratories of the Government and industry, as well as in the laboratories of private scientific and educational institutions, the talent of America is being marshaled in a united effort necessary to develop for our armed forces aircraft of greater performance and effectiveness than those of our enemies .... Thus is the foundation laid for achieving supremacy in the air.[1]

Just four years later, the members of the IAS celebrated their contributions to victory on honors night, January 28, 1946. The *New York Times* summed the matter up, commenting, "Wartime aviation's unsung valiants stood in the spotlight quietly and briefly last night at the Waldorf-Astoria Hotel's Grand Ballroom." Lieut. Gen. Jimmy Doolittle, a founding member, the guest of honor, and keynote speaker, applauded the efforts of his fellow engineers. "The men who flew the planes and the agencies which built them have gained a portion of their well-deserved recognition," he noted, "but the men who have designed them have, on the whole, gone unheralded."

The aircraft that won the war, he noted, "... as well as the better ones soon to come, are the brain children of our American aeronautical engineers. I salute that outstanding group of men and am delighted that the Institute of the Aeronautical Sciences has again honored some of them here tonight." Doolittle concluded by cautioning his listeners and the nation not to become complacent with victory. "Our future security," he warned, "depends on the aggressive continuation of research and development." The members of the IAS could not have agreed more.[2]

The Big Change

For the first 15 years of its existence, the IAS was widely regarded as an "old boys club." Indeed, the organization was exclusively male until 1939. The IAS was founded and managed by a small group of engineers and industry leaders who defined themselves as elite professionals and who modeled their organization after a similarly restrictive club, the Royal Aeronautical Society (RAeS). They created a relatively undemocratic governing structure dominated by a small elected Council and a less than completely open process for electing the senior officers. The heavy emphasis on the history and culture of flight reflected the attitudes and interests of an elite gentlemen's club rather than those of a traditional technical society like the Society of Automotive Engineers (SAE) or the American Society of Mechanical Engineers (ASME).

Reuben Fleet, IAS President for 1944. (Courtesy AIAA.)

So long as the IAS remained fairly small, Lester Gardner, a single dedicated individual, had been able to manage day-to-day operations and provide overall direction and a long-range vision. Under those circumstances, the small circle of oligarchic founders were able to maintain control of the organization through World War II.

In the decade to come, however, the IAS would be forced to shift away from the original management style toward a more democratic and bureaucratic structure. One of the first steps was to recognize the increasing importance of the aviation industry on the Pacific Coast. In 1945, the IAS Administrative Committee, with the approval of the officers and Council, established a Los Angeles office and hired William Dudley as West Coast executive. With offices in the Pacific Aeronautical Library, which had been established before the war, Dudley was to cooperate with New York and the local Sections in supervising IAS activities in the region.[3]

There was a new constitution as well, adopted by the Council in January 1946. The governing Council was increased from 12 to 28 members, including the president, vice president, past president, treasurer, fifteen national councilors, and six locally nominated area councilors representing the Sections. The Council was to appoint an Executive Committee, including no less than five of their own members, who were empowered to act between meetings of the full Council. The organization would have an Administrative Committee of four salaried officers: the assistant to the president, executive vice president, secretary, and controller. Their primary duty was to assist a fifth paid officer, the director, in the administration of the Society. The Chairman of the Council was to be compensated, as well.[4]

The comfortable, collegial, and very personal management style effected by Lester Gardner was about to give way to administration by a growing cadre of paid, professional staff members. As one astute observer has noted, the administrative staff of most professional organizations is critically important because "... the bulk of the membership tends to be indifferent, if not resistant, to participation in administrative processes."[5] In the postwar era, the IAS would face difficult financial issues, sticky and persistent real estate problems, and the need to serve a growing number of widely dispersed members working in a much more complex industry. Better management was a necessity. "Running the IAS was like running any business," one former staffer explained, "and you need professional administrators to keep it running smoothly."[6]

A Home of Our Own

The Berwind Mansion, home of the IAS, 2 East 64th Street, New York. (Courtesy AIAA.)

By 1944, IAS headquarters was shoehorned into a block of offices on the 15th floor of the RCA Building facing the Avenue of the Americas, with additional offices housing the staff of the *Aeronautical Engineering Review* and the *Aeronautical Engineering Catalog* elsewhere in the building. The expansion of the publications program had required some additional space, but it was the growth of the Aeronautical Index, the library and, finally, the Aeronautical Archive that were principally responsible for the real growth. At its peak, the Aeronautical Index project alone employed more than 100 people. The Institute was literally coming apart at the seems.

Then there was the question of prestige. The RAeS had recently purchased a large and lovely town house at 4 Hamilton Place, overlooking London's Hyde Park. The organization had raised a fund of 100,000 pounds to acquire the property, refurbish the building, and furnish it. Suitably impressed, Gardner had published a lengthy article on the project, complete with photos of the interior spaces. The corresponding French and German organizations, the secretary noted, were housed in similarly impressive and well-appointed quarters.

When Rueben Fleet took office as president in 1944, Gardner found him to be a receptive listener. In a single year, Gardner explained to those attending the Honors Night Victory Dinner in 1945, "Major R. H. Fleet ... made our greatest dreams come true." Remarking that "the aeronautical engineers of this country should have buildings worthy of their profession," President Fleet began his campaign by raising $176,000 for an IAS building in San Diego. Major contributions came from his own company, Consolidated-Vultee, the Ryan Company, and several other local manufacturers.

Later that year he launched a second campaign to fund an even larger IAS building for Los Angeles. Lester Gardner reported that ex-IAS President Donald Douglas supplied one-third of the total $268,000 raised. Although the money was in hand, construction of the West Coast buildings would have to wait until wartime building restrictions were lifted.[7]

Finally, Fleet turned his attention to Gardner's primary concern, a New York headquarters building for the IAS. The two men originally planned to construct a new 10-story building at an estimated coast of $1,500,000. The site was just off Fifth Avenue, opposite the University Club, on land that they hoped the Rockefeller family would donate. In addition to the extraordinary price tag, it would take years to design and construct a headquarters, a process that could not begin until after the war. In any event, Gardner was calling attention to the European tradition of remodeling large townhouses to serve as the headquarters of learned societies, the RAeS building being his favorite case in point.[8]

A statue of Icarus and a selection of historic prints greeted visitors to the IAS headquarters. (Courtesy AIAA.)

It was an idea that Fleet could sell. He began by asking the leading East Coast aircraft and engine manufacturers and subcontractors to donate $100,000 each to a building fund. Grumman and Edo responded as requested. Curtiss, Fairchild, Republic, Bell, General Electric, Square D, Beech, and a dozen other companies donated $25,000 each. Some firms, notably United Aircraft, refused to contribute anything. In the end, Fleet and Gardner raised $324,000 in a remarkably short period of time.[9]

Working with the real estate specialists at the Equity Preservation Corporation, Gardner canvassed the city in search of a building that would meet the needs of the Institute. He finally focused on a lovely town house at 2 East 64th Street. The building had been constructed as the home of Edward J. Berwind in 1896. Gardner delighted in explaining to visitors that Berwind was one of the original investors in the Wright Company and a member of the Board of Directors. Built in the style of a Renaissance palazzo, the mansion ran for 30 feet along Fifth Avenue and 120 feet down 64th Street. Constructed of stone and brick, with steel beams laid five feet apart at each floor level, the building boasted 49 rooms, 29 closets, and 10 washrooms—19,000 square feet spread over six floors, a basement, and a subbasement. Those interior spaces were richly finished in tile, concrete, and hollow brick.[10]

The Berwind mansion had an appraised value of $300,000. As a result of negotiations with the Guaranty Trust representing the estate, the IAS was able to acquire the property at a cut-rate $100,000. An additional $120,000 went for modifications planned by architect Harry P. Jaenike to suit the specific needs of the Institute. That left roughly $104,000 for contingencies.

The building seemed perfect and the price was right, but there was a final problem to be overcome. The New York City Housing and Buildings Department rejected Jaenike's plans for the modifications. Zoning laws did not permit "scientific societies" to occupy buildings in a "residential district." Gardner immediately asked the advice of Captain Thomas Harten, an old friend who was serving as an aide to Mayor Fiorello LaGuardia. Harten asked Commissioner Wilson of the Housing and Building Department to meet with IAS officials. "Major Gardner attended the Plattsburg military training camp with me and is an old friend of the Mayor as well," he explained. "Any consideration you can show him will be appreciated by the Mayor and me."[11]

Commissioner Wilson informed the Major that both Mayor La Guardia and city planner Robert Moses "were very solicitous about keeping Fifth Avenue a beautiful street and they would not approve any waiving of restrictions." Mr. Jaenike's drawings indicated that the 30-odd rooms on the third, fourth, and fifth floors of the mansion were intended to become offices for the business operations of the Institute, and such activity was prohibited by the zoning regulations. It was a shame that the IAS was not a library, museum, or club, he concluded, because a building dedicated to those purposes would be allowed. When Gardner responded that the organization could be regarded as all of those things, Wilson suggested that he reapply for occupancy "... as a museum, library and purposes similar to that of a club." The altered application was resubmitted and approved on April 7, 1945. The transaction was complete by July 1945, and work on the Berwind mansion was soon underway.

A display of rare ceramics with an aeronautical theme from W. AM. Burden's Tissandier Collection. (Courtesy AIAA.)

The IAS Reading Room in Rockefeller Center was closed in December 1946 while library materials and other collections were packed and shipped to their new home. All of the collections that had spent the war years in storage at the Guggenheim estate were also moved to 2 64th Street. The new Reading Room was the first element of the IAS headquarters building to open to the public on February 4, 1946. The library was now reopened with new hours: 9:30 a.m. to 5:30 p.m., Monday through Saturday, with extended hours until 10:30 p.m. on Thursdays. The first executive council meeting was held in the new building on February 14, 1946. With the Rockefeller Center lease expiring on April 30, 1947, the movement of the remaining staff began soon after the annual meeting in late January 1947.

The finished building was unveiled to the membership in a special supplement to May 1947 issue of the *Aeronautical Engineering Review*. Visitors entered the headquarters on the 64th Street side through two large glass doors decorated from top to bottom with intricate wrought iron scrollwork. Facing the door was a large niche with a statue of Icarus, surrounded by framed images of the winged man dating back to the 18th century. They were the first of almost 700 prints and paintings that adorned the walls of the building, most of them from the Harry F. Guggenheim and Bella C. Landauer Collections. The wooden chest beneath the sculpture, like all of the antique furnishings and rugs in the building, was the gift of Sherman Fairchild.

The library reading room was at one end of the entrance hall. Prints by artist Frank Lemon illustrating great events in the history of civil aviation decorated the walls, along with illustrations of World War I French aircraft. Five of the original Berwind murals decorated the

The IAS library stacks. (Courtesy AIAA.)

ceiling. An electric dumbwaiter near the library service desk connected to the basement stacks, which housed some 30,000 volumes along with ranks of filing cabinets and library card cases filled with index cards.

Gardner regarded the Council Room, at the other end of the hall, as "... one of the handsomest rooms in the United States." The woodwork, he noted, was matched mahogany "fabricated in France in the style of the Empire period." Cases filled with the rarest volumes in the collection lined the wall. The 11 bound volumes of aviation pioneer Octave Chanute's correspondence and the 9 volumes of Hart O. Berg's Wright brother's material, the collection that had launched the Aeronautical Archive a decade before, were housed here. Other bookcases contained what was probably the finest collection of rare American and French aeronautica outside the Library of Congress. The Edwin C. Musick, Pulitzer, and James Gordon Bennett Balloon Trophies were also displayed here.[12]

The main stairway, flanked by a large and small elevator, led to the second floor, where Gardner installed his aeronautical museum. This area was initially closed to unaccompanied visitors because of concern for the security of the priceless and fragile items on display. The walls were covered with prints in various groupings, portraits of the first men and women to take to the sky, images of the earliest flights in France, cartoons and caricatures, and scenes from the aeronautical history of other nations. Included were the work of well-known artists such as Francisco Goya, George Cruickshank, Thomas Rowlandson, and Honoré Daumier.

Framed song sheets from Bella Landauer's collection of music with an aeronautical theme were hung together in one corner of the Great Room. There were large glass-fronted cases containing ceramic and glass items produced to celebrate the birth of flight in the closing years of the 18th century. Clocks, fans, and china decorated with colorful balloons filled another case. Also on display, or carefully packed away in the specially constructed cabinets that lined the walls, were additional examples of prints, wallpaper, fabric, trade cards, book plates, calendars, medals, coins, philatelic items, and Gardner's special collection of Christmas cards, all illustrating the history of flight. Then there were the framed relics, which included letters from George Washington, Thomas Jefferson, and Walt Whitman, all describing their experience with balloons. The Model Room, at the rear of the main gallery, contained more than 500 miniatures of historic and current aircraft.

Visitors proceeded up the stairs to the third floor and the offices of the director, executive vice president, and other officers of the Institute and the Aeronautical Archive. The decorative items on this level ranged from a collection of badges identifying workers at various British aircraft factories to framed pieces of fabric from famous aircraft. Three hundred photos of

Fellows, officers, and better known members of the Institute graced the walls of the main reception room.

The accounting and membership operations were housed on the fourth floor, and the 10 rooms on the fifth floor served as offices for the president of the Aeronautical Archives and the editorial staffs of the *Journal*, the *Aeronautical Engineering Review*, and the *Aeronautical Engineering Catalogue*. The top floor provided a common area for staff members, a dressing room for members, and storage space to supplement the main storage areas in the basement and subbasement.[13]

The model collection, the tip of the iceberg. (Courtesy AIAA.)

From top to bottom, this was precisely the sort of building that Lester Gardner had dreamed of. There was plenty of room for the men and women who kept the IAS and its operations moving in the right direction. More than that, however, it was a showcase for the collections that he had spent a decade amassing. How proud he must have been! He would have all too little time to enjoy it.

Changing of the Guard

On January 30, 1946, President Charles Colvin reported that a special committee of the Council was considering management changes that would "release Major Gardner from his many responsibilities." Gardner was 70 years old. He had administered the Institute for 14 years. No one was more responsible for its success, either as a technical society or as an organization preserving and celebrating the history and culture of aeronautics.[4]

Gardner was by no means eager to simply fade away. He planned to remain actively involved with projects that he had been supervising, including key IAS real estate ventures and the creation of a new Center for the Advanced Study of the Aeronautical Sciences as a cooperative program between the Institute, Columbia University, and the Massachusetts Institute of Technology (MIT). In addition, he hoped to continue serving as chairman of the Council, a paid position, and as president of the Aeronautical Archive. On May 26, the Council voted to allow the Major to take a full year's leave, beginning on August 7, 1946, or earlier if he wished. He did not wish. In fact, a good many questions remained unresolved in so far as he was concerned.

Gardner had played a role in selecting his successor, Captain Samuel Paul Johnston, U.S. Naval Reserve. A native of Pittsburgh, born on August 3, 1899, Johnston attended the

S. Paul Johnston (right) and Harry Guggenheim (left) greet James S. McDonnell. (Courtesy AIAA.)

Carnegie Institute, trained as an Army aviator during World War I, and graduated from MIT with a B.S. in mechanical engineering in 1921. He was employed by the Aluminum Company of America until 1929, when he joined the staff of *Aviation*, where he rose to the position of editor. In 1940, he moved to the National Advisory Committee for Aeronautics (NACA), where he headed the expanded aircraft research program. Two years later, he took command of the Washington, DC, office of the Curtiss–Wright Corporation. Entering the Navy in 1943, he served with the Naval Air Transport Service in the Pacific and as deputy director of the Aircraft Division of the Strategic Bombing Survey in Europe and Japan, 1945–1946.

More writer than engineer, Johnston had spent a decade editing the most important aviation magazine in the country, had turned out four books between 1940 and 1944, and had written articles for the *Encyclopedia Britannica Yearbook* and the *Britannica Junior*. He seemed a perfect match for a professional society with a very active publishing program. The Council voted to offer him the post on January 30, 1946, at a salary of $12,000. He formally accepted in March and began work on April 25.

It was a difficult time. The postwar slump in the aircraft industry was reflected in declining Institute revenues. Corporate membership contributions were tied to earnings, and lay-offs meant a drop in individual memberships and dues. Clearly, the wise management of IAS real estate holdings would be important in maintaining financial health.

During the war, the IAS had leased the Guggenheim estate at Sands Point to the U.S. Navy, which conducted a variety of research programs there. In the months immediately following V-J Day, the estate was scheduled to become the home of the Special Devices Branch of the Navy Office of Research and Inventions. Headed by long-time IAS leader Admiral Luis de Florez, the organization was in the business of developing training aides, from maneuverable kites used to hone the skills of antiaircraft gunners to complex crew simulators that ultimately led to the development of the first digital computers.

Early in 1946, the Council agreed to extend the lease until 1950, giving the Navy a three-year option to buy for $350,000, subject to approval of the State Supreme Court, as required by the terms of the original gift. This, the IAS Council argued, was "... putting it [the estate] fully to the use on aeronautical research originally contemplated when the late Florence Guggenheim presented the property to the Institute."[14]

A group of West Coast IAS pioneers gather in Los Angeles in April 1951: (left to right) F. Allen Cleveland, chairman, Los Angeles section; James H. "Dutch" Kindelberger, IAS President for 1950; John K. Northrop, IAS President for 1948; Admiral Richardson; E. C. Wells, IAS President for 1958; Clark B. Millikan, IAS President for 1937. (Courtesy AIAA.)

Gardner had participated in negotiating the original lease. The action brought money into the IAS coffers and met the terms of both the Guggenheim and Martin bequests. The option to sell the Sands Point property, however, was something else again. Gardner had already explained that Harry Guggenheim approved of renting the facility but would object to the sale of the property. He presented his own views in clear and unmistakable terms in a letter to the IAS Executive Committee on August 20, 1946. Gardner opened by announcing that he was not attending the next Executive Committee meeting because, he "... valued the friendship of all the members ... too much to risk having the wide difference of opinion between us affect that pleasant relationship of many years standing."[15]

The notion of offering an option on the estate to the Navy, he remarked, "... is absolutely contrary to the expectations of Mrs. Guggenheim ...." Rather than profiting from the donor's generosity, he argued, it would be better if the IAS were to simply give Hempstead House to the Navy for use as a research facility. Moreover, violating the Guggenheim agreement would also abrogate the contract with Glenn Martin to the effect that the Minta Martin Fund would be used to support research at the Guggenheim estate.

Finally, Gardner felt that he had been personally mistreated. When he had asked to continue as President of the Archives, leading members of the Council countered that he could only remain under Johnston's supervision. It was, he continued, "the unkindest rebuff of my life." Under the circumstances, Gardner announced that he was severing his connections to the Institute, had removed all of his personal belongings from the building, and did "... not wish to have any further connection with the Institute or the Archive."[15]

He meant it. He refused to chair the committee appointed to negotiate the future of the Minta Martin Fund with Glenn Martin. He refused to accept the highest honor that the IAS could offer when he was elected an Honorary Fellow in December 1946. When he began writing letters describing the situation to his many friends, the IAS Council was sufficiently worried to ask President Arthur Raymond to counter with letters of his own.[16]

Gardner would live for another 10 years, gathering in his fair share of retirement honors, from a Doctorate of Laws conferred by the Brooklyn Polytechnic Institute, to the Silver Stein Award presented by the MIT Club of New York, and the prestigious Guggenheim Medal in 1947 in recognition of his role in the creation of the IAS and a lifetime's contribution to aeronautics.

His breach with the new IAS leadership healed over time. Major Gardner finally accepted the distinction of Honorary Fellow in 1954 and unveiled a plaque at Columbia's Pupin Physics Laboratory commemorating the founding of the IAS. The leaders of the Institute did their best to support him, ensuring his comfort and well-being, particularly after the death of his wife Margaret in 1952. When he died at 5:15 a.m. on November 23, 1956, the IAS arranged his funeral at Arlington National Cemetery.[17]

From the Cold War to the Space Age

The postwar years were a time of growth and challenge for the IAS. The postwar slump that had resulted from the cancellation of contracts, the failure of a large market for private aircraft to emerge, and other economic dislocations gave way to increased government spending on Cold War weaponry. The general trend is apparent from a simple statistical survey of membership figures. In the 14 short years following the end of World War II, total IAS membership increased by 288%!

IAS Membership, 1945–1959[a]

Member type	1945	1959
Honorary Members	18	20
Honorary Fellows	18	33
Fellows	166	222
Associate Fellows	533	1,685
Members	1,704	6,976
Associate Members		6,782
Industrial Members	321	
Technical Members	3,264	
Affiliates	134	
Student Members	1,035	4,557
	7,193[b]	20,275[c]

[a]both figures for December 31
[b]Ref. 4
[c]Ref. 18

At the end of World War II, IAS members were distributed through 13 active Sections and 36 Student Branches located in colleges and universities across the nation. Fourteen years later, there were 35 Sections, an almost 300% increase. With 78 Branches, there was scarcely a creditable aeronautical engineering program in the nation that did not sponsor an IAS student organization.

In 1945, the national organization had cancelled most of its meetings to comply with a National Defense Transportation office request to cancel meetings attracting more than 50 out-of-town attendees. President Charles Colvin promised a full schedule of seven national meetings for the following year, however.

IAS National Meetings, 1946

Meeting	Location
Aircraft Propulsion, March 21–22, 1946	Cleveland, Ohio
Light Aircraft, June 13–14	Detroit, Michigan
Annual Summer Meeting, August 15–16	Los Angeles, California
Air Transport, October	Chicago, Illinois
Wright Brothers Lecture, December 17	Washington, DC
Honors Night Dinner, January 27, 1947	New York, New York
Annual Meeting, January 28–30	New York, New York

In 1958, IAS leaders planned or participated in 14 national and international meetings.

IAS Meetings, 1958

Meeting	Location
27th Annual Meeting, January 26–29, 1958	New York, New York
Flight Propulsion Meeting, March 5–6	Cleveland, Ohio
5th Nuclear Conference, April 5–10	Cleveland, Ohio
IAS/IRE National Aeronautic Electronics Meeting, May 4–6	Dayton, Ohio
Heat Transfer and Fluid Mechanics Institute, June 22–23	Los Angeles, California
IAS/ARS/AIEE/ISA Telemetering Conference, May 25–27	Baltimore, Maryland
Summer Meeting, June 16–19	Los Angeles, California
National Specialist Meeting (Classified), August 20–21	San Diego, California
Midwestern Conference On Solid and Fluid Mechanics, September 9–11	Austin, Texas
IAS/UC Frontiers of Engineering, September 16–17	Los Angeles, California
7th IAS/RAeS Conference, October 5–16	New York, New York
Midwest Meeting, November 2–4	Wichita, Kansas
Turbine Powered Air Transport, November 17–19	San Francisco, California
Wright Brothers Lecture, December 17	Washington, DC

The era from the end of World War II to the dawn of the space age was a time of change as well as growth for the IAS. The era began with financial problems. In terms of profit and loss, the Institute ran $4,734 in the red for the period October 1, 1945, to September 30, 1946.[19] Nor did the immediate future look too bright. The cancellation of wartime contracts,

Dignitaries attending the joint IAS/RAeS conference, 1955: (front row, left to right) Sir Arnold Hall; Jerome Hunsaker; Theodore von Kármán; N. E. Roe; Robert Gross, President of the IAS for 1955; (back row, left to right): Charles Colvin, IAS President for 1945; Barry Laight; E. S. Moult; Charles J. McCarthy, IAS President for 1953; John. Leland Atwood, IAS President for 1954; P. B. Walker; and Lester D. Gardner. (Courtesy AIAA.)

the withdrawal of some wartime subcontractors from aviation, and a wave of postwar labor problems in the aircraft industry were among the factors that led 22 companies to drop their corporate membership. Even the income realized from those companies that remained dropped, because IAS annual assessments were based on sales.[20]

Soon after taking the reins, S. Paul Johnston began to reduce costs. He undertook a complete reorganization of the staff, reducing salary costs for fiscal 1947 by more than $7,000 over fiscal 1946, primarily through staff reductions. He conducted a broad study of job classifications and salary rates in the New York area and then redefined IAS jobs and implemented salary adjustments to reflect the going rate for similar work. Johnston was proud to point out that the process had led to increases in the clerical and secretarial salaries.[21]

He also paid considerable attention to reorganizing the all-important publications program, which was leaking money. In addition to the staff reductions, printing economies were put in place as well as the reduction "... of general operating expenses which would not effect the Institute services to members or the industry." The process of refining the organization and focusing on essential services would culminate in yet another major round of decisions in the decade to come.[22]

Johnston also had to face problems with IAS real estate holdings and the endowment. How was he to resolve the complex difficulties of managing the Guggenheim estate and the Minta Martin Fund? Florence Guggenheim had donated the site to the IAS as an aeronautical research facility. Glenn Martin had provided an endowment of $50,000 to fund research at the

Opening the IAS sponsored Goddard Rocket exhibit at the American Museum of Natural History, April 21, 1948: (left to right) James H. Doolittle, IAS President for 1940; Albert E. Parr, president of the museum; Mrs. Robert H. Goddard; Harry F. Guggenheim; Mrs. Roger Strauss; and Charles A. Lindbergh. (Courtesy AIAA.)

site. Yet it must have been clear from the outset that the IAS was not in the business of conducting aeronautical research.

Renting the property to the Navy as a site where the Special Devices Branch could conduct research and envelopment efforts was a stop-gap measure that met the terms of both gifts and provided some money to the Institute coffers. The director and members of the Executive Committee, however, preferred to escape the costs of managing and maintaining the buildings and grounds by selling the estate and creating an endowment. The attempt to do just that led to Lester Gardner's break with the IAS.

Ultimately, negotiators persuaded Harry Guggenheim to approve the sale. The Navy finally purchased the site in 1951 for $332,000. The money was invested as a restricted Guggenheim Fund, the interest from which was used to support U.S. participation in the International Council of the Aeronautical Sciences. In the spring of 1947, Glenn Martin agreed to the transformation of the Minta Martin Fund into an endowment that would support IAS student activities. Sherman Fairchild and others agreed that funds that they had contributed for special purposes before 1941 could now be used for the general activities of the Institute.[23]

The leaders of the IAS were always careful to avoid the appearance that major companies were shaping editorial or organizational policy. From the beginning, however, the fortunes of the IAS had been tied to those of the aviation industry. With the arrival of the Cold War and the vast increases in federal spending to support a military establishment armed with the finest high-technology weapons systems that American industry could provide, things were looking up for the U.S. aerospace manufacturers. That translated into increasing IAS corporate membership, a key to financial stability. In 1936, the year in which corporate memberships were introduced, the 26 firms that signed up contributed $14,400 to the IAS coffers. By 1962, the 107 Corporate Members paid dues totaling more than $178,000.[24]

As might be expected in such a period, the size and scale of the IAS budget steadily increased during the postwar years. During fiscal year 1950, IAS general income from all sources amounted to $178,524.37.[25] Nine years later, that figure had climbed to $926,684, an increase of more than 500 percent. If income and the scale of the budget climbed steadily during the 1950s, so did program activities and the cost of operations. By 1950, the Institute treasurer was once again able to close the books $2,779.49 in the black. The annual bottom line varied during the 1950s, from a high of $6725.28 in 1955 to a low of $895.48 in 1958.

Things began to change for the worse in 1959, when the Institute ended the year $673 in the red.[26] An expanded program of meetings and intensified Section and student activities, the treasurer remarked, "coupled with the never ceasing increases in material and labor costs, contribute to the continuing fiscal problem with which your society is faced."[27] For Johnston and his management team, each year was a struggle to keep income up and expenses down. By 1960, it was no longer clear that they were winning that struggle.[28]

The IAS also sought to grow by absorbing other organizations. Discussions of an affiliation with the American Helicopter Society (AHS) began in 1949. Under the terms of an agreement between the two organizations that went into effect on April 1, 1950, the AHS would retain its own identity, operate under its own constitution and bylaws, and elect its own officers. The IAS would provide offices at 2 64th Street and part-time secretarial services and would handle administrative details involving membership, dues, record keeping, elections, and all mailings. The Institute would also publish and distribute the annual *Proceedings* of the AHS rotary wing forum and the semiannual newsletter. In return, the IAS would receive a complete AHS financial account, all cash and assets on hand at the time of the agreement, and all AHS dues and income from publications for the life of the agreement. IAS leaders reached out to the American Rocket Society (ARS) in an unsuccessful attempt to reach a similar agreement.[29]

Public Service

The involvement of IAS leaders and staff members in the work of major national commissions and study panels during the postwar decade is an important indication of the prestige and respect accorded to the Institute. On October 28, 1946, the IAS was appointed to represent the interests of the aeronautical research community on the Aviation Industry Advisory Panel of the Air Coordinating Committee. Created by President Harry S. Truman on September 19, the committee included representatives of the State, War, Post Office, Navy, and Commerce Departments as well as the Civil Aeronautics Board and, on a nonvoting basis, the Bureau of the Budget. As its name suggested, it was intended to coordinate all of the aviation activities of the government. The IAS Council selected Frank Caldwell (IAS president, 1941), chief of research with United Aircraft, to represent it on the Advisory Panel.[30]

The advent of the jet engine, aircraft capable of spanning continents and oceans, the atomic bomb, and a new and unstable world situation prompted the President to create an Air Policy Commission on July 17, 1947, intended to consider current U.S. air policy and recommend changes. Chaired by Thomas Finletter, a prominent attorney with State Department connections, the Commission included four additional business leaders, none of whom had direct experience with the aviation industry. Wisely, they selected S. Paul Johnston to serve as executive director of the Finletter Commission.[31]

In that role, Johnston was responsible for preparing the Commission's final report, submitted to the White House in December 1947. It began by noting that the expected postwar boom in civil aircraft sales had not materialized. The group then outlined a set of problems that were unique to the aircraft industry:

- the airplane served the needs of both military and civil users;
- one customer, the government, normally purchased 80% of annual output;
- demand varied widely from year to year, with government requirements;
- the industry had difficulty maintaining steady production for cost savings and protection of the labor force;
- rapidly advancing technology translated into high costs and short product lifespan;
- extended design-manufacturing cycle; and
- production capacity exceeding demand.

Because of these qualities, and the critical importance of the industry to the defense of the nation, the Commission noted that the government could ill afford to allow the aviation industry to struggle with the laws of economics and the vicissitudes of commerce that plagued less-critical industries. "It may even be desirable," the Commission suggested, "to keep a few marginal manufacturers in business who might be forced out if the normal laws of supply and demand were allowed to operate."[32]

Johnston played the same role for the President's Airport Commission, chaired by IAS founder James Doolittle, in 1952.[33] With the advent of the space age and the need for the IAS to pay close attention to cutting-edge developments, Johnston once again played a key role, this time as a member of a 1957 advisory panel created by Nobel Laureate Edward Purcell of the President's Scientific Advisory Committee. In that position he wrote a key memorandum summing up the opinion of the panel with regard to the creation of a national space agency. As he had for the Finletter Commission, Johnston had once again used his own skills as a manager and writer to summarize the findings of an important advisory group in clear and concise terms that had a major impact on national policy.[34]

The West Coast

There must have been moments when S. Paul Johnston and his colleagues wondered if they were running an aeronautical institute or a real estate office. In addition to the problem of the Guggenheim estate, they had to face issues relating to the Western Division and the construction and operation of the San Diego and Los Angeles buildings.

When William Dudley died in 1947, James L. Straight was named manager of the Western Division. An ex-mayor of Boise, Idaho, Straight had been associated with West Coast aviation publications before the war. In 1942, he accepted a post as a coordinator with the Aircraft War Production Council, supervising some 30 technical committees involved in aspects of engineering, testing, and production. He developed a manual outlining aircraft inspection standards and directed the exchange of technical data and publications among aviation companies. In his new role, Straight would supervise all Institute activities on the West Coast—membership, meetings, Sections, and industry relations. He would also manage the personnel, funds, and services of the Pacific Aeronautical Library, where he had his offices, at 6715 Hollywood Boulevard. Two years later he was appointed assistant director of the IAS and was empowered to act on behalf of the director in all matters related to West Coast operations.[35]

Straight was heavily involved in work on the two California IAS buildings. The San Diego headquarters was constructed on a plot of land obtained under a 25-year lease from the San Diego Harbor Commission. The facility was dedicated on April 29, 1949, but the 140-member local Section had a difficult time making efficient use of the building. In August 1950, the IAS Council charged off $8,200 in accrued operational losses to a national surplus. Convair rented the building to house a section of its flight test division from August 1950 to February 1953. The $23,000 in rent received during this period was used to repay the Fleet Library Fund, which had loaned funds for construction, and to cover the operational costs and upkeep of the building when it returned to IAS hands.

It cost the San Diego Section $2,400 per year to maintain the building, most of which had to be absorbed by the national organization. The Section leaders made every effort to cut their losses by renting the facility for weddings, dances, and the meetings of other organizations. They thought they had found a more permanent solution in January 1958, when California Western University (CWU) took a five and one-half year lease on the building. The IAS Section could even continue to use the facility for meetings, while the CWU would bear all the expense of maintenance and repair. The arrangement threatened to collapse immediately, however, when the university failed to expand as quickly as expected. The IAS would have no choice but to resume the struggle to pay upkeep.

By 1947, construction costs in Los Angeles were soaring. The $268,000 building fund raised in 1944 would no longer cover the cost of the building. A new campaign raised an additional $100,000 from local industry to fund the construction of a modified building that had been stripped of some original features, including air conditioning, and expanded to include some income-producing rental space. The cost-cutting measures proved shortsighted. The national office eventually spent $25,000 to install air conditioning and other updates to the facility.

Completed in 1949, the IAS Western headquarters was located at 7660 Beverly Boulevard, Los Angeles. Initially, the building had three long-term tenants: the Pacific Aeronautical Library, funded by local aircraft manufacturers, and the Los Angeles offices of both the NACA and the Aircraft Industries Association. They would remain in place for more than a decade, until May 1959, when the newly created National Aeronautics and Space Administration (NASA) moved to larger quarters. Even so, the relatively low rent charged the three organizations did not cover maintenance costs. As in the case of the San Diego building, the Western headquarters was rented out for dances, television rehearsals, and wedding receptions. In order to remain competitive as a rental facility, the IAS installed kitchens and a bar and invested $6,800 in a liquor license.

Straight resigned as IAS assistant director early in September 1950 and was replaced by Ernest William Robischon. As a young man, "Robie," as he was known to a generation of IAS members, had built an impressive personal collection of aviation books, models, and stamps that would earn him a position as aeronautical librarian at Cal Tech. He held a variety of other positions with the university and served as founder, director, and consultant to the Pacific Aeronautical Library. During World War II he had served with U.S. Army Air Force Air Intelligence and established the Air Document Research Center in London, which organized captured German technical documents. After the war he had worked for both Cessna and Beech as a methods engineer.[36]

Soon after taking over the Western Division, Robischon created an aeronautical museum in the Los Angeles building, funded by the defunct Durand Republishing Company, for which he had served as executive secretary. The project enabled the IAS to save a bit more money by escaping certain Los Angeles city taxes. As a dedicated collector and historian of flight, Robie took his museum responsibilities very seriously. The project did not increase support for the building, however.

The Western Headquarters was a money pit. By 1957, in spite of everyone's best efforts, the building was running $2,700 in the red. The following year the loss had climbed to $4,700. The building lost $5,280 in the first six months of 1958! Clearly, the Los Angeles Section was fighting a losing battle as taxes and costs continued to outstrip income. Worse, the building was now too small to house the largest Section meetings. The Los Angeles IAS, which had numbered 1,055 members in 1947, had grown to 3,016 by 1959.

A pair of Los Angeles developers proposed to construct a new and much larger office building on the site in 1958. In return for the IAS contribution of the site to the project, the Institute would have a dedicated space in the new structure in perpetuity. The Council dithered until that opportunity had slipped away and then appropriated $10,000 to fund an engineering, architectural, and management survey of the options. No one, it seemed, could come up with a plan to deal with the white elephant that was the Los Angeles building.

International Relations

Surely no prewar aviation leader was more of an internationalist than Lester Gardner. An American member of the RAeS, he patterned his own Institute after the foreign organization that he so admired. Prior to 1939, he had done his best to encourage international contacts. The IAS exchanged publications with other aeronautical societies around the globe. IAS members attended European conferences, from the famous 1935 Volta conference on high-speed flight held in Rome to international gatherings of aeronautical journalists. "On this side," Paul Johnston reminded postwar members, "Institute headquarters became the first port of call for foreign visitors."[37]

In the spring of 1946, Sir Roy Feddon, a leading British propulsion specialist, suggested the potential value of a joint U.K.–U.S. postwar aeronautical conference. Paul Johnston, who agreed that it would be a fine thing to "... lean comfortably over our fences and swap yarns with our neighbors," opened discussions with J. L. Pritchard, his counterpart at the RAeS. It was a visit from Sir Frederick Handley-Page, president of the RaeS, that sparked the actual planning.

Held in London on September 2–6, 1947, the first joint conference included the presentation of 21 papers in concurrent sessions, with plenary gatherings to provide overviews of major topics: "Long-Range and Upper Atmosphere Aircraft," "Light Aircraft," and "Short-Haul and Medium-Range Aircraft." As might be expected of what were still, in many ways, a pair of old boy's clubs, much of the real business was conducted in informal meetings with the leadership of the RAeS and during the dinners, garden parties, receptions, and tours of every imaginable aeronautical facility.[38]

The meeting was a genuine success and became an annual event in the calendar of both organizations. Other international activities followed. The IAS had encouraged the growth of Sections in Ottawa, Montreal, and Toronto. As those Sections grew stronger, joint IAS–RAeS conferences in New York, London, and Ottawa led to the establishment of the Canadian Aeronautical Institute (CAI). The first joint IAS–CAI conference in Montreal followed soon thereafter.[39]

"Eventually," Johnston noted in 1947, "other aero-scientific bodies of other countries should be brought into the exchange program." At that point, he continued, "the scientific knowledge of all nations could be tossed into the common pot, and real progress in aviation would follow.

The airplane would quickly take its place as one of the greatest instruments for ensuring international peace and understanding that the world has ever seen."[37]

Fine words, but the next step in international cooperation would come from a Cold War organization – the North Atlantic Treaty Organization (NATO). At the invitation of Dr. Theodore von Kármán, chief scientific advisor to the U.S. Air Force and Chairman of the Advisory Group for Aerospace Research and Development (AGARD), the IAS had been participating in NATO activities since 1953. Those contacts and the ongoing IAS-RAeS relationship led a group of international scientists and engineers attending the IAS annual meeting in New York in January 1957 to plan for the creation of an international association of aeronautical societies.

The IAS would play the central role in getting the International Council of the Aeronautical Sciences (ICAS) off the ground. The creation and early work of ICAS was funded with money realized from the 1951 sale of the Guggenheim estate to the Navy. With the concurrence of the members, the IAS functioned as the secretariat for ICAS. Robert Dexter, the IAS secretary, took the lead in managing Institute participation and remained an important figure in ICAS for a number of years, including service as ICAS executive secretary. Hugh Dryden, a long-time IAS member and a key figure in the NACA/NASA, also took a serous interest in the foundation of ICAS.

ICAS would be the organization through which the aeronautical technical societies of the world could interact. As the draft statement of purpose suggested

> The general objective is to encourage free interchange of information in all areas with any bearing on the advancement of knowledge in all phases of aeronautics. This would be accomplished by a program of cooperation and collaboration among the organizing aeronautical and scientific societies of the world.[40]

Kármán served as the first president, with Maurice Roy, of the French organization ONERA, as the head of the executive committee. A general assembly made up of a single representative from each of the member societies set policy and appointed working committees.

Five hundred delegates from 25 nations (including 157 from the United States and 9 from the Soviet Union) gathered in Madrid on September 8–13, 1958, for the first International Congress of ICAS. By 2004, through the course of 24 successful biennial congresses, ICAS continued to meet the expectation of its founders by facilitating the interaction of aerospace engineers, scientists, and managers from around the globe.[41]

The IAS established a European office in Brussels in 1961 with a grant from the Arthur D. Little Foundation. The new office took the lead in managing the Institute's various international ventures. The organization forged yet another international link and indicated its determination to meet the needs of a changing industry when it affiliated with the International Astronautical Federation (IAF).[42]

The Library and "Project A"

In his first report to the Council, Paul Johnston remarked that "... the libraries of the Institute are potentially one of the greatest means of providing real services to the membership and also may develop new sources of revenue." Indeed, the IAS library was the busiest department in the organization. By January 1947, the combined holdings of the New York and Pacific Aeronautical Library totaled 32,256 volumes. Seventy-five new foreign periodicals were added to the 375 serial titles already being received. The reading room was closed for more than a month during the 1946 move. Even so, 319 people had paid 705 visits to the library that year, using 2,528 books and publications in the process. The staff had served a total of 3,306 users. They received 758 mail requests for information, responded to 536 telephone queries, and filled 248 orders for copies of 531 items. The Paul Kollsman Lending Library met the needs of 4,417 individuals.[43]

On October 7, 1946, even before most of the staff had moved into the new IAS headquarters building at 2 64th Street, the new library hosted an Industry Conference on Aeronautical Library Research Facilities. Fifty-two engineers and industry librarians from 33 organizations met to discuss mutual problems, including the difficulties of abstracting, classifying, and handling foreign language material. John A Sizer, representing the British Liaison Office at the Air Materiél Command (AMC), Wright Field, described the work of the British Technical Information Bureau of the Research and Technical Publications Division (RTP), the office responsible for the collection, analysis, and distribution of technical information. He was followed by Dr. Albert A. Arnhym, editor-in-chief of the Air Documents Division, with AMC.[44]

The problem was how to keep track of the exploding volume of technical literature. It was an issue to which the IAS had given considerable thought since 1935, when Lester Gardner had launched the Aeronautical Index. Just a year later, in January 1936, a new department was added to the *Journal of the Aeronautical Sciences*. Each issue would include a professionally prepared bibliography of the latest technical articles in the field, broken down by subject area. Initially, the index covered 33 periodicals in English, French, German, and Italian. The service grew until 1942, when the IAS launched its new publication *Aeronautical Engineering Review*, the principal mission of which was "... to provide Institute members with an up-to-date abstract and indexing service." Busy modern engineers, IAS leaders reasoned, "... have little time and but limited opportunity to scan the vast flood of books, magazines, reports, and other publications on aeronautical subjects that pour from the world's printing presses ...." The Institute could handle that task for them.[45]

By 1947, the *Aeronautical Engineering Review* had a staff of six professionals working under the supervision of Welman Shrader, Director of Publications, and a three-person technical advisory committee that included Director S. Paul Johnston, the NACA's Hugh Dryden, and Robert Dexter. In addition to their own indexing efforts, the IAS staff maintained a master card file of 107,800 articles, papers, and books indexed and abstracted by the NACA *Wartime Reports*, and the *Air Documents Index, Translations, Intelligence Reviews, Technical Reports, Summary Reports*, and *Interim Reports* produced by the Air Documents Division (Intelligence T-2) of the Army Air Forces AMC. It seemed to Paul Johnston that there ought to be a way to take better advantage of the indexing and abstracting work that was already underway.

Early in 1946, anxious to find ways to extend the utility and the profitability of the library operation, Johnston borrowed Ernest Robischon from the Army Air Forces. The director asked his consultant, who was already well known as an authority on aeronautical documentation, to study the IAS operation and offer recommendations for the future of the library program.

Robischon presented a comprehensive report late in 1946. Heading his list of recommendations was a suggestion that the IAS take the lead in the business of indexing and abstracting. The first step would be to develop a Standard Subject Heading List, a system of subject terms and categories to be used in classifying and cataloging aeronautical literature. Such a system could be adopted by all of the agencies currently involved in preparing bibliographies and indexes.[46] Recognizing that such an approach would enable him to join the strengths of the IAS publishing program and the library, Johnston lost little time in acting on the suggestion.

In February 1947, the director announced the successful transformation of what the staff had been calling Project A into Contract No. W 32—038 ac 17021 with the Army Air Forces. With funding from both the U.S. Army Air Force and U.S. Navy and the cooperation of the NACA, the Joint Research and Development Board, and the Special Libraries Association, the IAS would develop a Standard Aeronautical Index.[47]

Project Director Leslie Neville came to the IAS following a long career on the editorial staff of *Aviation*. Technical Assistant Harris Reeve and librarian Keith Brown were also members of the team, as was administrator Ruland Woodham. Within a few months, the project, which Johnston described as the most important undertaken that year, would employ a dozen people, including three in California and one in Dayton. They began by asking IAS members and users of the U.S. Air Force Air Documents Division to name their specialty area. The initial list of topics and categories was published in Wright Field's *Technical Data Digest*, with a request for reader comment. The result, a Standard Aeronautical Index List of Division Headings, was hammered out at an IAS headquarters meeting on October 24, 1947. The group recommended 49 main topics ranging from aerodynamics to wind tunnels and laboratories.[48]

The contract, extended into 1948, produced a thesaurus of terms and list of categories and subcategories that would be used by all aeronautical abstractors in the United States. The IAS leadership continued to regard the business of indexing, abstracting, and reviewing as one of their most important contributions to members and to the profession. Each month roughly 1,000 "periodicals, reports, house organs, books, catalogs, and releases in all languages" passed over the desks of IAS staffers. The resulting abstracts appeared in the *Aeronautical Engineering Review* and, beginning in 1948, in an annual compilation, the *Aeronautical Engineering Index*, published each year until 1958.[49]

In 1954, Theodore von Kármán indicated the need for an international abstracting service that "would designate to the men interested in research and development what they should procure and read of the foreign literature ...."[50] The Air Force invited the IAS to collaborate with AGARD, a newly organized NATO organization, on matters of aeronautical documentation. With funding from the U.S. Air Force Office of Scientific Research and Development, the IAS created International Aerospace Abstracts, accelerated reviews of the worldwide literature that appeared in each issue of *Aeronautical Engineering Review* beginning in December 1955. To streamline the process, the Institute was named the U.S. recipient of all reports from NATO counties. During 1957, a total of 6,770 abstracts appeared in IAS publications.[51]

A study of member reaction to IAS publications undertaken between 1958 and 1960 held some surprises for the staff. Only 10 percent of readers were strongly in favor of continuing the abstracts and reviews as a part of the magazine, while practically all readers wanted more feature content. As a result, the publication was retitled *Aero/Space Engineering* beginning with the May 1960 issue. The section of the magazine bringing readers up to date on news of the Institute and its members would now appear as a separate newsletter, *IAS News*. A new publication, *International Aerospace Abstracts*, funded by NASA, would be devoted to reviews and abstracts. At the same time, the Institute had applied for a National Science Foundation (NSF) grant to fund the transformation of the indexing and abstracting work into a subscription service. The organization would remain in the documentation business for many years to come, but it would increasingly be viewed primarily as a business operation.[52]

The End of History

Lester Gardner had amassed one of the world's finest collections of art and objects relating to the history of flight, and it continued to grow after his sudden departure. The leading benefactors, individuals like Harry F. Guggenheim and Bella C. Landauer, continued to add to the historical treasures they had already deposited with the IAS. Leading figures in aviation donated models, photographs, medals, trophies, awards, and aeronautical curiosities.

In May 1947, William A.M. Burden (IAS president, 1949) deposited the enormous collection of historical aeronautica gathered by two generations of the Tissandier family. There were tables, chairs, clocks, fans, snuffboxes, wallpaper, fabric, porcelain, and prints, all richly decorated with scenes from the history of ballooning. The collection was given pride of place in a section of the New York building known as the Burden-Tissandier room. The following year, the executors of the Orville Wright Estate presented 28 very special items to the IAS, including the stopwatch that had timed the first flights at Kitty Hawk.[53,54]

Paul Johnston, who had written extensively on the early years of aviation and who would one day head the National Air and Space Museum, shared Gardner's enthusiasm for the history of flight. During the course of his own research, Johnston discovered the papers of the English aeronautical pioneer William Samuel Henson in a New Jersey attic and persuaded the owners to donate the collection to the IAS Archive.

In the fall of 1948, he hired Claudine Lutz as curator of Archives, secretary to the Archives Committee, and curator of the New York collections, with general responsibilities for the growing history programs of the IAS. Educated in France and Switzerland, she attended Sarah Lawrence and graduated from Barnard with a degree in art history and literature. She had served as a Women's Army Corps (WAC) and had a deep interest in the history of flight. Her first project was to design and install a temporary exhibition, the "First Century of Flight," in the headquarters entrance hall. Using models and images, she walked visitors from the prehistory of aviation to the era of supersonic flight, the helicopter, and guided missile.

Encouraged by their first foray into exhibit design, Johnston and Miss Lutz made a tour of Northeastern and Mid-Atlantic museums to study display techniques, from the local giants—the American Museum of Natural History, the Metropolitan Museum of Art, the Museum of Modern Art, and the New York Historical Society—to more distant venues like Philadelphia's Franklin Institute and the Carnegie Institute of Pittsburgh. The resulting new look for the IAS Museum involved weeding out duplicate items, relocating cases, and improving the chronology.

The central reception room on the second floor was transformed into a modern exhibition gallery, complete with well-lighted wall panels and glass cases designed for changing displays. The first exhibit, mounted in the early summer of 1949, showed the best 18th century aeronautical prints in the collection to best advantage. As one commentator remarked, "The pale gray wall panels and gray-framed cases, enlivened by touches of coral and deep blue were noted with pleasure."[55] Other areas had also been refurbished in time for visiting members of the Aeronautical Society of Great Britain attending a joint IAS conference on May 22–29, 1949, to "... trace step-by-step progress toward modern flight by making one complete circuit of the model room."[56]

S. Paul Johnston and Claudine Lutz married following the death of Catherine Johnston in 1948. Mary Morgan Hamilton took over as curator of the museum in 1949, after three years of experience managing the public spaces of the Wings Club. In addition to managing the exhibition program, she doubled as secretary for the AHS, an IAS affiliate. In 1950, Miss Hamilton began working with Esther Goddard to develop a permanent exhibition chronicling the work of her husband, Dr. Robert Hutchings Goddard.

The American rocket pioneer was fast emerging as a space-age folk hero. Following his death in August 1945, Esther Goddard devoted herself to ensuring that his work would not be forgotten. With funds from the Guggenheim Foundation, which had supported Goddard's work during his lifetime, she combed his papers for additional ideas and inventions that could be patented. At the same time, she identified surviving Goddard rockets and equipment, some of which were incorporated into a large exhibition that opened at the American Museum of Natural History on April 21, 1948.

Following a national tour that took the exhibit to Washington, DC; Chicago; and Los Angeles, Mrs. Goddard worked with Mary Hamilton to plan a permanent installation in the IAS Museum, where it opened on December 1, 1950. Esther Goddard, who had donated the original Goddard items, was on hand along with Harry Guggenheim; Laurance Rockefeller; Smithsonian Secretary Charles Abbott; President Howard Jefferson of Clark University; and representatives of the IAS, the ARS, Reaction Motors, Inc. (RMI), and Aerojet-General.

Entrance to the museum was originally restricted, especially with regard to the valuable material on the second floor. Following the redesign and reopening of the museum, the IAS opened its doors to the public weekdays from 9 a.m. to 5 p.m. Special efforts were made to attract young people. There were tours for high school groups and special youth exhibits for the Industrial Museum of Jersey City and the Children's Room of the New York Public Library.

Ernest Robischon opened his own William F. Durand Museum in the Los Angeles building in the fall of 1951, soon after taking over the Western Division. He explained that the museum exhibits would be "designed not to portray history for history's sake, but to show how history can be used as a tool for the present and future." Appropriate topics, Robischon suggested, would include the development of air foils, landing gear, and fighter aircraft.

Over the next few years, local newspapers chronicled a series of temporary exhibitions mounted in the building, including a particularly impressive display of 73 models of modern aircraft, all in the same scale, prepared by Lloyd Jones, the proprietor of a Van Nuys camera and hobby shop. "At present," the *Los Angeles Times* noted, "the IAS is considering a plan to have Lloyd build a complete history of design evolution in the tiny models . . . ."[57]

With the approach of the 50th anniversary of flight, Robischon realized his dream of acquiring a full-scale model of the world's first airplane for the Durand Museum. His campaign, launched in the fall of 1952, attracted the support of 24 companies. John K. Northrop and James McKinley agreed that students at the Northrop Institute of Technology (NIT) would build the wings and the Los Angeles Trade-Technical Junior College (LATTJC) and the Pacific Airmotive Corporation constructed the elevator and rudder. NIT and LATTJC students did most of the final assembly and rigging. Marquardt Aircraft, with the assistance of Charles Taylor, who, fifty years before, had been the Wright brothers machinist, produced a wood and metal replica of the original four-cylinder engine.

"As time went by, strange bits and pieces began to turn up around the Los Angeles building," the *Aeronautical Engineering Review* reported. "One might find a wooden rib or two standing in a corner, or a strange looking gas tank under a desk."[58] When Theodore von Kármán asked Robie to visit Europe in February 1953 to advise on AGARD documentation, F. A. Cleveland of Lockheed Aircraft stepped in to coordinate the Wright Flyer project. Former IAS President Jimmy Doolittle, who was heading the national organization to commemorate the 50th anniversary of powered flight, dedicated the only full-scale model of the 1903 Wright airplane before a crowd of 500 people gathered at the W. F. Durand Museum in July 1953.[59]

For a professional engineering society, the IAS had made an enormous investment in the history of flight. By the 50th anniversary celebration, the IAS was operating a world-class aeronautical library, one of the world's leading archives of material related to the history of flight, and two aviation museums. "Our wartime experiences have shown," William A. M. Burden explained, "the potential value of such collections, and the special knowledge that collectors possess, to the intelligence aspects of our national defense." By the fall of 1948, the leadership of the Institute had decided to take full advantage of that strength. President John K. Northrop, who was supporting the growth of an historical archive at the NIT, appointed an Archives Committee to supervise matters of flight history within the Institute.

The high level of the members of the new committee certainly augured well for the future of history in the Institute. Chairman William Burden, the single most significant donor to the IAS collections, would take office as president of the Institute in 1949. The other members of the group included Executive Director S. Paul Johnston, founder Grover Loening, and two past presidents, Preston R. Bassett (1947) and Reuben Fleet (1944). The dozen remaining members included such leading collectors as Bella Landauer, E. Hildes-Heim, and Richard C. Gimbell; John J. Ide, the NACA's European representative; Arthur G. Renstrom of the Library of Congress; and aviation pioneer C. S. Jones. Claudine Lutz and Ernest Robischon represented the staff members involved in history programs.

The Council also created a new grade of member, Historical Associate, for individuals with a serious interest in aeronautical history or collecting who might not qualify for IAS membership in the technical categories. Members entering the organization through the new grade would not have a vote, but their $10 dues payment would entitle them to receive four issues of *Aeronautica*, a new quarterly publication that would carry historical articles, information on aeronautical collecting, and relevant news items. As part of the program, the Council appropriated $4,500 to support the new grade, which would include a renewed effort to catalog objects in the IAS collection.[60]

In 1948, the Historical Associates were invited to a special evening with Charles Dollfuss, chief curator of the Musée de l'Aeronautique and one of the world's authorities on the history of flight. The visiting French historian was treated to a tour of the collections and recounted his adventures as an aeronaut and collector. The new program was scarcely an overwhelming success. By the end of 1949 there were only 55 active Historical Associate members and fourteen additional Historical members already enrolled in other grades. "Expanded membership is our greatest need," Claudine Johnston explained. "We believe that interest in this field can be developed, but will require more time and effort."[61]

The new program was not all that well received, however. S. Paul Johnston closed the fifth issue of *Aeronautica* with a letter noting that although the Council and the staff had not expected the new membership grade to pay for itself in a single year, they had hoped for a much more positive response. The leadership had agreed to continue the program for another year, with some cuts. The museum would remain open from 9 a.m. to 5 p.m., although the curator would now be a part-time volunteer. Her assistant would continue the effort to catalog the collection but would have additional responsibilities. He asked the members to save the program by helping to expand the membership.[62]

Jack Straight, Western Division manager, returned to Los Angeles following the annual meeting in January 1950 determined to encourage Historical Associates on the Pacific Coast. The idea immediately appealed to the members of the Aero Historical Society, a recently disbanded group of aviation history enthusiasts, most of whom had been employed at North American Aviation. John J. Sloan, a North American standards engineer, was elected executive chairman of the IAS Historical Branch. The group began mounting small exhibitions at the Western Division headquarters building and appearing on local television programs.

Enthusiasm was high in Los Angeles. Within a month or two of the organizational meeting on March 3, 1950, the group counted 47 members. There were outings to local aviation events and sites and lectures by visiting authorities, such as the June 27 address on the work of aeronautical pioneer Octave Chanute by Professor Pearl I. Young, a NACA veteran who was teaching physics at Pennsylvania State College, or films by aviation pioneer James Mattern.[63,64]

The members of the Los Angeles Historical Branch were more active than their New York colleagues and their interests were a bit different, as well. The East Coast group was dominated by connoisseurs and collectors of aeronautica. Acquisitions by the Aeronautical Archives and the museum tended to focus on the early decades of ballooning and the prehistory or very early history of aviation. Lester Gardner had begun the IAS collection as a means of emulating elite European organizations like the RAeS. For him, collecting 18th- and 19th-century items was a way of establishing the cultural importance of the field and the Institute. A sense of that elite point of view remained after his departure.

The history enthusiasts in Los Angeles were, for the most part, engineers or technicians employed in the industry who simply loved airplanes. They tended to collect photos or aircraft and seldom missed an opportunity to attend an air show or visit a flying field to see a rare old airplane. The History Branch generated considerably more grassroots support than the New York program.

Beginning with the first issue of 1952, responsibility for *Aeronautica* and leadership of the Historical Associates program were transferred to the Western Division staff. G. R. Lawler, a volunteer editor, turned out the first Los Angeles issue, after which Ernest Robischon edited the publication for a time. The last issue of *Aeronautica* (Vol. 8, Nos. 3 and 4) went into the mail in the fall of 1956.[65]

It was apparent to Johnston and the members of the Council that although the IAS was certainly in the information business, it was not in the history business. The History Associates program was phased out in 1956. Next to go were the historical collections, as Paul Johnston recalled in a 1968 letter to American Institute of Aeronautics and Astronautics (AIAA) Controller Joseph J. Maitan. "You and other old IAS hands will recall that . . . Lester Gardner had great ambitions to turn the IAS into a historical museum." By the mid-1950s, however, ". . . it became quite apparent to the Council and the Director that all IAS funds and capabilities had to be directed toward the professional needs of the membership and not toward building museums."[66]

Perhaps 10,000 of the total 23,000 square feet available at 2 West 64th Street had been devoted to the library, archive, museum, and collections storage. By 1959, with the increase in staff, membership services, and activities directly related to the business of an active and growing technical society, the director faced the need to reconfigure the space. The museum displays vanished after 1956. Initially, the most valuable objects were stored on the sixth floor.

Storage is expensive and the dispersal of the collections was soon underway. At Esther Goddard's request, some key items from the large display of her husband's material went to the Roswell Museum, where the professor had conducted his rocket tests in the 1930s.[67] Paul Edward Garber, curator of the Smithsonian's National Air Museum, apparently learned of the storage problems during conversations with an IAS employee and expressed an interest in the Institute's collections. As early as December 1954, he presented a well-thought-out proposal, spelling out how the transfer might be conducted and indicating how the objects, library material, and archival items would be integrated into the Smithsonian collections.[68]

The Institute quickly shipped a few duplicate books and some of the potentially dangerous nitrate motion picture film to the Smithsonian. By the end of 1959, the Council had decided that the historic materials donated to the IAS had been intended for display. Because they had no intention to re-establishing a museum, they finally began to ship some of their large cache of models to the Smithsonian.[69]

"At some point in 1960 or 1961," AIAA Controller Joseph Maitan recalled, the New York City fire inspector ordered the IAS to clean out its overcrowded storage areas in the attic, basement, and subbasement. More than 100 file cabinets containing the hundreds of thousands of index cards generated by the Works Projects Administration (WPA) personnel of the 1930s along with decades worth of new clippings, press releases, old book, and magazines were shipped to the Smithsonian and the Library of Congress at government expense.

Finally, in 1963, facing the need to rent more storage space, the Council approved the final dispersal of remaining historical items to the Library of Congress and the National Air Museum. Marvin W. McFarland, Chief of the Science and Technology Division of the Library of Congress, dispatched Arthur Renstrom, a library authority on rare aeronautical books, to assist the IAS in dispersing the items.[70]

Paul Garber literally drove truckloads of material to the Smithsonian in July 1963. His inventory of the IAS items taken "on loan" to the nation's capital included

- 400 cartons of books;
- 166 cartons of pamphlets, brochures, and biographical items;
- 10 cartons containing the Sherman Fairchild photo collection;
- 39 cartons of miscellaneous photographs;
- 74 cartons of motion picture film;
- 5 cartons of material from the Bella Landauer Collection;
- 16 cartons of *Aeronautica*;
- 18 cartons of model aircraft;
- 29 boxes of commemorative medals (the gold medals were later returned);
- 171 early aeronautical prints from the Harry F. Guggenheim Collection;
- 3 filing card file cabinets with 60 drawers for 3 × 5 index cards;
- 4 card file cabinets of 24 drawers each containing the catalog of captured German and Japanese documents;
- 2 packages of posters; and
- 4 personal watches belonging to the Wright brothers.[71]

A substantial number of books and objects not selected by the Smithsonian or the Library of Congress were shipped to other historical organizations: the Franklin Institute, the Roswell Museum and Art Center, the Connecticut Aeronautical Historical Society, and the Nassau County (New York) Museum. Scattered items over the years also went to aviation trade schools and to the Canadian Air Museum.[72]

Johnston was aware that the dispersal of the collections might create tax difficulties or upset donors. He discussed the matter with Internal Revenue Service (IRS) officials and contacted as many donors as he could. Considering the less-than-ideal registration procedures practiced by IAS staff members who always had other duties, the organization kept surprisingly good lists of the material. Those occasionally proved useful in later years, when smaller donors who had not been contacted demanded the return of items given to the IAS. In several cases, Johnston was able to locate such items in the Smithsonian collections and return them to the donors.

At one point, the Institute recalled all of the gold medals sent to the Smithsonian. On other occasions, Garber was able to return specific items requested by IAS donors, including a silver cigarette case. Initially, the collections were placed on long-term loan to the institutions involved. This situation created administrative problems for the both the Smithsonian and Library of Congress. On June 26, 1968, upon receiving a report that the reacquisition of the materials would cost in excess of $100,000, the AIAA Council ordered that the long-term loans be deeded to the organizations holding them.[73]

Not all of the IAS treasures went to leading libraries and museums, however. In November 1963, with the final dispersal underway, S. Paul Johnston invited select members of the IAS staff to choose a personal "souvenir" from the collections before they were removed from the building. Controller Joseph Maitan always regretted that he had been hospitalized at that time. Five years later, when he was handling the final legal transfer of ownership to the historical material, Maitan asked Johnston to retrieve "... the silver French wine-tasting cup with a balloon engraving, circa 1790, if Dr. Garber and Robischon [of the Smithsonian] ever find it." That, the controller noted, would be a worthy memento "... of my security guardianship of the Institute's collections for some 25 years, during which period only one volume 'disappeared,' to my knowledge." Such a cup is not in the Smithsonian collection today, but neither is there a record of its being returned to the IAS for presentation to Maitan.[74]

The most important collections were handled individually and with great care. The fate of the Guggenheim prints and Bella Landauer materials was discussed with the donors at length. Both collections went to the Smithsonian, which was then planning for a new National Air and Space Museum on the Mall. William A.M. Burden's balloon-decorated furnishings and aeronautica went into insured off-site storage in 1959. A member of the Smithsonian Board of Regents, Burden ultimately donated his treasures to the museum as well.[75]

Lester Gardner's dream, at least a part of it, was over. The IAS and the AIAA would function in austere and businesslike offices, unadorned with the aeronautical antiques. The streamlined professional society had no place for the archival and museum treasures that had been so much a part of the prewar IAS.

Still, because Gardner did collect objects, books, and manuscripts documenting the early history of flight, the American people are inestimably in his debt. Had he not brought those collections to the IAS, Paul Johnston would not have been able to distribute them to the most appropriate public institutions. Gardner's treasure survives today, safely housed on the rare book shelves of great national libraries, in the manuscript and archival collections of the Library of the Congress and the Smithsonian, and exhibited for 10 million visitors a year to see and enjoy in the galleries of the world's most visited museum. For that, posterity owes a considerable debt to two forgotten figures, Lester Durand Gardner and S. Paul Johnston.

References

[1] Franklin D. Roosevelt to Frank Caldwell, Jan. 23, 1942, AIAA Headquarters Files.

[2] "Air Engineers Get Praise for War Aid," *New York Times*, Jan. 29, 1946.

[3] Minutes of the Fourteenth Annual Meeting of the Institute of the Aeronautical Sciences, the President's Annual Report for 1945, the Administrative Committee Report for 1945, AIAA Headquarters Files.

[4] Minutes of the Meeting of the IAS Council, Jan. 30, 1946, AIAA Headquarters Files.

[5] Daniel Rich, "IAS: The Institute of the Aeronautical Sciences and the Institute of the Aerospace Sciences," *Astronautics & Aeronautics*, Feb. 1971, pp. 68–74.

[6] Rich, "IAS," p. 73.

[7] *The Institute of the Aeronautical Sciences and Aeronautical Library and Museum*, IAS pamphlet, 1946(?), author's collection; "Memorandum to Council, Background Material On IAS Real Estate Projects," 1959, AIAA Headquarters Files.

[8] R. H. Fleet to Ernest Breech, Aug. 8, 1945, document folder concerning 2 West 64th Street, AIAA Head quarters Files.

[9] Lester D. Gardner, "Gifts to the Institute," *Aeronautical Engineering Review*, March 1946, p. 32.

[10] *The Institute of the Aeronautical Sciences and Aeronautical Library and Museum*, IAS pamphlet, 1946.

[11] All information on the zoning problem is from an undated memorandum by Lester D. Gardner in a document folder concerning 2 West 64th Street, AIAA Headquarters Files.

[12] The term "aeronautica" refers to rare books relating to the history of flight.

[13] The description of the interior spaces of the IAS Headquarters building at 2 West 64th Street is drawn from *The Institute of the Aeronautical Sciences and Aeronautical Library and Museum*, IAS pamphlet, 1946.

[14] Minutes of the Meeting of the ARS Board of Directors, Jan. 30, 1946, AIAA Headquarters Files.

[15] Lester D. Gardner to Members of the Executive Committee, Aug. 20, 1946, AIAA Headquarters Files.

[16] Minutes of the IAS Council, Dec. 4, 1946, AIAA Headquarters Files.

[17] Files pertaining to Lester Gardner and his funeral and estate, AIAA Headquarters Files; at his death, Gardner willed $2,406.15 worth of books and other museum objects to the IAS.

[18] S. Paul Johnston, "Director's Report, 1959" *Aero/Space Engineering*, March 1960, p. 17.

[19] IAS Comparative Operating Statement, Director's Report, Jan. 29, 1945, Minutes of the Meetings of the IAS Council, AIAA Headquarters Files.

[20] Minutes of the Meeting of the Executive Committee of the IAS Council, May 13, 1946, p. 4.

[21] Director's Report, presented to the Annual Meeting of the IAS Council, Jan. 29, 1947, AIAA Headquarters Files.

[22] Minutes of the Meeting of the IAS Executive Committee, Sept. 24, 1946, AIAA Headquarters Files.

[23] Memorandum to Council, Background Material on IAS Real Estate Projects, dated 1959, in AIAA Head-quarters Files.

[24] Daniel Rich, *AIAA at 50: Yesterday/Today/Tomorrow*, AIAA, New York, 1982, p. 24.

[25] "IAS Statement for the fiscal year October 1, 1949 to September 30, 1950," *Aeronautical Engineering Review*, Oct. 1950, p. 15.

[26] "IAS Statement for the fiscal year October 1, 1958 to September 30, 1959, *Aero/Space Engineering*, March 1960, p. 16.

[27] "Treasurer's Report – 1959," *Aero/Space Engineering*, March 1960, p. 17.

[28] An overview of each fiscal year is to be found in the annual reports published in the *Aeronautical Engineering Review*.

[29] IAS to Alexander Klemin, President of the American Helicopter Society, March 3, 1950, in the IAS Annual Report—1949, in AIAA Headquarters Files.

[30] "I.A.S. on Advisory Panel," *Aeronautical Engineering Review*, Dec. 1946.

[31] For information on the Finletter Commission, see John Rae, *Climb to Greatness: The American Aircraft Industry, 1920–1960*, MIT Press, Cambridge, MA, 1968, p. 193.

[32] Rae, *Climb to Greatness*, p. 194.

[33] For further information on S. Paul Johnston's role on the Finletter and Airport Commissions, see The Papers of S. Paul Johnston, Box 1, Folders 9 and 11, Smithsonian Institution Archives.

[34] http://www.hq.nasa.gov/office/pao/History/monograph10/nasabrth.html. Cited February 19, 2005.

[35] "Straight Appointed I.A.S. Western Manager," *Aeronautical Engineering Review*, Jan. 1947, p. 5.

[36] Robischon Appointed I.A.S. Western Manager, *Aeronautical Engineering Review*, Oct. 1950, p. 5.

[37] S. Paul Johnston, "Hands Across the Sea," *Aeronautical Engineering Review*, 1947, p. 21.

[38] "First I.A.S.-R.A.S. Conference Held in London, Sept. 2–6," *Aeronautical Engineering Review*, 1947, p. 5.

[39] S. Paul Johnston, "International Angles," *Aeronautical Engineering Review*, Sept. 1954, p. 22.

[40] "Draft Proposal for ICAS," Jan. 29, 1957, Hugh Dryden Papers, Box 2.55, File "ICAS, 1958–1965," Archives and Special Collections, Johns Hopkins Univ. Libraries.

[41] Subsequent meetings were held in Zurich, Switzerland, Sept. 12–16, 1960; Stockholm, Aug. 27–31, 1962; Paris, Aug. 24–28, 1964; London.

[42] Rich, *AIAA at 50*, p. 23.

[43] S. Paul Johnston, Director's Report, Jan. 29, 1947, AIAA Headquarters Files, p. 6.

[44] "Industry Librarians Confer," *Aeronautical Engineering Review*, Nov. 1946.

[45] S. Paul Johnston, "What About the Review?" *Aeronautical Engineering Review*, Jan. 1947, p. 21.

[46] Ernest Robischon, "Suggested Projects For The Institute of Aeronautical Sciences," Jan. 1, 1947, AIAA Headquarters Files.

[47] S. Paul Johnston, "Standard Aeronautical Index," *Aeronautical Engineering Review*, March 1947, p. 21.

[48] "S.A.I. Completes First Phase; Division Headings Determined," *Aeronautical Engineering Review*, Dec. 1947, p. 7.

[49] *Aeronautical Engineering Index, 1947*, IAS, New York, 1948.

[50] "A New IAS Service", *Aeronautical Engineering Review*, Oct. 1955, p. 24.

[51] S. Paul Johnston, "Director's Report," *Aeronautical Engineering Review*, March 1958, p. 21.

[52] S. Paul Johnston., "Publishing Program," *Aero/Space Engineering*, May 1960, p. 15.

[53] "The Burden-Tissandier Collection," *Aeronautical Engineering Review*, May 1947.

[54] Gardner, *The Aeronautical Archives*, p. 2.

[55] "I.A.S. Museum Reopens," *Aeronautica*, July 1949, p. 1.

[56] "New Look," *Aeronautica*, April 1949, p. 1.

[57] Marvin Miles, "Unique Models Put on Display," *Los Angeles Times*, July 1, 1956, p. A14.

[58] "Wright Replica," *Aeronautical Engineering Review*, July 1953, p. 23; it would not be the only full-scale model for long. In 1960, the National Capital Section built a 1903 the airplane and donated it to the Wright Brothers National Memorial, Kill Devil Hills, NC. In 2003, members of a Los Angeles-based AIAA team built a full-scale model and wind tunnel and tested it at NASA's Ames Research Center. The same group built another static model of the world's first airplane that toured the nation as part of the AIAA Evolution of Flight program.

[59] "Aerial Leaders Dedicate Replica of Wright Plane," *Los Angeles Times*, July 16, 1953, p. 4.

[60] "Historical Associate Grade of Membership Established by I.A.S.," *Aeronautica*, Jan. 1949, p. 1.

[61] Claudine L. Johnston, "Curator's report," *Aeronautica*, Jan.–March 1950, p. 3.

[62] S. Paul Johnston, "A Letter to the Historical Associate Members," *Aeronautica*, Jan.–March 1950, p. 4.

[63] "L. A. Historical Branch Addressed by Professor Young," *Aeronautica*, July–Sept. 1950.

[64] "Jimmy Mattern, Flyer, To Show 1933 Films," *Los Angeles Times*, June 17, 1955, p. 24.

[65] A complete run of *Aeronautica* can be found in the Library of the National Air and Space Museum.

[66] S. Paul Johnston to Joseph J. Maitan, March 12, 1968, AIAA Headquarters Files.

[67] Esther Goddard to S. Paul Johnston, Dec. 15, 1956; Marjorie Rodé, Secretary to S. Paul Johnston, Dec. 19, 1956; David Gebhart, Director of the Roswell Museum, to Rodé, Oct. 14, 1957, AIAA Headquarters Files.

[68] Joseph J. Maitan, AIAA Controller, Memorandum to AIAA President Floyd Thompson, May 7, 1968, AIAA Headquarters Files.

[69] Joseph J. Maitan to Floyd Thompson, May 7, 1968.

[70] Disposition of Collections at 2 East 64th Street, Folder of materials relating to the distribution of IAS collections, AIAA Headquarters Files.

[71] Inventory of AIAA Material on Loan to the National Air and Space Museum/Smithsonian Institution, undated, folder of materials relating to the distribution of IAS collections, AIAA Headquarters Files.

[72] "AIAA Aeronautica on Indefinite Loan to the Institutions Listed Below," undated list, AIAA Headquarters Files.

[73] Minutes of the Meeting of the AIAA Board of Directors, June 26, 1968, AIAA Headquarters Files.

[74] Joseph J. Maitan to S. Paul Johnston, Jan. 16, 1968, folder of materials relating to the distribution of IAS collections, AIAA Headquarters Files.

[75] The IAS kept a handful of historic objects. There was a rare Curtiss-Wright engine, a pair of aircraft engine models by the jeweler Cartier, and a dozen hand-crafted airplane models, including a diorama of the 1903 Wright aircraft being prepared for flight.

Chapter 8
The Merger and Beyond: IAS, ARS, AIAA, 1957–1980

AIAA

Space Flight Report to the Nation (SFRN)

Since 1892, sculptor Gaetano Russo's statue of the discoverer of America, perched precariously atop a narrow, 70-foot pillar, had dominated New York's Columbus Circle. In the second week of October 1961, the bright red tip of a Mercury-Redstone rocket, the symbol of a new age of exploration, rose 10 feet above Columbus's famous cap. The bright lights of the marquee in front of the New York Coliseum, bounded by West 58th Street, Broadway, and Columbus Avenue, announced that the American Rocket Society (ARS) was presenting a SFRN.

Space flight was a subject much on everyone's mind. The Soviets had launched *Sputnik I* just four years before. Since that time, the United States had been locked in a much-heralded space race with the Russians—a race that we seemed to be losing. Although the United States had scored a series of space achievements in 1960—including the launch of the first meteorological, navigation, communication, and reconnaissance satellites—the Soviets had delivered another major blow to their rivals with the launch of the first human being into earth orbit on April 12, 1961. American astronaut Alan Shepard made the first U.S. suborbital flight on May 5, 1961. Just 20 days later, President John F. Kennedy issued his ringing call for America to land human beings on the moon and return them safely to earth by the end of the decade.

The ARS SFRN, as *Astronautics* magazine insisted, offered everyone from veteran engineers to the kid down the block an opportunity to glimpse the exciting future that awaited human beings in space. As Chairman of the SFRN Space Flight Panels, Dr. Jerry Grey of the Princeton University Forrestal Research Center had spent a full year supervising the development of the technical program with the assistance and advice of George Gerard, ARS National Program Chairman, and the members of the ARS technical panels. A total of 12,800 professionals attended those technical sessions, which began on Monday, October 9 and continued through the morning of Friday, October 13. In all, 250 papers were delivered during the course of 50 individual sessions.

Having learned of the latest developments in their own technical specialties, engineers could get some notion of the big picture by attending the plenary sessions offering a preview of future vehicles, missions, and thoughts on the impact of space flight on life and culture as well as overviews of the space programs of the United States and the Soviet Union. In addition, there was an education day, with special programs for all students from doctoral candidates to junior high schoolers.

The presentations came from the biggest names in science, engineering, government, and industry. The list of luncheon speakers and panel leaders alone included Wernher von Braun, Director of the George C. Marshall Space Flight Center and the chair of the ARS committee that planned each phase of the extravaganza; James Webb, administrator of the National Aeronautics and Space Administration (NASA); Lt. Gen. Bernard Schriever, head of the U.S. Air Force missile program; Simon Ramo of TRW; James Doolittle; and New York Senator Jacob Javitts.

The marquee of the New York Coliseum announces the ARS *Space Flight Report to the Nation.* (Courtesy AIAA.)

The audience was international. The Twelfth Congress of the International Astronautical Federation (IAF) had just concluded in Washington, DC. Many, if not most, of the attendees had traveled to New York for the SFRN. The pioneers of space flight were there as well, including Hermann Oberth, Willy Ley, and Esther Goddard.

Public attention was focused on the exhibition that sprawled over the bottom three floors of the coliseum. The planners offered each of the 142 exhibitors an opportunity to spread their wares across both sides of the wide aisles. There were displays on propulsion systems, electronics, and earth satellites. Hughes Aircraft prepared a full-scale model of their proposal for a lunar landing vehicle. McDonnell Aircraft sent a Mercury spacecraft, complete with escape tower. B. F. Goodrich provided a spacesuit worn by Mercury astronauts. North American Aviation had the largest display including a scaled-down model of the X-15 research aircraft, a full-scale mockup of the giant, 1.5 million-pound thrust F-1 rocket engine that would ultimately boost the first human beings to the moon, and the boat-tail, or working end, of what was still known as the *Saturn 1,* with its cluster of eight H-1 rocket nozzles.

A *Newsweek* reporter noted that "the most popular exhibits were the 'moon dogs'-remotely controlled instruments, which are designed to rove the lunar surface in advance of man."

The commentator for the *New Yorker* agreed:

> Our own feet were put to the test shortly after we arrived [at the exhibit], when a large plastic ball, with wires attached, all but tripped us as it bounced down the aisle. We followed it to the RCA booth. "Sorry," said the operative, switching off a control panel as the ball hid under a table. "I was only merely demonstrating one means of remote-control locomotion on the moon. There are other systems besides rolling," he went on, directing our attention to another table which was covered with small four-and-six-legged mechanical monsters. "These are crawlers," he said, and turned them all on at once. Slowly and ponderously, amid much clacking of machinery, they walked about. Two of them collided . . . .[1,2]

Dr. Harold Richey, ARS President for 1961 (left) and Dr. William Pickering greet Vice President and Mrs. Lyndon Johnson during the *Space Flight Report to the Nation.* (Courtesy AIAA.)

The exhibition floors were open only to conference attendees from 9 a.m. to 5 p.m., Monday through Friday. The general public was admitted ($1.50 for adults, $.75 for children under 16) from 5 p.m. to 10 p.m. each day and all day Saturday and Sunday. Some 42,000 visitors walked through the turnstiles that week. "The program has only two objectives," remarked a *New York Times* reporter. "To educate the scientists and engineers outside the space industry to the opportunities that the Space Age has opened to them, and to educate the public, which will be asked to contribute billions to the program, on what it means to the nation and the world."[3]

Vice President Lyndon Johnson, the Kennedy administration point man on matters relating to space, put the matter in starker terms when he addressed the 1,500 individuals attending the SFRN banquet at the Waldorf-Astoria ballroom. "If I could get but one message to you it would be simply this," he remarked. "The future of his country and the welfare of the free world depend upon our success in space. There is not room in this country for any but a fully cooperative, urgently motivated, all-out effort toward space leadership."[4]

The ARS at 31

The SFRN said a great deal about the ARS in its 31st year. Before 1956, the ARS staff had been shoehorned into two small cubbyholes in the Engineering Societies Building and had held its annual meeting in conjunction with the American Society of Mechanical Engineers (ASME) because the Board feared that it could not attract sufficient attendance on its own. Just six years later, it was a self-confident and aggressive organization that was willing to take the risk of renting four floors of the New York Coliseum to mount a combination meeting/ exposition/media event on such a scale.

"You can imagine what kind of staff workload was involved in organizing an IAF Congress one week and then the largest space meeting ever held to that date the week afterwards," Jim Harford explained. Indeed, the staff had pulled off a management triumph. It had involved

Harold Ritchey, ARS President for 1961; James Harford; General James Gavin; and Kurt Stehling at an ARS event. (Courtesy AIAA.)

recruiting talented members of the organization, like Jerry Grey, and providing the support that they needed to craft a stellar program. Others had worked to convince industry managers to support the organization and the effort with funds as well as world-class exhibitory. Finally, the staff and board members had to convince industry and military leaders, Federal officials, and public figures from the highest levels of government that the SFRN was a golden opportunity to get their message across. The result was a triumph, a landmark in the history of the ARS.[5]

The Society passed yet another milestone in the weeks following the SFRN. Lovick Osteen Hayman, Jr., a 25-year-old engineer who was employed at the Langley Research Center while working on his graduate degree at Old Dominion University, became the 20,000th member of the organization. On January 1, 1956, the ARS had 3,500 members. Six years later, the number would stand at 20,500, an increase of well over 500 percent! The 45 Corporate members of 1956 had grown to 170. The 30 Sections had increased to 53. There had not been a formal student program in 1956. By 1962, there were 45 Student Chapters. The ARS had formed its first six technical committees in 1956: solid rockets, propellants and combustion, ramjets, liquid rockets, instrumentation and guidance, and space flight.[6] By 1962, the number of Technical Committees (TCs) had climbed to 19.[7]

Nor was ARS student activity confined to colleges and universities. By 1959, the number of accidents and injuries to young people involved in "amateur" rocketry was on the rise. Led by Jack Schafer, the Southern California Section prepared a report that took a very strong stand against amateur rocketry. Several ARS Sections held workshops to assist teachers in developing classroom curricula on rocketry and space flight. The national office published two short guides to careers in astronautics. John Newbauer of the New York staff prepared a book of carefully designed classroom experiments that were safe, practical, instructive, and even a bit entertaining.[8]

L. Eugene Root, IAS President for 1962; and William Pickering, ARS President for 1962, discuss details of the merger outside a Phoenix hotel, 1962. (Courtesy AIAA.)

The only fly in the ointment was the fact that the organization continued to struggle with shoestring financing. "The Society is a nonprofit institution," President Harold Ritchey (1961) admitted to the members. "Its vitality is exemplified by the fact that there is always more to be done than there are financial means for the doing. Fortunately, we were born to be poor!" Indeed, the ARS had finished in the red for three of the last six years of its existence (1957, 1960, and 1961). Although the SFRN had demonstrated that the staff and board could fund important projects when required, there were threatening financial clouds on the horizon.[9]

The Institute of the Aeronautical Sciences (IAS) Contemplates the Future

If things were looking up for the ARS, the IAS faced a far more uncertain future. By 1958, expenses rose above income, as they had in the immediate postwar years. The problem continued to grow through 1960 and 1961, including unexpected losses on the West Coast operation. In an effort to restore the balance, the IAS increased dues early in 1961. Although income did rise a bit, membership, which had climbed steadily through the 1950s, dropped to 15,572 by September 30, 1962.

Thanks to the rich endowments acquired by the Institute during the years before the war, however, it remained a very well heeled organization. In 1954 those restricted funds amounted to $570,352.26. Just five years later, they had grown to $1,247,205. These funds did not contribute to the yearly calculation of profit and loss, but the return from the investments continued to fund the specific programs for which they were intended. The ARS, on the other hand, literally lived from hand to mouth.

By January 1962, the ARS had 162 Corporate members, the IAS only 107. The ARS had 55 Sections coast to coast, compared to only 38 for the IAS. The ARS had outgrown the IAS by most measures. By 1961, the ARS would have 20,500 members in round numbers. The IAS could count only 19,000, down from 20,120 the year before. The rocketeers might be

Allan Emil, Martin Summerfield (last ARS President, 1963), Robert R. Dexter, Harold Luskin, and James Harford sign the documents merging the IAS and the ARS, Wings Club, New York, December 7, 1962. (Courtesy AIAA.)

expected to have more members at Aerojet, the Army Ballistic Missile Agency, and Hercules, a company that produced solid-propellant rockets, but they also led the IAS at Bell, Bell Labs, Bendix, General Motors, Honeywell, General Electric, Hughes, Martin, and North American and United Aircraft. The ARS even outdrew the IAS from within the ranks of the Air Force.

ARS advertising revenues had peaked in the late 1950s at $465,000 and were steady at $365,000 in 1961. That year, IAS advertising income fell to a record low of only $123,000. Worse, there was every reason to believe that the gap would continue to grow wider for the IAS.[10]

The IAS was doing its best to reverse the trend. In one key area, they followed the lead of the ARS. In 1958, the Institute established technical committees covering various engineering specialties. Like the Sections, the TCs would become a critically important avenue through which individual members could influence the life of the organization.

S. Paul Johnston had also fundamentally redirected the organization toward a more streamlined, focused, and efficient operation since the mid-1950s. He had effectively divested the Institute of its historical and museum functions. The nation's finest collection of model aircraft were now little more than office decorations and the historic materials themselves were shipped south to the Smithsonian and the Library of Congress, with duplicates and seconds going to other institutions better prepared to care for them.

The second floor of the IAS headquarters, once the core of the museum area, was transformed into a modernized library. By the early 1960s, the precious historical volumes that Lester Gardner had collected joined the objects placed on long-term loan to suitable libraries. What remained were the current items and technical materials. There was a real hope that the abstracting and indexing effort, relocated to new quarters on the first floor, could become a subscription service. The printing operation was consolidated in the basement. Office functions found a home on the third floor. In some cases, the fixtures and furnishings of which Gardner had been so proud paid for the modernization of the building. The great crystal chandelier that hung in the Council Room, for example, was sold to cover the costs of rewiring the antiquated structure.

The staff was streamlined, as well. Experienced managers were brought on board to oversee key functions—publications and corporate, member, and technical services. In other cases, IAS veterans moved into new positions. With the History Associate program dead and the Durand Museum closed, for example, Ernest Robischon accepted responsibility for arranging national meetings and for student services. Western, Central, and Eastern managers assisted the Sections. Controller Joseph Maitan watched over fiscal affairs, and Secretary Robert Dexter supervised the various service departments. S. Paul Johnston stood at the tip of the pyramid.[11]

Although the changes did result in some savings and made for a more efficient organization, they did not address the fundamental problem. Aeronautics, once a word that inspired visions of up-to-the minute technology, suddenly sounded archaic. By the mid-1950s, anxious to stake a claim on the future, Air Force Chief of Staff Thomas D. White described air and space as "an indivisible field of operations."[12] The leaders of the Aircraft Industries Association renamed their organization the Aerospace Industries Association in 1959, the year in which the term "aerospace" ("... an operationally indivisible medium consisting of the total expanse beyond the earth's surface") found a place in the official U.S. Air Force lexicon.[13]

"As readily as missiles become operationally suitable," an Air Force source commented in 1957, "they will be placed into units either to completely or partially substitute for manned aircraft ...."[14] There were those who wondered about the future of an industry that had always depended on the sale of winged vehicles. "What has put it in jeopardy is the change that missiles have brought to the industry," *Time* magazine reported:

> They not only promise the end of the manned military bombers and fighters, but have brought such other lightning changes that huge projects, calling for hundreds of millions of dollars, can be obsolete overnight. To meet the challenge, the plane and engine makers are well aware that their industry must undergo the fastest and most radical change in its history—or die.[15]

The most dramatic change in the century-long history of the industry, the shift from aviation to aerospace, occurred in a single extraordinary decade, 1954–1964. In 1954, missile production amounted to only 9% of total industry sales. A decade later, the dollar value of missile and space production stood at 49% of the industry total! The mid-1960s marked the peak spending years for both the Apollo lunar program and the development of new families of nuclear-tipped land-based and submarine-launched guided missiles. By the 1970s and 1980s, missile and space production stabilized at 25%—35% of total industry sales.

At the end of 1960, officials of the Glenn L. Martin Company, the oldest surviving U.S. aircraft manufacturer, announced that they were leaving the airplane business. Following a 1961 merger with the American–Marietta Corporation, a chemical firm, Martin–Marietta built an enviable reputation as a manufacturer of missiles, launch vehicles, and electronic systems.

Other veteran firms, including Boeing, Lockheed, Martin–Marietta, General Dynamics, McDonnell, Fairchild, and North American, were soon earning a significant portion of their income from the production of missiles and spacecraft. A host of new companies swarmed into the field: the Aerospace Corporation; TRW, Inc.; LTV Corporation; and Space

Technology Laboratories. Electronics firms like Raytheon, Hughes, and Sperry were not only critical to the effort, they were often prime contractors.

The IAS had little choice but to enter the space age. The first step came in the fall of 1957 in the immediate wake of *Sputnik*, when the Council approved a rather strange alteration to the IAS logo. The new trademark, as S. Paul Johnston called it, looked for all the world as if the traditional symbol of laminar flow over a symmetrical airfoil was caught in a gunsight. "Some people see in it a system of mathematical coordinates or a set of conventional centerlines," he noted. "Others see a missile or rocket (from the south end going north). Still others have said, 'Ah, *Sputnik*'! Whatever the reaction, he concluded, "the intent is to indicate that our interests extend into areas far beyond the confines of classical aerodynamics ...."[16]

The announcement of the new logo was accompanied by an editorial in which Johnston argued that the interests of IAS members did not stop at the edge of the atmosphere. "We see no basis for the argument that there is any essential difference between Aeronautics and Astronautics." The Institute, he concluded, would represent the interests of its members in flight at all altitudes, "from zero to infinity." To underscore that notion, the *Journal of the Aeronautical Sciences* and the *Aeronautical Engineering Review* became, in May 1958, the *Journal of the Aero/Space Sciences* and *Aero/Space Engineering Review*. Two years later, on October 27, 1960, IAS General Counsel Allan D. Emil officially registered a new name for the organization with the State of New York. Henceforth the IAS would be the Institute of the Aerospace Sciences.[17]

In 1957, the cover of only one monthly issue of the *Aeronautical Engineering Review* had a space theme, a drawing of the Vanguard satellite. By 1960, fully one-half of the covers portrayed spacecraft or guided missiles. That year the IAS devoted a complete issue to manned space stations. For the next three years, at least half of the covers were devoted to space. Flipping through the pages of the *Review* during these years it is apparent that advertisers were focused on space as well. The change was more than skin deep. Both the *Journal* and the *Review* were now carrying many articles on space flight.

The response to the shift toward cutting-edge technology was positive. In May 1958, 1,000 members attended an IAS guided missile meeting. Later that summer, the San Diego Section expected 200 members to register for a local meeting on space technology. They had to scramble when 650 individuals signed up.[18,19]

The Merger

By the late 1950s the IAS and the ARS had evolved into organizations that were headed for a collision. They were in competition for members, corporate support, and recognition as the principal technical society representing professionals at the cutting edge of the aerospace enterprise. The ARS, which was still growing, could afford to ignore the situation for a time. The IAS could not.

In December 1960, S. Paul Johnston prepared a memorandum summarizing the problems posed by multiple organizations attempting to represent technical professionals in a single industry. "Under such conditions," he remarked, "there is little wonder that the Aerospace Industries are putting up loud and continuous complaints of 'too many societies,' 'too many meetings,' 'too many demands.'" In short, there was no reason why industry leaders should choose to support two organizations when one could provide all of the professional services required.[20]

The problem grew worse as both the IAS and ARS sought increased advertising to expand and spruce up their competing magazines, *Aero/Space Engineering* and *Astronautics*.

First Meeting of the AIAA Executive Committee, 1963: (left to right) R. Dixon Speas; Robert Gross, IAS President for 1955; Andy Haley; Allan Emil; L. Eugene Root; W. H. Pickering, first President of the AIAA; Martin Summerfield; Harold Luskin; Charles Tilgner; Howard Seifert; S. Paul Johnston; and James Harford. (Courtesy AIAA.)

To complicate matters, burgeoning interest in space flight and missilery led to the introduction of new commercial competitors like *Missiles & Rockets*. Then there was the American Astronautical Society's *Journal of Astronautics*. Even the West Texas–New Mexico section of the ARS created problems inside the organization when they attempted to expand their local publication into a national magazine.

In July 1958, IAS President Ed Wells appointed a long-range Planning Committee to develop recommendations that would guide the Council as they directed the course of the Institute into the future. A discussion of the problem with Harry Guggenheim in the spring of 1959 led to a Guggenheim Foundation study of Technical Meetings in the Flight Sciences, which only confirmed the seriousness of the problem. With all of that in mind, Johnston prepared a memorandum laying out three broad options for the Council and the Planning Committee.

The first possibility was to create a National Council of the Aerospace Sciences, a sort of "United Nations" forum in which all societies would come together to make mutually binding decisions regarding programming and national meetings. Although all of the organizations would maintain their own sovereignty and traditions, the decisions of the Council could "... reduce the demands for technical and financial support against industry which result from irresponsible competition."

At the opposite end of the spectrum was the possibility of creating a quasi-governmental National Academy of the Aerospace Sciences. In this case, "all prior interests, loyalties, commitments ... could be submerged and all existing assets pooled for the common welfare." There would be divisional units devoted to specific fields, but all activity would be funneled through a Board of Governors functioning for the whole organization.[20]

Option three called for a straightforward merger of the IAS and the ARS. In spite of what can only be described as gut-level opposition to such an event, this was the approach that made the best sense. It followed the pattern established in 1957, when the Institute of Radio Engineers and the American Institute of Electrical Engineers merged to form the Institute of Electrical and Electronic Engineers (IEEE) to solve a very similar problem in their industry. The IAS and ARS were already beginning to collaborate. In November 1960, a joint committee of the two organizations had met in Los Angeles to begin planning a collaborative space program for the IAS summer meeting the following year.[21]

If the forces of history and national policy seemed to be moving the IAS and ARS together, it would still not be a match made in heaven. The basic problem standing in the way of such a merger would be the instinctive competitive stance that the two organizations had adopted toward one another. The leaders of the ARS tended to see the IAS as a stodgy organization that retained too much of the elitist spirit with which it had begun and was less than fully responsive to the needs of its members.

The leaders of the IAS did see themselves as an elite organization that brought real estate holdings and a rich endowment to the table. They feared "a dilution of membership standards" and opposed national membership drives, preferring that "membership in the IAS should grow because of more and better services supplied to members." IAS traditionalists must surely have had the SFRN in mind when they remarked that "trade association activities, particularly the organization of commercially oriented equipment exhibits are not considered to fall properly within the sphere of activities of a professional society." They may well have been thinking about the various ARS policy initiatives during 1954–1958 when they expressed opposition to any "... efforts to influence programs of the Congress or of the Defense Department or other government agencies."[22]

Finally, there were those within the IAS who charged the ARS with fiscal irresponsibility, "... in trying to offer too many services without having the necessary income to cover their obligations." There was, they claimed, "no sign that the ARS finance group is ready to adopt the type of sound fiscal policy under which the IAS has successfully operated." Executive Secretary Jim Harford would have countered by arguing that the ARS was turning out a better magazine, was doing a better job with business operations, and offered better service to members and the public.[23]

Harford believed that the success of the SFRN was a key factor precipitating the merger. IAS President L. Eugene Root (1962), he argues, recognized that given the visibility and influence of the ARS, the time had come to unite the groups. Harford became aware that Root and William Pickering, the Director of the California Institute of Technology's (Cal Tech) Jet Propulsion Laboratory (JPL) and the incoming president of the ARS, were quietly discussing the issue, "knowing full well that Jim Harford was not going to be in favor of this, and I sure wasn't."

Harford never hesitated to voice his opinion. Writing to Pickering on January 2, 1962, he argued that "there is a clear conflict in principles between the ARS and the IAS which ... will require that the ARS make serious compromises in matters that are very important." He believed "that there will be hundreds—perhaps thousands—of dissidents who feel as I do that one society ought to concern itself with space alone." The merger might provide the AAS with a golden opportunity to market itself as *the* American space flight organization. He pointed to organizational differences that might be a problem and underscored IAS opposition to ARS involvement in public policy and the inclusion of industrial/technical exhibitions at meetings.[24]

At the same time, Harford was a realist who had no doubt which way the wind was blowing. He knew that Root, who was president of Lockheed Missiles and Space, spoke with

Jerry Grey (left) and Norman Augustine, AIAA President for 1983 testify in 1978 hearings before Senators Wendell Ford and Harrison Schmitt. (Courtesy AIAA.)

the voice of the industry. Rather than standing in strong opposition to a merger, he supported his friend Pickering, recognized the strength of the forces favoring consolidation, and promised to continue operating as a team player if things did not go in the direction he preferred. "If the two Boards decide to merge," he concluded, "I and the staff will aggressively try to win jobs in the merged organization. What's more, we'll try to dedicate ourselves as much to commercial supersonic transports and air line management as we do to space vehicles."[25]

William Pickering of the ARS and IAS President General Donald Putt (1960) began discussing the relationship between their two organizations late in 1960. At the invitation of IAS President H. Guyford Stever (1961), Pickering attended an IAS Council luncheon at Palo Alto, California, on Saturday, November 4, 1961. Following a general discussion of issues, Pickering and Stever agreed to establish a special study group made up of four members from each society to continue the discussions and report to the IAS Council and the ARS Board at the earliest possible date. The group reported favorably on January 21, 1962. The IAS and ARS then created a 10-person Consolidation Steering Committee made up of the officers of both organizations. They also established five working groups responsible for dealing with a series of specific problem areas: constitution and bylaws, organization, member services, finances, and ways and means.[26]

While the working groups were hammering out details, the IAS dispatched Secretary Robert Dexter to offices provided by Eugene Root at Lockheed's Sunnyvale, California, facility. He represented Johnston in coordinating and approving the work of the committees. Both organizations provided their representatives with budget studies and draft management plans suggesting how various functions would be melded and funded.

Who would head the new organization? Jim Harford remembers that he and Paul Johnston were invited to a joint meeting in Phoenix on April 5 and 6, 1962. Both men described the approach that they would take if placed in command of the new organization. Ultimately, the decision was to allow Johnston, who had agreed to retire just 18 months after the new organization was launched, to serve as the first executive director. Harford would be named deputy executive director.[24]

The Phoenix meeting seems to have been the place where the forces in opposition to the merger came to a head. As one commentator noted, Gene Root had to deal with IAS members who feared that he was "giving away the store," while Pickering had to answer ARS adherents who feared they would be forced to "mix with those airplane guys." Four decades later, the echo of those arguments can still be heard among American Institute of Aeronautics and Astronautics (AIAA) veterans who remember the merger.[27]

The business of resolving sticky issues relating to programs and the initial election procedures continued into early May, when the Principles of Consolidation were accepted by the leadership of both organizations. The document presented a detailed description of what the new Institute would be; how it would be organized, governed, and financed; and what services it would offer members. The compromises between the IAS and ARS ways of doing things were pointed out at every stage. Although it was complex, it provided the members of both groups with all of the information they required to make a decision.

The opening paragraph of the principles announced that the new organization would be known as the American Institute of Aeronautics and Astronautics. "This name carries our national identification, labels it as an organization for the promotion of learning, arts and sciences; and it identifies the primary subject areas of interest." In addition, members were informed, "its acronym, AIAA, is simple and dignified. Numerous other names were exhaustively debated in a critical search for the compromise proposed here."[28]

Only once in the next 40 years would a change of name be seriously considered. The November 1982 issue of *Astronautics & Aeronautics* carried an article suggesting that the AIAA be given a more streamlined name, the American Aerospace Institute. The Board approved an even simpler name, the Aerospace Society of America, and invited comment from the members. The response covered the spectrum of possibilities. Here are just a few:

- I would rather return to IAS Institute of Aerospace Sciences
- I'll resign first!
- About as much class as the title "Flat Earth Society"
- ASA is popularly known as the Acoustic Society of America
- ASA refers to film—highly disapprove
- Not Again!
- American Statistical Society, AIAA is unique
- ASA-American Standards Association
- NO! NO! NO!
- We are trying to get aeronautics back, why take it out of AIAA?
- Change the logo too![29]

When put to a vote of the members, the change was defeated by a narrow margin.[30,31]

The principles were in the mail to all nonstudent members of both organizations by early June. Enclosed with the printed copy of the document was a postcard on which the member was asked to endorse or reject the principles and to offer any comment that he or she might wish to make—and they did have comments to make. Irwin Hersey, editor of *Astronautics*, published dozens of letters in the July and August issues of the magazine. He remarked that 96% of the members who had responded were in favor of the consolidation. Three percent disapproved, and one percent failed to express an opinion.[32]

If the letters from the members are a fair indicator of opinion across the board, Hersey's breakdown seems about right. At the same time, scanning an assortment of the letters provides evidence of opposition to what one member described as, "the shot gun wedding." Leon Green, Jr., a member of both the ARS and IAS, regretted the loss of the "more spirited

ARS," but generally regarded "... the tempering influence of the infusion of cooler IAS blood as valuable ...." Robert W. Bussard, on the other hand, was a hard-liner who believed that "... the proposed merger ... would be a national tragedy—for out of it the ARS mode of life would be destroyed ...." Jerry Grey commented that he would be "extremely sorry to see the ARS, or the subsequent merged organization, cease its efforts at public information and active participation in national affairs." In the years to come, he would work to ensure that those elements of the ARS tradition were carried over to the AIAA.[33–35]

With the acceptance of the Principles of Consolidation by the membership, a schedule for the rest of the process took shape. The legal merger package would be discussed and approved by the ARS Board in Cleveland on July 19 and by the IAS Council in New York on September 10. At a joint meeting in New York on October 4, 5, and 6, the leaders of the two organizations would approve the constitution, sign the articles of consolidation, and discuss staff and budget issues. The ballot approving consolidation and electing the first AIAA officers would go to members by October 10. The results of the vote would be announced at an ARS meeting in Los Angeles on November 12 (Ref. 36).

Launching the AIAA

The AIAA officially began operations at midnight on January 31, 1963. The vote had been overwhelmingly in favor of the merger, but it was not unanimous. Of the 13,502 IAS members entitled to vote, 8,768 votes were cast, with 8,201 in favor of the measure and 567 against.[37,38]

If the decision to consolidate had been difficult, so was the transition to the new organization. "Practically all Gene [Root] and Bill [Pickering] did," explained one individual who went through the process, "was to dump the two staffs, journals, and committees in the middle and let them struggle for survival after the merger."[39] The first order of business was simply to discover how large the organization was.

The IAS and ARS had a combined membership of perhaps 43,500.[40] Because some 7,000 of those individuals had been members of both organizations, the AIAA began operations with an estimated 36,500 members, a figure confirmed by the first official count late in 1963. In any case, it was the largest aerospace society in the world.[41]

AIAA Membership, December 31, 1963

Member Type	Total
Fellows and Honorary Fellows	400
Associate Fellows	3400
Members	25,100
Associate Members	3,600
Students	4000
Total	36,500

The AIAA opened its doors with perhaps 180 Corporate members. By July 1, 1964, 47 of those had resigned, but 9 new sponsors had come aboard, bringing the total to 142.

Deputy Director Jim Harford and some of the old ARS staff remained in their offices at 500 Fifth Avenue, while Johnston, the IAS staffers, and some ARS folks were housed in the IAS building on 64th Street. Harford had the thankless but essential task of forging a new staff for the AIAA. It was a traumatic situation. The leaders of both organizations wanted to protect their professional staff to the extent possible. On the other hand, both groups had

Jim Harford meets with New York Mayor John Lindsay during a meeting on the potential of technology to effect positive change in urban areas. (Courtesy AIAA.)

planned meetings, published journals and magazines, and managed local Sections. One of the principal arguments for the merger had been the opportunity to end the duplication of effort and create a unified operation that was streamlined, efficient, and cost effective.

In the spring of 1962, the ARS had a staff of 49. The IAS employed a total of 67 people—56 in New York, 8 in Los Angeles, 2 in San Diego, and 1 in Europe. The staff of both the Pacific Aeronautical Library and the Aerospace Abstracts Office had been excluded from the count. Initial studies indicated that the new organization would require a staff of 109. Seven positions would be eliminated in the first round.[42]

Harford did his best to preserve the talent on the two staffs, shifting individuals from one specialty to another when possible. Confident in the staff he had built at the ARS, he put Irwin Hersey in command of the new magazine, *Astronautics & Aeronautics*. "We didn't need two meeting managers," he recalls. Walter Brunke of the ARS would replace Ernest Robischon. Geoffrey Potter took over the membership operation, and Harford himself would replace Wellman Schrader as the person responsible for corporate membership, which had been his specialty at the ARS. Marjorie Rodé, Johnston's secretary, would eventually assist Harford with the corporate duties.

During the second year, the organization hired a consulting firm to assist in planning for an efficient staff. With the support of President Courtland Perkins, the deputy executive continued his careful trimming, "not without pain," as one commentator remarked.[43] By July 1964, the staff was down to a lean 87 employees, not including the large number of specialists employed in the indexing operation, the Technical Information Service (TIS), which brought to total staff up to 180.[44]

The most significant of the early personnel decisions came early in 1964. After 18 years at the helm, S. Paul Johnston announced that he would be retiring in August, with plans to become the director of the Smithsonian's National Air and Space Museum.[45] Jim Harford thought that there had been an unwritten understanding that he would follow Johnston into the top job. With the moment at hand, that seemed much less certain. Harford learned that L. Eugene Root and R. Dixon Speas, the IAS/AIAA treasurer, favored replacing Johnston with a retired high-ranking military officer. During a few weeks of jockeying behind the scenes, Pickering and President Courtland Perkins (1964) came to Harford's defense. By the July Board meeting, he had been confirmed as the next executive director. It was a wise and critically important decision.[24]

Slowly but surely, the new organization was developing a personality of its own. The familiar blue and white AIAA logo was unveiled in February 1964. Members were assured that it was the result of "months of careful scrutiny of many designs." President Pickering suggested that it was "... chosen from among eight final entries, ... [and] symbolized the field of aeronautics and astronautics encompassed by AIAA activities." It was the work of Walter Dorwin Teague Associates, a firm with a long history in industrial design, whose senior partner and namesake was a longtime member of the IAS/AIAA.[46]

Sections, Branches, Technical Committees, and Meetings

At the time of the merger, Technical Committees were emerging as the primary building blocks of both the ARS and IAS. One of the most important meetings in the early history of the new organization came in San Francisco on January 5, 1963, when Harold Luskin, the new AIAA vice president for Technical Affairs, called together the AIAA technical directors and the Technical Activities Committee (TAC) to establish the functions of the new committees, recommend chairmen, and define the scope of each committee's activities.[38]

The AIAA started with 47 committees in 1963. By 2005, that number had grown to 69. The TCs included professionals in all areas of flight technology. They generated meeting sessions, papers, articles, reviews, continuing education programs, and award nominations. The TCs advised the Institute in their areas of expertise and assisted in developing and communicating public policy initiatives. For more than four decades, the Technical Committees have been the intellectual heart of the AIAA.

The local Sections were the other cornerstone of the new organization. In 1963, local ARS and IAS Sections were invited to meet with their counterparts and negotiate the creation of local AIAA organizations. When the dust had settled, there were 66 Sections arranged in 6 regions spread across the United States Ten Sections had 800 members or more:

Ten Largest AIAA Sections, January 1964

Section	Total Members
Los Angeles	6,800
New York	2,000
National Capital	2,000
San Francisco	1,300
New England	1,200
Long Island	1,000
Pacific Northwest	865
Alabama	850
Greater Philadelphia	840
Orange County, California	800

By 2005, there would be 64 Sections organized into 7 geographical regions, including 2 Sections in Australia.[47]

The ARS brought 2,400 Student members with them to the merger. The IAS contributed 3,200. Allowing for those who were members of both organizations, the AIAA opened its doors with perhaps 4,000 Student members.[47] Over the next 40 years, the 92 Student Branches of 1963 would grow to 160 located at universities scattered around the globe, including China, India, Turkey, Israel, Russia, and eight other foreign nations. Two high schools, one in New York and one in Philadelphia, also have Student Branches. Now, as in the beginning, the Sections and Branches are the primary contact between members and the national organization. They guarantee a more democratic approach to governing the Institute and provide members an opportunity to become involved in the governance of the Institute.

Technical meetings were the core program elements for both the IAS and ARS. During the first year of operation the new AIAA scheduled 22 national gatherings.

AIAA National Meetings, 1963

Meeting	Papers Number of Papers	Attendance[47]
Solid Rocket Propellants Philadelphia, Pennsylvania, January 30–February 1	35	623
National Propulsion Cleveland, Ohio, March 7–8	16	366
Electric Propulsion Colorado Springs, Colorado, March 11–13	69	347
Space Flight Testing Cocua Beach, Florida, March 18–19	49	846
Launch and Space Vehicle Shell Structures Palm Springs, California, April 1–3	42	496
Manned Space Flight Dallas, Texas, April 22–24	63	1,214
Hypersonic Ramjet White Oak, Maryland, April 23–25	36	393
Manned Space Laboratory Los Angeles, California, May 2	18	?
Reliability & Maintainability Washington, DC, May 6–8	16	518
AIAA Summer Meeting Los Angeles, California, June 17–20	202	2,678
Meteorological Support for Aerospace Testing Fort Collins, Colorado, July 24–25	24	220
Torpedo Propulsion Newport, Rhode Island, July 24–25	36	342
Guidance and Control Cambridge, Massachusetts, August 12–14	59	668
Simulation for Aerospace Flight Columbus, Ohio, August 26–28	47	430
Physics of Entry Cambridge, Massachusetts, August 26–28	33	461
Engineering Problems of Manned Interplanetary Exploration Palo Alto, California, September 30–October 1	19	389
9th Anglo-American Conference Cambridge, Massachusetts Montreal, Quebec, Canada, October 16–17	9	66
AIAA/ASD(AFSC) Vehicle Design and Propulsion Dayton, Ohio, November 4–6	44	667
AIAA/AFFTC/NASA Testing Manned Flight Systems Palo Alto, California, December 4–6	23	248
AIAA Heterogeneous Combustion Palm Beach, California, December 11–13	37	180
AIAA Aerospace Sciences New York, New York, January 20–23		

In March 1972, when veteran employee Ruth Locke was ill, the AIAA staff gathered for a group photo to cheer her up: (back row, left to right) Gerald Gilbert; Leon Auerbach; Jim Harford; Nelson Friedman; Leonard Rosenberg; Katherine Finn; Stanley Beitler; Walter Brunke; Lawrence Craner; John Newbauer; David Kaufman; Joseph Maitan; Robert Dexter; (middle row, left to right) Lorraine Johnson; Isidore Zysman; Valentine Shakun; Dorothy Hambach; Bobbi Chifos; Jo-Ann DeNigris; Eleanor Gray; (bottom row, left to right) Sophie Terlecki; Gorda Benjamin; Elizabeth Pfeiffer; Josephine Sosa; Annelise Ranze; and Ruth F. Byrans. (Courtesy AIAA.)

Publications

Few subjects had been more contentious during the period of the merger than that of publications. President Richard Horner (1965) put his finger on the problem when he pointed to the unique nature of the AIAA "No other professional society in the world represent[5] such a broad spectrum of disciplines . . . .," he noted. "Our membership represents a composite of diverse educational backgrounds—aeronautical engineering, biology, physics, electrical engineering, mathematics, astronomy, law, medicine."[43]

How could such diversity be covered in a journal? Nick Hoff of the Massachusetts Institute of Technology (MIT), represented a point of view that saw value in multiple journals covering special areas of interest. "I am a structures man," he noted. "My field is one of the most active in aerospace and ought to have its own journal. I don't want to have to pore over 200 pages of rocket propulsion and control systems to find five structural papers." Martin Summerfield, who would head the new publications program, responded that aerospace leaders like Hoff ". . . ought to have an interest in other fields, too. We are an interdisciplinary society, after all."[43]

Martin Summerfield, no longer serving on the staff as editor, became the first vice president for publications, then as now A. I. Hersey was staff director for publication services. They would work jointly with Dr. William Sears, the last editor of the IAS *Journal of Aerospace Sciences*, to ensure the continued quality of the AIAA product. In addition, they created a Board of Associate Editors, with members drawn from a variety of disciplines.

The new organization had little choice but to begin with a single broad technical publication, the *AIAA Journal*, which carried articles on a wide variety of disciplines. By 1964, because a sufficient backlog of manuscripts was built up, two slightly more specialized publications appeared, the *Journal of Aircraft* and the *Journal of Spacecraft and*

Over a quarter of a century after its founding, gatherings of the International Astronautical Federation continued to attract space leaders from both sides of the Iron Curtain. Here, cosmonauts Alexi Leonov and Vitali Sevastyanov pass the time of day with astronaut Russell Schweikart. at the 27th IAF Congress, Anaheim, California. (Courtesy AIAA.)

Rockets. By 1980, three additional titles had been added to the series: the *Journal of Hydronautics*, the *Journal of Guidance and Control*, and the *Journal of Energy*.

AIAA Journal Circulation, 1980

Journal	Member subscribers	Library Subscribers
Journal of Aircraft	2,883	1,221
Journal of Energy	1,866	747
Journal of Guidance and Control	2,243	766
Journal of Hydronautics	614	576
Journal of Spacecraft and Rockets	1,789	1,305

By 1980, all of the journals were losing money. Institute officials were seriously considering enlisting a series of specialized professional groups (respectively, the Society for Aeronautical Systems, the Spacecraft and Missile System Society, the Marine Systems and Technological Society, the Aerospace Control and Information Systems Society, and the Aerospace Energy Systems Society) to accept responsibility for their own journals, which the AIAA would continue to publish for them.

Much of the debate revolved around the *Journal of Hydronautics*, which had a very small but devoted group of adherents. At one point in the discussion, a wag suggested that they simply change the name of the organization to the AIP, the American Institute of Polynautics. Another countered that if they changed the title to the American Institute of Almost Anything, they could keep their accustomed acronym. As a result of the long discussion, both the *Journal of Energy* (1973–1980) and the *Journal of Hydronautics* (1967–1980) were discontinued that year.[43]

There were three more technical publications to come. The *Journal of Propulsion and Power* first appeared in 1985, followed by the *Journal of Thermophysics and Heat Transfer* in 1987, and the *Journal of Aerospace Computing, Information, and Communication* in 2004.

When does a discipline become so important that it deserves a journal of its own? As one might expect, that decision is the result of a good deal of internal give and take. In recent years, for example, there has been considerable pressure to establish a journal covering hypersonic flight. Even the proponents of such a publication have come to recognize that papers in that field can find a good home in any one of three existing AIAA journals, however. The establishment of a new disciplinary journal in 2004 occurred only after extended discussion. In the end, it would appear that Nick Hoff won his argument. Although the AIAA has attempted to keep the number of referred archived journals to a minimum, the needs of emerging fields had led to the growth of publications.[48]

Tangled advertising contracts and arguments over the title and format prevented the appearance of *Astronautics and Aerospace Engineering*, the new, "engineering magazine," until February 1963. It would serve the function of *Astronautics* and the *Aerospace Engineering Review* as a lighter, newsier publication that would keep members up to date with regard to the latest happenings in industry and the AIAA. The title changed to the catchier *Astronautics & Aeronautics* in 1964. In October 1982, Harford presented a five-year plan that would retitle publication *American Aerospace* and transform it from what was generally regarded as a "house organ" into a "dynamic and profitable magazine." After some discussion, the magazine became *Aerospace America* in 1984. *The AIAA Student Journal*, launched in 1963, continues to meet the special needs and interests of student branches.[49]

The ARS had launched its Progress in Astronautics and Rocketry book series in 1960 with a volume of papers on solid-propellant rocketry edited by Martin Summerfield. The AIAA inherited that series and changed the title to Progress in Astronautics and Aeronautics. *Electric Propulsion Development*, edited by Ernst Sthulinger (1963), was the first contribution to the new series. Still going strong more than four decades later, the series publishes both single-author manuscripts and edited compilations of papers.

Through its AIAA Education Series, the Institute is a major publisher of textbooks covering the entire range of aerospace science, technology, and management skills from *Advanced Classical Thermodynamics* to *Weaponeering*. Books copublished with the Aerospace Press include detailed engineering and scientific monographs, and a line of general interest books enables readers to explore the past, present, and future of the aerospace enterprise.

Honors and Awards

From the outset, the leaders of the IAS recognized the importance of an active honors and awards program to the health of the organization. Although it took a bit longer, the ARS had instituted its own awards program as part of its transformation into a professional society. At the time of the merger, combining the two awards programs had been easy enough. Since that time, the program has continued to grow. By 2005, AIAA offered a total of 81 distinct awards in 10 categories.

Some awards honor the great names in aerospace history:

- Chanute Flight Award
- Dryden Lectureship in Research Award
- Durand Lectureship for Public Service

- George M. Low Space Transportation Award
- Goddard Astronautics Award
- Hap Arnold Award for Excellence in Aeronautical Program Management
- Jeffries Aerospace Medicine and Life Sciences Research Award
- Lawrence Sperry Award
- Otto C. Winzen Lifetime Achievement Award
- Piper General Aviation Award
- Von Braun Award for Excellence in Space Program Management
- Von Kármán Lectureship in Astronautics
- Wright Brothers Lectureship in Aeronautics

Others honor great names in the history of the Institute:

- De Florez Prize for Flight Simulation
- Gardner–Lasser Aerospace History Literature Award
- Haley Astronautics Award
- Pendray Aerospace History Literature Award
- Summerfield Book Award
- Wyld Propulsion Award

Still others honor individuals known only to specialists:

- F. E. Newbold V/STOL Award
- Harry Staubs Precollege Outreach Award
- Losey Atmospheric Sciences Award
- Theodore W. Knacke Aerodynamic Decelerator Award
- Walter J. and Angeline Crichlow Trust Prize

There are awards for achievement in Aeroacoustics, Operations, Communications, Measurement, Software, Computer Aided Engineering and Manufacturing, Digital Avionics, Energy Systems, Flight Dynamics, Ground Testing, Information Systems, Multidisciplinary Design Optimization, Plasmadynamics, Structures, Survivability, Propulsion, Thermophysics, History, Education, International Cooperation, and Public Policy. Finally, there are the awards for service to the AIAA.

Are there too many awards? Most certainly not. Just ask any of the hundreds of individuals whose professional achievements have been recognized at an AIAA Honors luncheon or banquet. It is a tradition as old at the Institute, one that has lost none of its luster since the first awards were presented at a black-tie gathering in the Columbia University Faculty Club.

Location, Location, Location

The decision to sell the IAS building and bring the staff together in new quarters had been made during the consolidation discussions. Johnston and many of his staff loved their historic townhouse and the image of the organization that it projected. Even they had to admit, however, that the time had come for a move into new quarters where the combined staff could be accommodated in a larger space more appropriately configured as the office of a space-age organization.

By December 1963 all of the staff members had come together in new quarters in the Sperry Rand Building at 51st Street and the Avenue of the Americas. The AIAA Board authorized the sale of the old building at a meeting on July 1, 1964. The IAS had paid

$100,000 for the house in 1945. It sold to an anonymous buyer in the spring of 1965 for $425,000. Nevertheless, in the years to come, Jim Harford wondered how much more they could have gotten if they had held onto the place a while longer.[50]

The Institute faced far more serious real estate problems in California. The leadership was so anxious to escape the financial burden of the San Diego building that they took a loss on the sale. The local Section could not afford to maintain the structure and had resorted to renting it out in order to pay the costs. Finally, in the summer of 1965, with the approval of the AIAA Special Finance Committee and the Supreme Court of the State of New York, the building was sold to Lockheed Aircraft for $117,000. At the time, the structure and its contents were carried on the books at a value of $176,706. It was, at least, the last time that the Institute would lose money on the San Diego building.[51]

Finding someone to take the Western Headquarters building at 7660 Beverly Boulevard off their hands would be a good deal more difficult. At the time of the San Diego sale, the Los Angeles building was costing as much as $40,000 a year, with the prospect of replacing the air conditioning, plumbing, and the roof. By December 1966 the building had been on the market for more than a year and had elicited only one "disaster" bid that was below the minimum price established by the Board. By the close of fiscal 1967, more than half ($27,000) of the total $55,000 annual deficit was credited to the Los Angeles building. Another $22,000 was chalked up to losses incurred by the Pacific Aeronautical Library (PAL).

The library had been founded in 1941 by West Coast aircraft companies who continued to support it for four decades. By 1966, those large companies had developed their own technical libraries. It was still useful to smaller organizations, however, and the Institute did attempt to drum up fresh support, but fees continued to decline. In 1968, PAL, as it had always been knows, closed its doors. Its holdings were donated to the Northrop Institute of Technology (NIT). The AIAA claimed a loss of $6,055 as a result of the donation.[52]

The hope that an imminent sale of the Los Angeles building was in the offing continued through 1968 and 1969. Legal problems developed as AIAA officials worked with the aircraft companies that had originally donated the funds for the building to effect a sale. At the going rate of $10 per square foot, the estimated value of the structure was $1,366,000, just under the original $1,400,000 value. In 1970, Parke-Bernet Galleries signed a three-year lease with the AIAA with an option to buy. The property finally sold in 1974, apparently for a little more than $400,000.

The sale of the two western buildings and the dissolution of the PAL marked the end of a long era in the history of the IAS/AIAA. Lester Gardner's vision of the need for information resources and social centers and Reuben Fleet's determination to provide them were over. The AIAA was moving in entirely new directions, along with the industry that it served.

Fifth Anniversary Snapshot

The upcoming fifth anniversary of the AIAA, Harford suggested to the Board, would probably pass unnoticed. February 1, 1968, would be another workday for the 217 members of the Institute staff. Ninety-six of them worked in New York, 12 in Los Angeles, and 3 in London.[53]

The largest number of employees, 109, worked for the TIS under the supervision of Robert Dexter and John Glennan. It is likely that most members were unaware of the TIS operation. It was rooted in the 1930s with Lester Gardner's great indexing project and had picked up momentum in 1947 with a contract from the Air Force Office of Scientific Research for indexing and abstracting of international aerospace literature. NASA had begun funding the effort in 1962, with an annual $1,600,000 contract that was renewed each year until 1972.[54]

The TIS staff offered translations from 19 different languages and prepared abstracts from 1,200 international journals. In 1967, International Aerospace Abstracts reported on 33,000 documents in 24 twice-monthly issues. TIS captured 15,000 documents on microfiche cards, 378,000 copies of which were produced and distributed around the world. That year the TIS library loaned 18,000 documents to companies or government agencies and provided 13,800 photocopies.

Ruth Bryan supervised a staff of 15 editors and assistants who prepared 1,924 manuscripts for publication in 5 journals (including the *AIAA Student Journal.*). Their efforts produced 6,776 printed pages in 1967. John Newbauer and his 10 editors and assistants produced 12 annual issues of *Astronautics & Aeronautics*. Director of Advertising Sanford Wolfe, with his staff of two full salesmen and three part-timers, booked more than $300,000 in a very difficult year.

The meetings staff, led by Paul Burr, including Steven Braunstein, Walter Brunke, Bobbie Chifos, and Ruth Meyers, managed 15 meetings for 15,777 attendees at which 1,254 papers were presented. The same team supported the efforts of 36 TCs that had 100 meetings of their own. The exhibits arranged by Technical Display Director Lawrence Cramer helped to pay for the meetings.

Geoffrey Potter's team, which included Arthur Kortheuer, David Kaufman, and Robert Schuldt, assisted the 68 Sections and 106 Student Branches. They also arranged three regional Section officer gatherings and developed two weekend seminars on shell theory and optimal control theory for the new AIAA Professional Studies Series. Nelson Friedman managed the new electronic data processing system and Alfred Kildow managed the public relations effort. Leonard Rosenberg managed the New York office operation.

Joseph Ryan was executive secretary for Western operations, and Sandford N. Harris held down the AIAA bastion in London. Controller Joseph Maitan and his staff of nine kept track of the money coming and going, including the annual $1,620,000 in staff salaries. The AIAA was becoming a big business.

Cycles: Members and the Money

Jim Harford took the reins at a moment when things were looking up for the aerospace industry and the AIAA. In 1965, the Institute counted 37,931 members—33,328 Professional and Associate members and 4,603 Student members. Three years later, the last of the flush Apollo years, the total was up to 39, 992, with 7,141 Student Members.

Compared to what he had been used to at the ARS, the new organization had fairly deep pockets. The IAS brought a $1.3 million dollar general fund to the merger. The ARS came to the table with a slight deficit. To ensure fiscal responsibility in the new Institute, the organizers created a Finance Committee that supervised fiscal operations for the first five years.

Annual revenues represent another way of looking at the development of the Institute over the years. The $2.3 million figure of fiscal 1964 rose to $4,114,975 by 1969.[55] As early as fiscal 1964–1965, the AIAA finished the year $91,045 in the black.[56] The good times were coming to an end. By the close of 1970, the Institute faced a net operating deficit of $92,593.[57]

Membership was also falling. By 1971, with hard times in the industry, professional membership was down to 27,000, with 3,700 Student Members, down from 4,200 the year before. That year the number of Corporate Members, which stood at 131 in 1964, was down to 77. The decline reached its nadir in 1974, when the professional count stood at 23,467, with 4,083 students. In just five years, the Institute had lost one out of every four Professional Members.

All revenues were in decline. At its peak, the Institute had received up to $500,000 in advertising revenues alone. By 1970 that figure had plummeted to $208,000. Corporate dues, which had brought in $250,000 a few years before, dropped to $200,000. The income from a 1967 technical display in California brought in $249,000. A similar exhibit in Houston in 1970 produced $92,000.[58]

The reasons for the decline were not difficult to identify—an economic downturn coupled with a period of major business mergers. Still, it seemed somehow unfair. "1970 was a tough year for the country, an even tougher year for aerospace, and a very difficult year for AIAA," noted President Ronald Smelt (1969, 1970).[59] "I believe historians will record this era as incredible: that a profession which enabled us to reach the moon, and that provides, through its transport aircraft, the greatest positive item in our balance of trade, should be so completely rejected by the country."[58]

With membership dropping, the Los Angeles and Orange County Sections instituted employment seminars for their out-of-work members. The national office increased its efforts to pass information on available positions along to members. With the assistance of the U.S. Department of Labor, Vice President for Member Services Fred Bagby and Staff Director Geoffrey Potter established a program with the assistance of the Department of Labor offering employment counseling workshops for out-of-work engineers. Outreach efforts in an era of declining travel budgets included a series of short courses that could be presented to local Sections either live or on tape.

As income continued to drop, the National Board reluctantly increased dues in 1971. In order to mitigate any personal hardship, however, dues for any member unable to find professional employment was lowered to $10 per year.[60]

Membership climbed back toward its old levels as the recession eased, reaching 37,366 members in 1993 on its way to a post-1970s peak of some 39,000 in 1994. A decade later, in 2003, the AIAA roster was back down to 30,072 Professional members and 4,181 Student members. Try as it might, the Institute had a difficult time breaking the 40,000-member ceiling.[61]

With the decade of the 1970s drawing to a close, money problems seemed to be easing as well. By 1979, the general fund had reached $3.561 million. As AIAA General Manager Al Pryes demonstrated, however, there were other ways to look at the financial health of the organization. He began by translating all figures into 1972 dollars and then graphing revenues and expenses (with and without contracts, investments, and contributions) 1963–1983. The result showed the expected rise in the 1960s, followed by a plunge beneath the break-even line in the early 1970s, and a slow recovery by the early 1980s.[62]

What it did not show, Pryes explained, was that AIAA had spent (in 1983 dollars) a total of $6 million in interest income to pay program costs. The result, given poor market conditions, was that the organization's portfolio had fallen from a value (in 1972 dollars) of $2.99 million in 1964 to $1.63 million in 1981. The AIAA began limiting the practice in 1980 and halted it completely in 1983. The lesson, President Joseph Gavin suggested, was clear. "Fiscal matters often fail to get a high level of attention in a technical society." A failure to pay attention to Pryes report, he predicted, "will only lead to larger and more difficult financial problems in the years ahead."[62]

Banding Together

Since 1931, when ARS President David Lasser had dreamed of establishing an International Interplanetary Commission to link the space flight societies of the world in an international effort to conquer space, the leaders of the ARS and IAS had reached out to cooperate with other engineers at home and abroad. As the aerospace industry began to dip into the recession of the early 1970s, the old desire to band together in common cause reappeared in the AIAA.

Both Presidents Raymond Bisplinghoff (1966) and Robert Seamans (1969) had raised the issue. No one was to take it more seriously, however, than President Ronald Smelt (1969–1970). At a Board meeting on July 30, 1969, he suggested that the AIAA take the lead in creating a federation of U.S. engineering societies. Such an organization, he argued, would "provide for efficient use of services, avoid overlap in meeting coverage, inhibit the proliferation of technical societies and give a stronger engineering voice rather than the fragmented one that is represented by many individual societies."[63]

Raymond Bisplinghoff responded that although such a step might require the AIAA to give up its identity, it would result in economies of scale and, through the pooling of resources and staffs, might even allow everyone involved to reduce dues. It might even be possible to create such an organization while allowing the member organizations to retain some measure of their own governance and to issue their own publications and hold their own meetings. The federal organization would simply coordinate and make final judgments when necessary. Apparently somewhat taken aback, the Board at least approved the creation of a study group to be appointed by the president, so long as "... establishing this committee does not preclude member objections to the establishment of a Federation."

At a Board meeting on January 21, 1970, Edgar Cortright (president 1976), who had spearheaded the preparation of a report on federation, argued in favor of such a step, pointing to the advantages of lower costs, smaller staff size, the ability to take advantage of economies of scale, and the virtues of central planning. The first step would be to discuss the matter with other large engineering societies and then to obtain the endorsement of the National Academy of Engineering and the President's Scientific Advisor. At that point, the group could hire a management consulting firm to assist in the process and create a "Tiger Team" of representatives from the various organizations who would hammer out the details. The Board suggested that Cortright discuss the subject with the National Academy and the Engineering Joint Committee.[64]

As he approached the end of his second term as president, Ronald Smelt remarked that his inability to inspire a movement to create a federation of engineering societies was one of his "great failures." He praised Ed Cortright and his task force and noted that he himself had written a letter proposing the idea to the President of the National Academy of Engineering. "But the task is a monumental one," he concluded, "and I do not hold out great hopes for an early Federation."[65]

If the AIAA was unable to create the "one big society" of which Ron Smelt dreamed, they were able to join with other groups to achieve much good for their members. In 1971, the AIAA joined with the Institute of Electrical and Electronics Engineers (IEEE), the National Society of Professional Engineers (NSPE), the ASME, the American Institute of Chemical Engineers (AICHE), and the American Institute of Mining Engineers to create a Joint Societies Employment Activity Committee (JSEAC). The purpose of the new group was to assist out-of-work members with employment counseling and job location services.[65]

JSEAC led to the creation of the Coordinating Committee of Engineering Society Presidents (CCESP), which gave way to the Association for Cooperation in Engineering (ACE) in 1974, with AIAA as a charter member. ACE struggled along for five or six years relying on volunteer services provided by the member organizations, but it was never strong enough to be effective in any sphere. Finally, in 1980, a consortium of societies formed the American Association of Engineering Societies (AAES). The old drive to band together that had animated early leaders had dissipated, however. In spite of the strong arguments of Daniel Fink (president, 1974), the AIAA Board voted against affiliation.

Speaking Out

Appropriate involvement in public policy was nothing new for the AIAA. The leaders of the IAS had offered advice to the White House and the Congress on aviation issues. A generation later, Paul Johnston had served as executive director for both the Finletter Commission and President Truman's Airport Commission. The ARS had urged the government to create a satellite program and offered their thoughts on a national space agency to anyone in a position of power who would listen.

During the years immediately following the merger, the leaders of the AIAA were too busy organizing themselves and dealing with problems to pay a great deal of attention to public policy. That began to change at a luncheon in June 1966, when Courtland Perkins (president, 1964) suggested that the Institute consider involvement in public policy. That afternoon, Technical Vice President Robert W. Bussard proposed to the Board that the time had come to "make the Institute a national force" and approve the principle of AIAA involvement in national issues.[66]

After considerable nervous discussion, the Board agreed to the creation of a working group to plan a public policy forum. At the December meeting, Alfred Eggers, who had been appointed to chair the planning group, called for the creation of an AIAA Forum Committee on Aerospace Technology and Society. With the approval of the Board, Eggers was instructed to continue developing a "forum for discussion of all divergent views on matters of national interest and to provide the means of making the results of such a forum known to the members and the general public."[67]

In his farewell report to the Board as president of AIAA, Raymond Bisplinghoff underscored the importance of the initiative:

> Improving public understanding of the profession and its contributions could well be the least appreciated objective of our constitution. For this reason, little has been done. One of the dilemmas of modern times is that science and technology can do a multitude of things—many of which the community may not necessarily desire and so will not support. As professionals, we have tended to preoccupy ourselves with the potential of science and technology rather than the expressed needs of the community. A responsible professional society of the future must find the wisdom among its members to relate its technologies to human needs.[68]

The forums began at the Anaheim meeting in 1967 and would involve public figures like Jonas Salk, Margaret Meade, Norman Cousins, labor leaders, members of Congress, writers, and other public figures. The next step came in 1968 when President Fred Thompson appointed a task force to study the impact of Research and Development policy on civil aviation. It was the first time that the Institute had launched its own study of a public policy issue and crafted recommendations for policy makers. In 1969, 20 AIAA Technical Committee members working under the direction of Starr Colby prepared a report titled: "The Post-Apollo Space Program: An AIAA View." Two years later, the Institute decided to produce a report on technological and economic issues swirling around the decision to support the construction of an American Supersonic Transport (SST). It was the project that brought Dr. Jerry Grey into the policy business.

Jerry Grey had come out of the Navy and earned his B.S. in mechanical engineering and M.S. in engineering physics in 1949 from Cornell; and a Ph.D. in aeronautics and mathematics in 1952 from Cal Tech. After working as an engineer with the Marquardt Company (1951–1952), he accepted a teaching post at Princeton University, where he remained until 1967.

Grey had been general chairman for the ARS SFRN in 1961 and had served as a member of the AIAA Board and vice president for publications since 1966. In 1971, when his consulting firm, the Greyrad Corporation, fell victim to hard times in the aerospace industry, Jim Harford invited him to join the AIAA staff as administrator of technical activities. In that role, he took the Institute's public relations activities in hand.

When it became apparent that the SST program was in jeopardy, and with it thousands of jobs, he set to work on an assessment of the technical and economic impact of the program, which was published as an AIAA report, "The Supersonic Transport: A Factual Basis for Decision." Grey and Holt Ashley, who had chaired the study panel, were invited to testify before Congress on behalf of the Institute. It was the first time that had happened. Delivered just a month before the Congress decided to scuttle the SST program, however, the first report got the AIAA public policy effort off to a slow start.[69]

Discussions now swirled around the possibility of formalizing what had so far been a somewhat haphazard approach to public policy. All of the old IAS arguments about the impropriety of professional society involvement in political matters were heard once again. Jerry Grey recalls that many members feared that speaking out on public issues might "... compromise the business prospects of our corporate members by interfering in the government's (their customer's) decisions and attitudes, questioned how AIAA could represent the different views and business prospects of the different corporations."[69]

Finally, in July 1971, the Board approved a public policy initiative. President Martin Goland issued a cautious statement intended to guide future policy pronouncements, "Inform but do not persuade."[70] The organization would move forward under that somewhat limp banner. There was to be a new post, vice president for public policy, with TRW's Richard DeLaurer at the helm. Jerry Grey was named staff administrator of public policy, a post that he would hold until Harford asked him to take on the duties of publisher of *Astronautics & Aeronautics* in 1983.

The initial project was a study of space shuttle development, "New Space Transportation Systems—An AIAA Assessment." The team, led by J. Preston Layton, with staff support from Jerry Grey, struggled for a year and a half to prepare the report. The essential problem, Grey remembered, was "... how could the committee—all of whose members depended for their jobs either on NASA programs or on companies bidding for them—generate an assessment whose conclusions might not agree with NASA's 'party line'?[71]

It remains an essential problem when considering difficult questions involving science and technology. The people who are best prepared by education and experience to comment on such questions may depend on the answers for their livelihood. Grey, for example, recalled the "... hours Layton and I spent together, drafting and redrafting sections of the report, only to elicit howls of rage from a committee member concerned about what his company management would do to him."

With the first draft complete, Preston Layton had to sell it to the AIAA Board, "mostly high-up aerospace executives and government agency heads." Jerry Grey remembers Layton, "defending, parrying, explaining, and not infrequently pounding the big table to make his point." The members of the committee were able to overcome those difficulties and produce an honest report that supported the basic NASA shuttle compromise, while pointing to the difficulties of calculating the operational costs of the program without knowing how many flights could be scheduled.[72]

The success of the space shuttle study pointed the way to the future of public policy within the AIAA. Any remaining doubts about the value of the program vanished when Norman Augustine (president, 1983), already one of the most respected executives in the aerospace business and destined for even bigger things, signed on as vice-president for public policy. He was, Jerry Grey thought, "in good part responsible for the high esteem most AIAA members now have for Public Policy."[69]

The evolving program involved establishing an improved AIAA publicity machine as a means of ensuring that the Institute's opinions were made known to legislators and the public. Those efforts combined with Congressional testimony and contacts with key government officials were generally successful and established the AIAA as an important voice on matters of national science and technology policy.

Grey is quite right in suggesting that the transformation of the AIAA from "a purely technical group that spoke only to itself to a nationally respected voice in the aerospace community," represents one of the great shifts in the history of the organization. The members of the Institute have employed that voice on behalf of some of the most imaginative ideas of the past half century, from innovative aircraft design and the potential of new communications and navigation satellite systems to potential for even more visionary projects to effect revolutionary change.[69]

References

[1] "Rocket Rendezvous," *Newsweek*, Oct. 23, 1961, p. 58.

[2] "Away You Go," *The New Yorker*, 1961, p. 44.

[3] "The Age of Space," *New York Times*, Oct. 10, 1961, p. 42.

[4] "SFRN Hears Call for Full Community Participation to Achieve Space Goals," *Astronautics*, Dec. 1961, p. 23.

[5] James Harford, "The American Rocket Society, 1953–1963: A Memoir," IAF-88-605, author's collection, p. 14, 1988.

[6] Harford, "The American Rocket Society," p. 14.

[7] Harold W. Ritchey, "Six Years," *Astronautics*, Dec. 1961, p. 19.

[8] John P. Stapp, "Community of Interest," *Astronautics*, June 1959.

[9] Undated chart labeled "ARS Five Year Statistics," James Harford File of Items Relating to the Merger, AIAA Headquarters Files.

[10] A letter to the members of the IAS, Jan. 26, 1962, 4, AIAA Headquarters files; James Harford Notes, Harford File of Items Relating to the Merger, AIAA Headquarters Files.

[11] "A New Look for the Staff of the IAS," *Aero/Space Engineering*, June 1960, pp. 18–19.

[12] Frank W. Jennings, *Air Power History*, Spring 2001, p. 48.

[13] *Interim Aerospace Terminology Reference*, Air Force Pamphlet No. l 11-1-4, Oct. 30, 1959, in Jennings, *Air Power Hisotory*, p. 48.

[14] "The Guided Missile," *The Air Force Reservist*, Dec. 1957, *The History of the American Aircraft Industry—An Anthology*, edited by G.R. Simonson, MIT Press, Cambridge, MA, 1968, p. 229.

[15] "Aviation," *Time*, Sept. 14, 1959, p. 92; also in *The History of the American Aircraft Industry—An Anthology*, edited by G.R. Simonson, MIT Press, Cambridge, MA, 1968, p. 229.

[16] S. Paul Johnston, "Zero to Infinity," *Aeronautical Engineering Review*, Dec. 1957, p. 24.

[17] Official change of name registration with the State of New York, AIAA Headquarters Files.

[18] "1000 Members Attend Guided Missiles Meeting," *Aero/Space Engineering*, Aug. 1958, p. 13.

[19] "200 Expected, 650 Show at 'Local' Space Meeting," *Aero/Space Engineering*, Oct. 1958, p. 23.

[20] S. Paul Johnston, "Memorandum to Long Range Planning Committee," Dec. 15, 1960, AIAA Headquarters Files.

[21] S. Paul Johnston, "Progress Report to the IAS Council," undated 1960, AIAA Headquarters Files.

[22] Supplemental Minutes to the Meeting of the Council of the Aerospace Sciences, Nov. 4–5, 1961, AIAA Headquarters Files, p. 4.

[23] Arthur W. Gilmore to L. Eugene Root, July 31, 1962, AIAA Headquarters Files.

[24] Oral history interview with James Harford, Aug. 4, 2004, author's collection.

[25] James Harford to William Pickering, Jan. 2, 1962, James Harford Notes Relating to the Merger, AIAA Headquarters Files.

[26] L. Eugene Root, Letter to IAS Members, Jan. 26, 1962, AIAA Headquarters Files.

[27] Daniel Rich, *AIAA at 50: Yesterday, Today and Tomorrow*, AIAA, New York, 1982, p. 27.

[28] "Principles of Consolidation," printed version, AIAA Headquarters Files.
A list of other proposed names for the organization is contained in a file of merger notes in the AIAA Headquarters Files. They include:

AAA	American Aerospace Association
AAI	American Aerospace Institute
AAIS	Association of Aerospace and Interplanetary Sciences
AAS	Aerospace Society of America
AIAS	Association of Interplanetary and Aerospace Sciences
AIS	Institute of Interplanetary Sciences
AISS	American Institute of Space Sciences
ASAS	American Society of Aerospace Sciences
ASASE	American Society of Aerospace Scientists and Engineers
IAA	Interplanetary and Aerospace Association
IASE	Institute of Aerospace Scientists and Engineers
IRAS	Institute of Rocket and Astronautical Sciences
NARAS	National Academy of Rocketry and the Aerospace Sciences
SAIS	Society of Aerospace and Interplanetary Sciences
SAS	Society of Aerospace Sciences
SASE	Society of Aerospace Scientists and Engineers
SASS	Society of Aeronautical and Space Sciences
SIAS	Society of Interplanetary and Aerospace Sciences

[29] "1983 AIAA Ballot, Constitutional Amendment, Name Change to Aerospace Society of America," Minutes of the AIAA Board of Directors Meeting, May 13, 1983, AIAA Headquarters Files.

[30] "Should AIAA Become the American Aerospace Institute?" *Astronautics & Aeronautics*, Nov. 1982.

[31] "Still the American Institute of Aeronautics and Astronautics," *Astronautics & Aeronautics*, July/Aug. 1983, pp. B4-5.

[32] Irwin Hersey, "More From Members on ARS-IAS Consolidation," *Astronautics*, Aug. 1962, p. 82.

[33] Jerry Grey, letter *Astronautics*, July 1962, p. 74.

[34] Leon Green, Jr., letter *Astronautics*, Aug. 1962, p. 83.

[35] Robert Bussard, letter *Astronautics*, Aug. 1962, p. 89.

[36] Consolidation Memo, June 18, 1962.

[37] *This is IAS, IAS Facts and Figures*, IAS, New York, Jan. 1962.

[38] "AIAA Countdown," *Astronautics and Aerospace Engineering*, Feb. 1963, p. 122.

[39] James J. Harford, "The American Institute of Aeronautics and Astronautics, 1963–1993," a paper presented at the Space History Symposium of the History of Science and Technology, Moscow, Russia, May 12, 1993.

[40] The first accurate membership count provided in the minutes of the 1964 annual meeting clearly indicates that the figure of an initial combined membership of 46,000 provided by Daniel Rich and copied by other writers is clearly too large.

[41] Annual business meeting of the AIAA, Jan. 26, 1964, AIAA Headquarters Files.

[42] "Personnel Analysis for AIAA," no date, AIAA Headquarters Files.

[43] Rich, *AIAA at 50*, p. 28.

[44] Robischon found a new position as librarian/archivist at the National Air and Space Museum, returning to his first love, aeronautical history and documentation. For the total employment figure, see Rich, *AIAA at 50*, p. 28.

[45] S. Paul Johnston moved from the AIAA to become director of the National Air and Space Museum from 1964 to 1969. Much of his work there involved planning for the construction of a new museum on the National Mall. Frustrated by his inability to move Congress to fund the project, he retired from the Smithsonian in 1969 and settled in Easton, Maryland, where he died on August 9, 1985. The Smithsonian Institution Archive has a small collection of Johnston's papers relating for the most part to his service on the Finletter and Truman Airport Commissions. The AIAA Headquarters Files hold some pre-IAS publications of Johnston's and his desk diary covering a period in the 1950s.

[46] "Institute's Official Emblem Selected," *Astronautics & Aeronautics*, Feb. 1964, p. 78.

[47] AIAA Annual Report, Jan. 27, 1964.

[48] I am indebted to my colleague Dr. John Anderson for his informed judgments and priceless insight on matters relating to AIAA publication policy.

[49] Minutes of the Meeting of the AIAA Board of Directors, Oct. 21, 1982, AIAA Headquarters Files.

[50] Minutes of the Meeting of the AIAA Board of Directors, March 5, 1964, AIAA Headquarters Files.

[51] Treasurer's Report, AIAA Board of Directors Meeting, July 29, 1965, pp. 2–3.

[52] Statement of Income and Expenses, Annual Report, 1969, AIAA Headquarters Files, p. 4.

[53] Unless otherwise noted, this material is entirely drawn from James J. Harford, "1967 Report of the Executive Director," Jan. 23, 1968, AIAA Headquarters Files.

[54] Minutes of the AIAA Board of Directors Meeting, Jan. 10, 1973, p. 3, AIAA Headquarters Files. In FY 1972, the contract was reduced to $1,350,000.

[55] Condensed balance sheet and operating statement, AIAA, Fiscal Year October 1, 1968 to September 30, 1969.

[56] Treasurer's Report, Annual Business Meeting, Jan. 25, 1966, p. 17, AIAA Headquarters Files.

[57] "Treasurer," 1970 Annual Report, AIAA Headquarters Files.

[58] Ronald Smelt, "AIAA Annual Report, 1970," *Astronautics & Aeronautics*, March 1971, p. 52.

[59] Smelt is the only AIAA President to have served two terms, having stepped into office in 1969 when President Robert Seamans resigned after only three months to become Secretary of the Air Force and then having been elected for his own term in 1970.

[60] Minutes of the Meeting of the AIAA Board of Directors, Jan. 22, 1971, p. 2.

[61] All membership numbers in this chapter are drawn from the official statistics provided in the minutes of the annual meetings, AIAA Headquarters Files.

[62] Joseph Gavin, Jr., "Eighteen Year Financial Analysis of the AIAA," *Astronautics & Aeronautics*, April 1982, pp. 223–24.

[63] President's Report, AIAA Board of Directors Meeting, July 30, 1969, p. 8, AIAA Headquarters Files.

[64] "Federation of Societies," Minutes of the Meeting of the AIAA Board of Directors, Jan. 21, 1970, p. 14, AIAA Headquarters Files.

[65] Rich, *AIAA at 50*, p. 32.

[66] Minutes of the Meeting of the AIAA Board of Directors, June 29, 1966, p. 11, AIAA Headquarters Files.

[67] Minutes of the Meeting of the AIAA Board of Directors, Dec. 1, 1966, p. 14, AIAA Headquarters Files.

[68] Raymond Bisplinghoff, Report of the President, AIAA Annual Meeting, Jan. 1967, pp. 15–16, AIAA Headquarters Files.

[69] Jerry Grey, e-mail to the author, February 11, 2005.

[70] AIAA and Public Affairs—Policy Guidelines," Exhibit 7, Minutes of the Board of Directors Meeting, July 14, 1971, AIAA Headquarters Files.

[71] Jerry Grey, *Enterprise*, William Morrow, New York, 1979, pp. 86–87.

[72] Grey, *Enterprise*, p. 87.

AIAA Comes to Washington

May 2, 1988, was a big day in the history of the Institute. The Aerospace Building was a sleek, triangular structure, 10 stories tall, with the narrow rounded point fronting on L'Enfant Promenade. If the name of the building was not a dead giveaway as to the business of the top floor tenants, the line of blue runway lights leading from the sidewalk to the front door should have sealed the matter.

The Air Force band played that morning. Jim Harford, President Lawrence Adams, and Incoming President William Ballhaus made brief remarks, after which Thomas Jones, CEO of the Northrop Corporation and the man who had headed the fund-raising effort, cut the red ribbon. With the ceremony complete, the 300 guests were invited in for refreshments, music, and a tour. Frederick C. Durant, III, who, as president of the ARS in 1953 had hired Jim Harford, displayed the finest pieces of his personal collection of space art on the walls of the new facility. It was a day long in coming.

Visitors to Jim Harford's 10th floor office were treated to a spectacular view. The Smithsonian castle, the Washington Monument, and the White House were off to the right, the Potomac River snaked along to the left, and Washington National Airport and the high-rise towers of Crystal City were in the distance.

The story of the building began at a luncheon meeting of the Board on May 26, 1982, when President Joseph Gavin noted that one of the biggest decisions facing the Institute was what to do when the lease on the headquarters at 1290 Avenue of the Americas expired in April 1983. The rent was scheduled to rise from $9 to an unacceptable $40 per square foot. The time had come to search for new quarters.[1]

The first thought was to develop a new Science and Technology Network (STN) building in cooperation with other engineering societies. The American Association of Engineering Societies (AAES) was interested in the project, as were the Institute of Electrical and Electronics Engineers (IEEE), the American Institute of Physics (AIP), and the United Engineering Trustees (UET). Because the UET was required by its charter to remain in New York, the group was exploring possible sites with local real estate brokers. The Institute Development Committee suggested the appropriation of no more than $50,000 to support the effort. At the same time, staff members were discussing the possibility of a move to Washington, DC.[2]

In May 1982, Seymour Zeiberg, chair of the newly created Headquarters Relocation Committee, reported to the Board. The time and money required to design and construct a new building seemed to put that option out of reach. The cooperative STN project, he noted, was moving very slowly. He recommended that the Institute move independently and presented four basic options.

1) less prestigious New York rental space at $20 per square foot;
2) rental space in downtown Washington, DC, at $20 per square foot;
3) rental space in suburban Washington at $16 per foot; and
4) rental space in suburban Washington with an option to purchase at the end of a five-year fund-raising campaign.

Jim Harford introduces members of the Board to plans for the Aerospace Building. (Courtesy AIAA.)

The advantages of a site in the nation's capital were clear. It made good sense for the Institute to be headquartered in Washington, where the staff and members would have easier access to Congress, the military services, Federal agencies like the National Aeronautics and Space Administration (NASA) and the Federal Aviation Administration (FAA), and the headquarters of other national aerospace organizations, including both the National Aeronautic Association and the Aerospace Industries Association (AIA). Nor did the board have to be reminded that many large aerospace manufacturers had established offices in the District of Columbia or the nearby suburbs.

Of course, there would be a downside to such an historic shift. The cost of moving, including the relocation of staff members, might run as high as $1–2 million. A survey of staff members indicated that perhaps only 10 percent would be willing to move in any case. A shift to Washington would thus represent a significant break in corporate memory and tradition. The move would certainly force the Institute to reduce its $4.4 million investment reserves. The Board voted 10 to 5 against the Washington option. President Michael Yarymovych instructed the staff to continue the search for reasonably priced rental space in New York while planning for a larger presence in Washington.[1]

The dream of a building shared by a large number of societies was hard to let go. In January 1983, Harford brought his counterparts in other New York-based societies together with the UET, city and state officials, and representatives of a real estate development firm to discuss the possibility of constructing a high-tech complex on two East River piers in the South Street Seaport area. The notion was for the city to provide the acreage at low cost and for the developers to build the structure for a mix of profit and nonprofit clients. The participating societies would share common facilities, including meeting rooms, television studios, computer systems, libraries, and even typesetting and publishing equipment. In the meantime, the Institute would have to move out of their Rockefeller Center space to temporary and much less expensive quarters at 1633 Broadway by May 28.[3]

The logic of Washington would not die. By the early fall of 1983, all of the issues were back on the board. At a meeting in Washington on August 31, John McLucas, chair of a newly established Building Committee, noted that he was comparing costs between New York and Washington. Past president Yarymovych suggested that the issue was not resolved and that the battle lines were drawn between New York and Washington. The question, he suggested, was "where should the visible flag of the aerospace field be located." Other Board members pointed to the fact that Washington was an aerospace center. Michael Robinson suggested that "... we consider our image and identity and identify with the government." President Norman Augustine, who had said little during the session, asked how many Board members had recently attended meetings in Washington, New York, and Los Angeles. Washington and Los Angeles tied, with New York coming in a distant third.[4]

The turning point came at a Board meeting on May 24, 1984. Harford was discouraged by his inability, after a three-year effort, to move his colleagues in other societies any closer to a serious agreement to jointly develop what he was now calling a World Technology Center. Jerry Grey spoke for a group of worried staffers, pointing out that a relocation to the New York suburbs would preserve office continuity and minimize the disruption of service that would be occasioned by a move to Washington. Nelson Friedman, American Institute of Aeronautics and Astronautics (AIAA) secretary and administrator of facilities and services, had researched the options in both cities. New construction in New York was prohibitive because of land and labor costs. The cost of an older building that could be restored in an acceptable area would also be very high and not attractive to potential funders.

On the basis of the information on hand, the Board finally reached a decision. Unless conditions changed, the 83 positions represented on the New York staff would be shifted to a new headquarters building in Washington, DC, within the next two years. If possible, they would try to obtain an equity position in a new building. The 88 staff members of the Technical Information Service (TIS) would continue at their present location, 555 West 57th Street. That group could not be shifted to Washington because New York offered the rich mix of languages required by the translation operation. Affected staff members would be provided with fair and equitable relocation or severance pay. Norm Augustine, chair of the President's Advisory Committee (PAC), agreed to head a Headquarters Relocation Task Force—to include the Advisory Committee, Board volunteers, and staff representatives—that would present recommendations as to a firm course of action.[5]

The group met in July, and Augustine presented their recommendations at a Board meeting on August 16, 1984. They suggested pursuing two possibilities simultaneously until it was clear which option was superior.

The first project was the prospect of a limited partnership opportunity in new construction, an Aerospace Building, to be built at 370 L'Enfant Promenade in Southwest Washington. The project was actually the result of some creative thinking on the part of AIAA planners, who discovered that the District of Columbia owned a triangular piece of land below the L'Enfant Plaza development. At the time, it was being used as a parking lot for city vehicles. In an effort to attract the AIAA to town, officials in the DC Development Authority were willing to give the Institute very favorable terms on a ground lease for the property. With that in hand, Institute officials approached real estate developer D. Kenneth Patton, who put together a limited partnership to support construction of the building. In exchange for the ground lease, AIAA would receive 13-percent equity in the building and a very favorable rate (a $5 per square foot discount) on the entire top floor.[6]

If the deal to create the Aerospace Building fell apart, Augustine and the headquarters relocation team offered an alternate possibility, the chance to purchase a building on Duke Street in Alexandria, Virginia. The Board approved both the PAC plan and Jim Harford's

AIAA leaders make their mark on a beam headed for the structure of the Aerospace Building. (Courtesy AIAA.)

relocation and severance policy for staff members. In October 1984, recognizing the need to move quickly, the Board delegated the authority to negotiate a deal on its behalf to the PAC and the Executive Committee.[7]

Developer Ken Patton, who was putting the Aerospace Building deal together, brought a parade of architects, builders, and Washington city officials to an AIAA Board meeting on April 12, 1985. Although schedules had slipped and financial commitments were not yet in place, the deal was becoming sweeter for the Institute. The building had grown larger (from 300,000 to 380,000 square feet), while AIAA's commitment to lease had been reduced from 40,000 to 37,000 square feet. The Institute would still have 13-percent equity in the building as well as a large block of guaranteed parking spaces, electricity at cost, and the right of first refusal to purchase the entire building if and when it was sold. The Board learned that the building was being constructed as a tax shelter, so a sale was likely as the value increased over the first 10 years. The members of the PAC and the Executive Committee voted to accept the deal on October 16, 1985 (Ref. 8).

Groundbreaking came less than two months later, on December 10, 1985. Two hundred officials and guests gathered in a gaily striped blue and white tent to witness the ceremony. The highlight came when Thomas V. Jones, the CEO of Northrop and chair of the committee that would raise the $5 million required to fund the AIAA expenses related to the building and the move, walked to the podium. The first commitments were in hand, he announced, $500,000 from Boeing and $211,000 from Northrop. The rest of the money, he explained, would come as the result of a "multi-tiered, multi-pronged approach appealing for support from the aerospace industry and the members."

"Fair share" targets had been developed for corporate sponsors based on a percentage of their 1984 earnings. The top fifteen manufacturers, with $80 billion in sales, were expected to cough up $4 million, 80% of the total, with the rest coming from smaller firms and individuals. All contributors would be honored on a plaque to be mounted in the lobby of the building. Norm Augustine thought that the plaque would be a big selling point with the members. "Join the gang," he suggested. "Someday you can bring your kids to the Air and Space Museum, then walk two blocks to the Aerospace Center and show them you helped make it possible."[9,10]

On July 10, Jones informed the Board that commitments had been received for $3 million of the $5 million goal. "Mr. Jones is performing heroically," Harford noted. "Going one-on-one with the CEOs of target donor corporations." Augustine, Alan Lovelace, and John McLucas pitched in as well. Big contributions did arrive early on. Rockwell led the way with a $239,000 pledge, followed by General Dynamics with $220,000, Hughes with $211,000, Martin–Marietta with $282,000, and LTV with $90,000.[11,12]

When the total had stalled at only $3.3 million that fall, Harford called for an all-out effort to bring in individual contributions. "Early November will see a second appeal mailed," he noted, "and a third one will be scheduled." Things were looking even more serious by February 19, 1987, with no more money on hand and only six months until the scheduled move-in date. Equipment purchases budgeted for the last two million dollars would have to be proposed as individual items.

The opening date slipped until the following spring, and money did begin to flow once again, some $4.7 million by mid-November 1987. Still, these were not easy times. The AIAA stock portfolio took a 16% loss in the stock market downturn. Taking full advantage of all of its resources, the Institute auctioned off the last of Lester Gardner's historical treasures.

John P. V. Heinmuller was only 20 years old in 1912 when Emile Wittnauer, the Swiss watch maker, appointed him his U.S. representative. Over the next few decades Heinmuller climbed the ladder of success to become president of the Longines-Wittnauer Watch Company. Along the way he became an aviation enthusiast, chief timer for the National Aeronautic Association, a writer on aviation history, and the leading American collector of philatelic items relating to flight and air mail. In January 1945 he presented the 73 volumes of his air post collection to the Aeronautical Archives.

Because of its value, the Heinmuller collection had not been dispersed with the rest of the Aeronautical Archive. By the early 1980s, the time seemed ripe to sell. The history technical committee chaired by Fred Durant urged a cautious approach and the involvement of professionals. Two appraisals of the collection set its value at more than $130,000. When finally auctioned by John A. Fox of Floral Park, New York, the collection brought $195,000 into the Institute of Aeronautical Sciences (IAS) treasury.[13]

The final move to New York began on Friday, August 28, 1987. Jim Harford promised that the AIAA would be open for business in its new location the following Monday morning. Only the 10th floor of the building, where AIAA headquarters was located, was ready for occupancy. Construction continued on the rest of the building. Until the front entrance was finished, visitors had to enter at the rear of the building on the D Street side. When Jim Harford finally reported that the building fund had topped the $5 million mark on September 2, 1988, four months after the opening, the Board of Directors immediately toasted the moment with champagne.

Changing of the Guard

Only 25 of the 90 non-TIS staff members chose to relocate from New York to Washington. They were joined by half a dozen employees from the Washington and Los Angeles offices, although the Los Angeles operation was not closed. AIAA paid a high price for the move in the loss of veteran employees. Walter Brunke and his deputy, Bobbi Chifos, had arranged hundreds of meetings since coming to work at the American Rocket Society (ARS) with Harford in the early 1950s. David Kauffman had directed student and educational programs for more than 20 years.

The man at the top was going, as well. Jim Harford had put in 35 years with ARS and AIAA. He had no intention of giving up his house in Princeton, New Jersey, from which he had commuted into New York for years. The board had agreed to offset his housing costs in Washington, where he would stay long enough to get the new headquarters up and running and to assist his successor through the difficult early months. Then he would retire. "There are," he explained, "other things I want to do."

On February 12, 1988, President Larry Adams and President-Elect William Ballhaus appointed John McLucas, past president of AIAA (1984), former Secretary of the Air Force, and FAA administrator to chair the search committee for Harford's replacement. The other members of the committee would include past presidents Norman Augustine, Allen Fuhs, and Michael Yarymovych and three current board members, Michael Francis, Richard Kline, and Richard Peterson.[14]

After surveying the field and soliciting nominations, the McLucas Committee whittled the field down to 15 candidates for interviews and then down to five finalists. Two of the individuals were Congressional staff members specializing in aerospace affairs. One was in the marketing department at Honeywell, and another was a civilian Air Force employee. The fifth individual, Cort Louis Durocher, was clearly the strongest candidate.

A native of Houghton, Michigan, Durocher earned a B.S. in aeronautical engineering from the Air Force Academy in 1968 and an M.S. in astronautical engineering, guidance, and control from the Massachusetts Institute of Technology (MIT) in 1977. Following a successful flying career in the Air Force he had held a variety of high-level positions in the U.S. Air Force Space Division, winding up as chief of the Advanced Systems Division with responsibility for planning future Department of Defense space programs. Retiring from the military, he took a job as program manager of defense initiatives with Hughes Aircraft.

Durocher had been on the job for only two weeks when he was invited to interview for the executive director's job. Jerry Grey, an acquaintance whom he had asked for career advice before taking the Hughes job, had called him to Harford's attention. Both men were impressed with Durocher's record as an engineer as well as his considerable AIAA experience. He had joined the organization as a graduate student in 1975. The Institute had published his M.A. thesis. He had been an active member of Technical Committees and had helped to found the Space Transportation Technical Committee. Only 41 years-old, Durocher was already an Associate Fellow.

After two rounds of interviews with Ballhaus, Augustine, and McLucas, Durocher was informed that he was the committee's choice. By no means certain that he wanted to give up his existing career, the young engineer spoke to a top Hughes executive and AIAA member who urged him to take the offer, promising that he could keep his badge and return to his old job without loss of seniority if things did not work out. As Durocher recalls, he sent his badge back after a year at AIAA.[15]

Durocher began work as executive secretary on October 1, 1988. As he notes, no two men could have been more different than Jim Harford and himself and yet have gotten along so well. They were indeed different. Harford was a genial guy with a ready smile and an encouraging word, although he could certainly be tough when that was required, as when he was charged with merging the ARS and IAS staffs. It would have been hard to find an AIAA staffer who genuinely disliked him.

From the outset, the AIAA and its predecessor organizations have emphasized support for student engineers. (Courtesy AIAA.)

Jerry Grey, who knew and liked both men, notes that Durocher's approach was "... much more like a corporate CEO style than Jim's more casual, person-to-person hands-on interaction with the staff and the volunteer Board and committees." Durocher puts it more succinctly: "Harford was an intellectual. I was an engineer." Far more important than their differences, however, was the fact that they got along well together, which made for a good transition.[16]

Cort Durocher's first goal after coming aboard was to bring AIAA fully into the computer age. When he arrived, the office boasted four "dumb" terminals linked to a mainframe. The terminals were on the desks of the lowest paid staff members, data entry personnel. Information moved slowly. In his first year on the job, it took four months to close the books.

The new executive secretary was determined to establish a system that would put information at the fingertips of all key employees and ease the business of making decisions. Using a portion of the $5 million fund raised to equip the new headquarters, he purchased 100 desktop computers that could be networked. Because no one seemed certain what sort of machines to buy, the organization decided to let the users decide. They bought 50 Macintosh machines and 50 IBM personal computers. Staffers would let them know soon enough which machine they preferred. (The Macs came out on top.) Ken Berlin, the Institute's first computer specialist, came on board to supervise the Management Information System.[15]

Hard times were returning to the industry and, as in the early 1970s, AIAA felt the pinch. After five consecutive years in the black, the fiscal year ending on September 30, 1988, closed with a $132,000 operating budget shortfall. "We were in a spiral down," Durocher recalls of his early years with the organization. Convinced that "paying attention to our financials" was critically important, Durocher approached President Bastian "Buzz" Hello (1989), explaining that he was a hardworking guy with two engineering degrees but not a lot of experience in finance or business management. Dispatched to Northwestern University, Durocher enrolled in a business management graduate certification program and returned to Washington anxious to "... put management and financial structures in place to weather the bad times."[15]

There were problems in several of the eight areas reporting to Durocher. He decided to focus on them one at a time, beginning with technical activities, which had "hemorrhaged money" that first year, producing a $520,000 variance. He put a new manager in charge and underscored his focus on budget and finance by bringing David Quackenbush on board as treasurer and controller. The process of focusing on one division at a time would continue for three years. By the end of that time, he had "... totally changed every product line we had."[15]

The Library and TIS: The End of an Era

Another financial blow came with the loss of income from the NASA contract supporting the TIS. As things had developed, books, journals, papers, and other items flowed into the TIS office where they were indexed and abstracted. The original published items were then turned over to the AIAA library, which was now primarily a repository for abstracted material. The great bulk of the historic library had long since been dispersed to the Smithsonian, the Library of Congress, and other research libraries.

The TIS office published its material in *International Aerospace Abstracts*. In addition, electronic copies were posted on NASA's Scientific and Technical Information Program (STIP) Aerospace Database. At its peak, the annual NASA contract paid $3–4 million per year. In 1992, government contracts provided the largest portion of the Institute's income (23%). The NASA contract was the largest on the AIAA books.[17]

International Aerospace Abstracts published its one millionth abstract in 1989. That year, TIS abstracted 45,000 articles. Then things began to go awry. In the early 1990s, the head of the NASA STIP decided that the abstracting operation should be opened to bid. When the winning bidder could not meet the terms of the contract, NASA turned to AIAA's TIS once again. Now, however, the contract called for abstracts on a per-item basis, an operation that would be far less lucrative than the old arrangement. It was the beginning of the end.

AIAA management did its best to keep costs down. In June 1994, the 88 TIS employees who had remained in New York were moved from their 57th Street offices to less expensive workspace on John Street.[18] The AIAA library had continued to operate, supplying copies of articles and other materials upon request for a fee. In the spring of 1994, AIAA signed a joint venture agreement with Dynamic Information, Inc. The agreement established a new operation, Aeroplus Dispatch, that would service orders for articles and other information from the AIAA library, which was shipped in toto to Burlingame, California. The venture was short lived. When the business could not support itself, the AIAA donated the entire library collection to the Linda Hall Library, a Kansas City, Missouri, research library specializing in science and technology collections.

TIS continued to struggle along on only a small portion of the old NASA contract. In 1995–1996, the first full year under the new arrangement, AIAA contract revenue fell to only 3% of the total. When Cort Durocher asked Treasurer Dave Quackenbush to appraise the activity, it was clear that the program would now begin losing money, something that AIAA could not afford. Quackenbush was able to sell the entire TIS operation to Cambridge Scientific Abstracts (CSA) for $1 million. CSA, which did abstracting across the board in science, technology, and medicine, an especially lucrative field, could take advantage of economies of scale and make money with the operation. For AIAA, the result was a one-time infusion of cash and the end of an era that had begun with Lester Gardner and his Aeronautical Index more than six decades earlier.[15]

Sheila Widnall, former Secretary of the Air Force and AIAA President for 2000. (Courtesy AIAA.)

Hard Choices for the Sake of the Future

In 1992, just four years after AIAA had relocated to the Aerospace Building, the majority owner filed for bankruptcy and the building went into foreclosure. Under the terms of the agreement, AIAA had an opportunity to buy the building. Recognizing that such an action would put the Institute in the real estate business, the Board rejected the purchase. In the end, however, AIAA did find itself with a 33.25% equity in the building.

Even so, AIAA was now paying roughly one million dollars a year in rent. Beyond that, Durocher, who had hired a local firm to provide a 36-hour crash course in the basics of commercial real estate for members of the staff, recognized that the building was a "loser." The Aerospace Building was a lovely structure, in a good location, with an underground garage and a great view from the upper floors. There were restaurants and shopping just across the street at L'Enfant Plaza. For all of its advantages, however, the building had not attracted the assortment of high-end corporate tenants that would have spelled success. Instead, it was filling up with government agencies. The AIAA itself was a problem, sitting on the prime space on the top floor and paying a cut-rate rent.[15]

As President A. Thomas Young (1995) remarked, the changing dynamics of the defense industry, "the merging of leading aerospace companies, loss of aerospace engineers and scientists due to downsizing, diminishing undergraduate and graduate student enrollment," demanded that AIAA leaders consider, "whatever action necessary to insure the organization's vitality into the next century." In August 1993, the Board created an ad hoc committee to study options for reducing their occupancy costs. The committee included Young, past President Malcolm Currie (1994), incoming President Skip Fletcher, and Vice President of Finance Sidney Pauls. The group recommended that locating to less expensive rental space in the Washington suburbs and subletting the high-value 10th floor space in the Aerospace Building would offer maximum financial security and flexibility.[19]

An aggressive marketing effort led to a U.S. Postal Service lease for two-thirds of the AIAA office space. The rest went to the staff of *Air & Space Smithsonian* magazine. The AIAA staff moved to Reston, a planned community 20 miles southwest of Washington, DC, where it would take a 10-year lease on the fifth floor of a building at 1801 Alexander Bell Drive that also housed the American Society of Civil Engineers. The Institute opened its doors at the new location on January 29, 1996. With their occupancy cost reduced to $300,000 per year, the organization saved an annual $700,000, so long as the Aerospace Building office space was rented. President Young assured the members that the relocation "was a positive step to effectively shape our organization for the new century."[20]

Visitors stepping off the elevators and walking through the glass doors into AIAA's Reston headquarters are confronted by several large metal wall plaques. They bear the names of all those corporations and individuals who contributed to the $5 million fund that supported the move from New York to Washington. They look far too large for the blue wall on which they are mounted. They were intended, after all, for the much larger lobby wall of the Aerospace Building.

The move to Reston was only the first in a series of steps taken to put AIAA on a firm financial footing that would enable it to better withstand economic and industry fluctuations. Like all such efforts it was painful, but it was also one of the great successes of the Durocher years. The Institute entered that era struggling to keep from losing financial ground. The treasurer had closed the books on 1988, Durocher's first year in command, more than $130,000 in the red. Fifteen years later, the Institute was ending its fiscal year $93,000 in the black.

The size of the endowment is a more important and far more impressive measure of financial stability. In 1979, total investments in the endowment amounted to $3,564,000. At the end of 2003, that basic endowment portfolio was worth $19.8 million. A second portfolio for the AIAA Foundation totaled $6.6 million. A pension fund account contained a further $6.5 million. Durocher and his team had established a solid financial foundation on which to build for the future of the Institute.[21]

Education, Public Policy, and Meetings: Business as Usual, and Then Some

As with any well-run organization, shifts in programmatic emphasis within AIAA are usually reflected in changes in institutional structure. When AIAA was launched in 1964, for example, education was the responsibility of the Vice President for Section Affairs, who had oversight over Student Branches. Not until 1971 did a Vice-President for Education join his colleagues in charge of Publications, Sections, and Technical Affairs. Through its strong support for the Accreditation Board for Engineering and Technology (ABET), AIAA was obviously a force to be reckoned with in higher education. By 1986, AIAA volunteers were serving on accrediting panels surveying 10 of the 54 university aerospace engineering programs annually.

The AIAA student program followed the pattern established by the IAS and ARS, offering special programs, a dedicated *AIAA Student Journal*, and a series of honors and awards. Over time, Branch activities came to mirror Section activities. One thousand student engineers participated in seven AIAA Regional University Conferences held in March and April 1983. There was a separate system of student awards, capped by an AIAA Industry Scholarship Program launched in 1976. Ten years later, the program offered 23 $1,000 awards for sophomores, juniors, and seniors. The nation's leading aerospace manufacturers supported the program: Boeing, Northrop, Lockheed, General Electric, General Dynamics, Vought, Honeywell, and Comsat.[22]

By 1972, with enrollment in university engineering programs in decline, President Allen Puckett pointed to the importance of interesting high school students in science and technology. He encouraged AIAA Sections and Student Branches to "... offer counseling services, provide lectures, assist in curriculum development and otherwise help high school students to become knowledgeable about the disciplines that relate to aerospace engineering training."[23]

Early the next year, the constitution was amended to extend student membership to "... persons interested in aeronautics or astronautics whose primary activity is study at recognized colleges, universities or secondary schools offering curricula and studies acceptable to the Institute."[24] Special efforts were made to encourage the creation of high school groups where there were established sections or branches to nurture and encourage the youngsters. Section leaders were encouraged to work with other local groups: the Civil Air Patrol, Boy Scouts, National Association of Rocketry, and National Association of Physics Teachers. The plan was to identify 40 or so of the 1,000 high schools in the nation that emphasized science and technology.[24]

In 1993, the Institute launched yet another experiment, a one-year pilot program for a new class of Educator Members. "Pre-college" teachers could become AIAA members, receiving both *Aerospace America* and the *AIAA Student Journal* as well as the opportunity to participate in Section activities. The program was modified by 1995, with the teachers being identified as Educator Associates. A Precollege Outreach Committee (PCOC) established four subcommittees devoted to resources, teacher workshops, technical writing, and electronics/communications.

The PCOC presented Educator Achievement Awards to Educator Associates and offered $200 grants to teachers to support the acquisition of supplies and publications to be used in teaching mathematics and science. In true AIAA fashion, the PCOC held its first regional conference, Space 2000, in Long Beach, California. There were exhibits of educational materials, student projects, and four astronauts to inspire the 200 invited teachers and students.

If AIAA has recognized the need to attract youngsters into engineering careers, it has also paid increasing attention to the need for continuing educational opportunities for members. By the 1980s, the Institute was offering a wide range of short courses, tutorials, workshops, and home study courses. In addition, there were management courses on finance, law, business operations, and manufacturing processes.

By 1993, AIAA offered 40 professional development courses, including 5 home study correspondence courses, 33 seminars, and in-house-produced programs at corporate facilities. Nearly half of the 773 students registered for these courses in 1992 were nonmembers, 69 percent of whom subsequently jointed the Institute. Thirty-one per cent of the home study students were foreign nationals. Joint courses planned and prepared with other institutions were launched in 1993, including Computational Aeroacoustics in Aerodynamics, offered at Brussels, Belgium, in cooperation with the von Kármán Institute.[25]

Since the mid-1960s, involvement in public policy has remained a major concern of the AIAA. In 1986, members testified eight times before various Congressional Committees and were invited to comment on everything from the NASA budget to plans for a National Aerospace Plane. The public policy staff produced a variety of products, from full-scale reports on major issues (five in 1989) based on months of research and thought, to workshops that provided AIAA experts an opportunity to discuss problems, to position papers, shorter documents offering preliminary information and analysis.

Public policy leaders like Jerry Grey emerged as nationally recognized authorities on a wide variety of space issues, from down-to-earth subjects like metric conversion to the possibility of space colonies and interplanetary voyages. The office also worked to provide individual members with access to their own Congressional representatives and other policy makers. Congressional Visits Day, inaugurated in 1994, enabled more than 100 members from 25 states and several Technical Committees to meet and discuss issues of concern with Senators, Representatives, and key staffers. That year, the Public Policy Committee unveiled a strategic plan designed to focus the energies of the organization on critical issues in civil aeronautics, space flight, and national defense.

Publications and meetings remain the bread and butter of the AIAA. The Institute continues the tradition of the IAS and ARS, offering a full range of general meetings that draw large numbers of members supplemented by smaller, very focused gatherings. On occasion, AIAA pulls out all the stops in emulation of the 1961 ARS Space Flight Report to the Nation (SFRN). The First World Space Congress was just such a gathering.

Held in Washington, DC, on August 28–September 5, 1992, the Congress was jointly sponsored by the Committee on Space Research (COSPAR), the International Astronautical Federation (IAF), and AIAA. Attracting 10,000 scientists, engineers, industrial leaders, government officials, and academics from around the world, it was the largest gathering of space professionals in history. A decade later, the Second World Space Congress was staged in Houston. The goal, Cort Durocher announced, was to "outline a vision for the next quarter century of space exploration."[26]

By the early years of the 21st century, AIAA leaders began to meld their years of experience in planning technical meetings with their long-standing determination to offer expert advice on national issues in science and technology. The Ballistic Missile Defense Agency (MDA) was created in 1999 to deploy a technological system to protect the nation from a limited missile attack. Lieutenant General Roland T. Kadish, U.S. Air Force, decided that the agency could afford to sponsor two major conferences a year that would bring together scientists and engineers involved in the program. Cort Durocher won the contract to organize that aspect of the MDA effort. In similar fashion, NASA leaders were persuaded to contract with the AIAA to organize special meetings related to their efforts to meet President Bush's challenge to return Americans to the Moon and Mars.[15]

The AIAA Foundation

In their 1994–1995 annual report to the members, the Board announced that it was "proceeding with setting up a Foundation to endow additional scholarships, design competitions, and outreach activities to attract and motivate youth to technological professions."[27] Cort Durocher was thinking bigger than that. He was troubled by the fact that the business of raising funds to support the industry scholarships and some other educational programs was an ongoing process. Why not, he wondered, create a separate AIAA Foundation endowment that would fund the entire awards and education programs in perpetuity?

He persuaded Norman Augustine and five other past presidents to join a nonprofit, tax-exempt AIAA Foundation Board. He promised Augustine that he would never have to contribute a dime or attend a meeting. In the end, the project so appealed to Augustine that he donated the royalties from his best-selling book, *Augustine's Laws*, to the Foundation.

Durocher envisioned a program that would have two "legs." The first would encompass all AIAA awards. Since the establishment of the Sylvanus Albert Reed Award in 1934, the AIAA and its predecessors had accumulated considerable endowment funds dedicated to honoring important contributors to flight technology in the name of one pioneer or another.

The Foundation would include all of those funds and any others contributed to support the honors and awards program in the future. The other "leg" would cover AIAA educational activities, from the industrial scholarships to student conferences, the K-12 program, and other efforts.

The Foundation was an enormous success. The original $6 million endowment grew to $8.7 million by 2005. The programs that it supported grew, as well. The Foundation revived interest in the honors and awards program, attracting new contributors. Not long after the project was launched, Durocher met with Walter Crichlow, a long-time member who wanted to discuss, in general terms, establishing a new award. Two days later, there was a call from Merrill–Lynch informing AIAA that Mr. Crichlow had signed over $660,000 in stock to the Institute. The result was the Walter J. and Angeline Crichlow Prize, presented every four years "... for excellence in aerospace materials, structural design, structural analysis, or structural dynamics." The honorarium—$100,000 (Ref. 15).

Now the AIAA education program came of age. In 2003, the Foundation provided $150,000 in scholarships. Thirty students won $2,000 AIAA Foundation Undergraduate Scholarships, 10 received $5,000 graduate awards, and a lucky 4 won $10,000 Orville and Wilbur Wright Graduate Awards. Through the Pre-College Program, a single school could receive up to $1,000 in grants for the purchase of supplies to support the teaching of math and sciences. Grants totaling $20,000 went to individual classrooms serving 15,000 students. Nine hundred and fifty K-12 teachers had signed up as Educator Associate members. The AIAA Foundation and the new educational initiatives that came with it represented one of the real achievements of the Durocher years.

Evolution of Flight

An appreciation for the story of Wilbur and Orville Wright had been a tradition since the foundation of the IAS. Orville Wright was the first IAS Honorary Fellow. The annual Wright Brothers lecture was one of the early IAS celebratory events. For his part, Orville Wright's gifts to Lester Gardner, from pieces of the wood and fabric from the world's first airplane to the stop watches used to time the first flight, had become the treasures of the early Institute collection.

On the occasion of the 50th anniversary of powered flight, Ernest Robischon had engineered the creation of a full-scale model of the 1903 airplane, even involving Charles Taylor, who had built the engine for the world's first airplane. The machine had been the centerpiece of the Durand Museum in the IAS Los Angeles building. When the museum was disbanded, the airplane went to the San Diego Aerospace Museum.

Ten years later, John Attinello of the Institute for Defense Analysis and Hal Andrews of the Bureau of Naval Weapons led a team of National Capital Section members who produced a full-scale model of the world's first airplane for presentation to Wright Brothers National Memorial. The elevator, rudder, propellers, landing skids, and other parts were produced in the basements and garages of team members. The Pratt & Whitney Company constructed the four-cylinder engine, but without the pistons or crankshaft so that it could never be started. The wings were built at the Patuxent River Naval Air Station (NAS). The team did the final assembly in a warehouse near Washington National Airport.

AIAA President William Pickering presented the airplane to Conrad Wirth, director of the National Park Service, at the Wright Brothers National Memorial at Kill Devil Hills, North Carolina, on December 16, 1963. Major General Marvin Demler, President of the

National Capital Section, Attinello, Andrews, and 400 others looked on with pride. The airplane, which the National Capital Section refurbished 30 years later, was the centerpiece of the Flight Room at the Wright Brothers National Memorial, where millions of visitors would enjoy and learn from it. It remains on exhibit at the park today, the most historic full-scale model of the world's first airplane.[28]

On the evening of February 22, 1978, an arsonist burned the San Diego Aerospace Museum to the ground and the original IAS Wright Flyer with it. Uncertain what to do with the insurance money resulting from the loss, Los Angeles Section officials invited members to suggest ideas. Dr. Fred E.C. Culick, a California Institute of Technology aerodynamicist, suggested that he would spend the money to build and fly a full-scale model of the 1903 machine and use it to gather flight data on the world's first airplane. They gave him the money.

From the beginning to the end, the volunteer project took shape under the leadership of Culick and Section members Jack Cherne, Howard Marx, and others. The first step was to conduct wind tunnel tests of 1/6 and 1/8 scale models of the aircraft. The data obtained provided the grist for one dozen articles and any number of public presentations on the program. Next, in 1995 the team constructed a full-scale model of the airplane and tested it in the National Full-Scale Aerodynamics Complex at the NASA Ames Research Center. With the additional information resulting from these tests, Culick and the others were able to more fully understand the aerodynamics of the machine and make solid judgments as to its flying characteristics.

Having fulfilled its technical purpose, the airplane was loaned to the Flight Deck Museum of the Western-Pacific Regional Office of the FAA in Hawthorne, California, where, as one Los Angeles Section member noted, "it would be available to the kids." The members of the evolving team then went back to work on a second full-scale 1903 airplane that they could fly.[29]

AIAA national leaders had also noted the approach of the centennial of powered flight. Cort Durocher credits President Edward "Pete" Aldridge, Jr. (1997) with the decision to invest heavily in celebrating the centennial. "St. Pete," Durocher says with a smile, had served as CEO of the Aerospace Corporation, President of McDonnell Douglas Electronic Systems Company, and 16th Secretary of the Air Force. He immediately saw the advantages of participating, even leading, the commemoration of a century of wings. Aerospace had ridden an economic roller coaster during the 1970s, 1980s, and 1990s. Here was a golden opportunity to call the attention of the nation to the extent that flight had shaped the history of the 20th century. Who better to plan the celebration than AIAA?

They would call the campaign the Evolution of Flight. "We took that concept, marketed it and branded it," Durocher explains. The Board allocated $3.7 million for the project, the largest investment in a single program in the history of the Institute. They then raised another $2 million from corporate sponsors: Boeing, GE Aircraft Engines, Lockheed Martin, Northrop Grumman, Parker Hannifin, Pratt & Whitney, Raytheon, Rockwell Collins, and Snecma. Merrie Scott provided staff supervision, and Kim Grant oversaw the educational aspects of the program. The board recruited an advisory committee of stellar aerospace figures: Neil Armstrong, the first man to set foot on another world; Scott Crossfield, the first man to fly twice the speed of sound; Patty Wagstaff, three times U.S. national aerobatic champion; and AIAA Fellow Richard Culpepper, who chaired the Evolution of Flight Committee.[30]

Although it was not specifically an element of the Evolution of Flight program, the AIAA also launched an effort to mark historic aerospace sites. Developed by Anthony M. Springer, an historically conscious engineer who chaired the History Technical Committee, the program invited members to nominate spots where the great events in aerospace history had occurred. Each site selected would be honored with a metal plaque and an informative pamphlet. Approved in 2000, the program had honored 25 sites by the end of 2004, from the spot in Annonay, France, where the first balloon took to the sky in June 1783 to the *Apollo 11* landing site on the Sea of Tranquility. That plaque is being held in safe storage until returning astronauts can be place it on the lunar surface.

The events of the centennial year were coordinated by a Federal Centennial of Flight Commission and a First Flight Centennial Federal Advisory Board. State commissions in Ohio and North Carolina and local organizations across the nation provided a full schedule of events. The AIAA program ran through all of it like a thread, helping to bind the individual efforts into a single memorable commemoration. Some of the leading projects included:

- Four million people in almost a dozen cities visited a large AIAA traveling exhibition, with the full-scale 1903 machine constructed by members of the Los Angeles Section as its centerpiece.
- The AIAA Foundation established the Orville and Wilbur Wright Graduate Fellowships, which provided $10,000 scholarships to students of flight science and technology.
- Sixty Learn-to-Fly scholarships went to student pilots between the ages of 16 and 25 years of age.
- A Distinguished Lecture Series sent authorities on flight science, technology, and history to sections and branches in every corner of the nation.
- Student art contests gave youngsters an opportunity to share their visions of the future of flight.
- Commemorative publications ranged from a chronology of the history of flight to definitive volumes on the development of aerodynamics and the history of rocketry.
- Five million people around the world visited the Evolution of Flight Web site, where they found everything from interactive descriptions of the principles of flight to the contributions of nations around the globe to flight science and technology.

The Class of 2003 program deserves special notice. In 1999, entering high school students scheduled to graduate in 2003 were invited to enter an essay contest. Over the next 3 years, the 20 lucky winners would have an opportunity to savor some unique experiences, including

- the chance to plan a trip to the moon at the Washington, DC, Challenger Center
- five days at the Advanced Space Academy of the U.S. Space and Rocket Center in Huntsville, Alabama
- meetings with Boeing officials and VIP behind the scenes tours of company factories and test facilities
- a visit to Colorado Springs, Colorado where the young AIAA Ambassadors bunked with U.S. Air Force Academy cadets, and to Denver for an opportunity to fly the simulators at the United Airlines Training Academy.
- time spent at the Experimental Aircraft Association Advanced Aviation Leadership Academy and at world-famous AirVenture 2001
- a visit to Naval aviation, museum, and university programs in the San Diego, California area
- six days behind the scenes at Cape Canaveral, Florida
- attendance at the second World Space Congress in Houston, Texas
- a never to be forgotten opportunity to join thousands of other flight enthusiasts from around the world in the place where it all began—Dayton, Ohio

In addition to the national program, 36 AIAA Sections and Student Branches pitched in with an array of imaginative projects. There were events commemorating the Wright brothers:

- The AIAA Cape Canaveral Section worked with Student Branches to construct a ¼ scale model of the Wright Flyer and accompanying presentations on the invention of the airplane.
- The AIAA Connecticut Section wrote a play that brought the story of the Wrights to life and performed it during the appearance of the AIAA 1903 Wright Flyer tour in their area.
- The AIAA Hampton Roads Section constructed a series of accurate replicas of the Wright wind tunnel and prepared educational materials highlighting the process of invention for secondary school students.
- An Educator Associate member of the Oklahoma Section built a radio-controlled model of the Wright Flyer and demonstrated it in classrooms and science camps.
- Members of the San Diego Section worked with a group of K-12 students who built 5/8 scale replicas of the 1899 Wright kite and flew them at a special picnic.

Other Sections developed and offered educational programs, curriculum materials, exhibitions, art contests, and competitions of every stripe. There was nothing quite like it in the history of the ARS or IAS. All over the nation members of the Institute extended themselves to ensure that all Americans would better appreciate the heritage of flight and have a good time in the process.

A Centennial Snapshot, July 14–17, 2003

Dayton, Ohio, was the center of the aerospace universe for 17 days in July 2005. Visitors could see the places where the history of flight was made at historic sites and museums scattered around town. They could literally walk in the footsteps of the inventors of the airplane by touring the West Dayton neighborhood where Wilbur and Orville Wright had lived and worked. The world's first practical airplane was preserved at Carillon Park, less than fifteen minutes away. At Huffman Prairie, eight miles east of Dayton, you could squint your eyes a bit and imagine that you could still see that airplane circling over the grazing cattle. From Huffman Prairie it was five miles to the sprawling Air Force Museum at old Wright Field, where researchers had been probing the outer edges of the aviation envelop since 1928.

The 600,000 enthusiasts who participated in the Inventing Flight celebration, July 3–20, 2005, scarcely had time to catch their collective breaths. Eight thousand of them gathered in the city's downtown ballpark to enjoy what was described as an Olympic-style opening ceremony, complete with 400 costumed dancers, singers, and performers and ending with fireworks. The stars of the evening were two Ohioans whose fame rivaled that of the Wright brothers, Neil Armstrong and John H. Glenn.

The high point of the Fourth of July was an appearance by President George W. Bush at the Air Force Museum. Events were scheduled chock-a-block in the days that followed. There were to be balloon races, an International Blimp meet featuring a record numbers of airships in one place, a Black Cultural Festival, and a fantastic air show spread over two weekends. You could treat the family to an afternoon at Celebration Center, an island in the Miami River where educational exhibits rubbed shoulders with food kiosks and an assortment of rides and games for the kids. Visitors flocked to the west side of town, where they could chat with archaeologists digging for evidence of the famous Wright bicycle shop that once stood there or pass the time of day with Wilbur and Orville, their younger sister Kate, or their friend, the poet Paul Dunbar, all portrayed by young actors participating in the Time Flies street theater program.

It would be difficult to imagine a more fitting conclusion to the commemorative activities than the ceremony conducted at Woodland Cemetery on the final Sunday morning. The first American to orbit the earth and the first man to walk on another world joined members of the Wright family laying wreaths on the graves of the two brothers who set the air age in motion. "Their bodily remains are buried here," John Glenn noted, "but ... their spirit ... lives on in the life of every person young or old who is inspired, by their example, to imagine, to dream, to do, to move the world ahead."[31]

For the world of aviation, the AIAA/ICAS International Air and Space Symposium was the centerpiece of the events in Dayton. There had been bigger meetings, but few that were quite so impressive. The event drew 1,300 participants from 20 nations. Attendees could choose from a total of 500 technical presentations offered in as many as 25 parallel sessions on each of 4 days. The exhibit area of the Dayton Convention Center was crowded with displays prepared by 80 exhibitors.

"To appreciate the scope of the international aerospace symposium," reporter Tim Gaffney noted, "consider this: two of the twelve men who walked on the moon were in a Tuesday session ... ." Senator Harrison Schmitt, the second from the last man to walk on the moon, was a member of the panel discussing the impact of government policy on the history of air and space technology. His Senate colleague, John Glenn, was also on the panel, as was Scott Crossfield, the first man to fly twice the speed of sound. A few minutes after the session began, Chairman Jerry Grey noted that Edwin "Buzz" Aldrin, the second man to set foot on the Moon, had just walked into the meeting room. It was that kind of week. The leadership of the nation's aerospace community, from industry leaders to astronauts to working engineers, were in Dayton to celebrate a century of flight and consider solutions to problems that block the way to the future.[32]

There was really nothing so extraordinary about the AIAA gathering in Dayton. The organization had been attracting the great names in aerospace to such meetings for more than 70 years. It would be difficult to think of a major contributor to the American aerospace enterprise who had not been involved with the ARS, IAS, or AIAA.

Consider, for example, the individuals who had served as president of those organizations. The list of leading scientists and engineers includes Jerome Hunsaker, Clark Millikan, Frank Caldwell, James Doolittle, Arthur Raymond, Edward Wells, James Wyld, Martin Summerfield, Hall Hibbard, Raymond Bisplinghoff, Robert Truax, John Paul Stapp, and William Pickering. Charles Lawrance; Donald Douglas; Glenn Martin; Reuben Fleet; John Northrop; J. H. Kindelberger; John Leland Atwood; Robert Gross; L. Eugene Root; Joseph Gavin, Jr.; Edward "Pete" Aldridge; and Norman Augustine were among the industry leaders who took the helm of AIAA or one of its predecessors. Then there were the government officials and military leaders, men and women like George Lewis; Hugh Dryden; Donald Putt; Robert Seamans, Jr.; George Mueller; Joseph Shea; and Sheila Widnall.

The very first issue of the *Journal of the Aeronautical Sciences* carried articles by Theodore von Kármán, Max Munk, and John Stack. Since that time, the publications of the IAS, ARS, and AIAA have introduced some of the great ideas in the wide range of disciplines that make up aerospace science and engineering. For more than seven decades, local and national meetings of those organizations have provided a forum for introducing new discoveries, offering fresh perspectives, discussing critical issues, and building personal relationships.

AIAA's Honors Night Banquet was one of many commemorative events marking the 100th anniversary of powered flight. Special guests pictured with members of the AIAA staff were Neil Armstrong, John Glenn, Amanda Wright Lane (great grandniece of the Wright brothers), Patty Wagstaff, Stephen Wright (great grandnephew of the Wright brothers), John Travolta, Alan Mulally, and Scott Crossfield.

Participation in those activities has come to define the aerospace profession in the United States.

Would Wilbur and Orville Wright be pleased to know that the citizens of their home town, their nation, and the world were honoring the centennial of their achievement? Probably so. Never great ones for ceremony or fanfare, they might have wondered at the need for the fireworks, the 400 performers at the kickoff ceremony, and the rides and games at Celebration Central. They would certainly have appreciated the Dayton Air Show, however, as well as the care with which their 1905 airplane had been preserved and the vast array of aircraft on display at the Air Force Museum, each and every one of which traced its lineage back to the craft they had flown from the sands of Kill Devil Hills.

The two men who set all of this in motion were engineers, after all. Although neither of them had ever spent a day in a college classroom, they had discovered the engineering process for themselves on the way to the invention of the airplane. Nothing delighted them more than wrestling with the most difficult technical challenges. It gave them a reason to get up in the morning. "Isn't it astonishing," Orville Wright wrote to a friend in the late spring of 1903, "that all of these secrets have been preserved for so many years just so that we could discover them!"[33]

One thing you can be sure of, they would have been found among the crowd of professional technologists attending the working sessions of the AIAA's International Air and Space Symposium. They would have felt right at home and approved of what had become of the organization founded by Orville's friend Lester Gardner seven decades before. Nothing would have delighted them more than to be among a large group of engineers and scientists drawn from around the world to commemorate their achievement by laying the foundation for tomorrow.

The AIAA would soon be transitioning into a new era. On January 18, 2005, the Board of Directors announced that Robert S. Dickman, a retired U.S. Air Force Major General who had been serving as Deputy for Military Space in the Office of the Secretary of the Air Force, would be leading the AIAA into the future. Cort Derocher was departing after 16 very active and successful years on the job. The new executive director was inheriting an organization that had never been in a stronger fiscal position or offered a broader range of programs supporting the members, the industry, and the nation. AIAA had met the needs of its members, the industry, and the nation for three-quarters of a century. It was an organization that could take great pride in its achievements. Like the Wright brothers themselves, however, the members of AIAA prefer looking to the future to dwelling on the past.

References

[1] "Board Votes Against Headquarters Move to Washington, *Astronautics & Aeronautics*, July–Aug. 1982, p. 85.

[2] Minutes of the AIAA Board of Directors Meeting, Jan. 1, 1982, AIAA Headquarters Files.

[3] Minutes of the AIAA Board of Directors Meeting, Jan. 1, 1983, May 13, 1983, AIAA Headquarters Files.

[4] Minutes of the AIAA Board of Directors Meeting, Aug. 31, 1983, p. 8, AIAA Headquarters Files.

[5] Minutes of the Meeting of the AIAA Board of Directors, May 24, 1984, pp. 3–5, AIAA Headquarters Files.

[6] Cort Durocher outlined the details of the deal for the author in an oral history interview.

[7] Minutes of the Meeting of the AIAA Board of Directors, Aug. 16, 1984, p. 205, AIAA Headquarters Files.

[8] Minutes of the Meeting of the AIAA Board of Directors, Feb. 28, 1985, p. 2; April 12, 1985, p. 4; Oct. 16, 1985, p. 4, AIAA Headquarters Files.

[9] Minutes of the Meeting of the AIAA Board of Directors, Feb. 13, 1986, p. 7, AIAA Headquarters Files.

[10] "Augustine Asks Members to Contribute to the Fund Drive," *Astronautics & Aeronautics*, May 1987, p. B4.

[11] "Headquarters Fund Off to a Good Start," *Astronautics & Aeronautics*, April 1986, p. B4.

[12] Minutes of the Meeting of the AIAA Board of Directors, July 10, 1986, p. 7, AIAA Headquarters Files.

[13] Minutes of the Meeting of the AIAA Board of Directors, Nov. 12, 1987, p. 3, AIAA Headquarters Files. When Jim Harford reported that the Building Fund had topped the $5 million mark on September 2, 1988, the Board of Directors immediately toasted the moment with champagne.

[14] Minutes of the Meeting of the AIAA Board of Directors, Feb. 12, 1988, p. 1, AIAA Headquarters Files; see a file of documents on the Heinmuller sale in the AIAA Headquarters Files.

[15] Oral history interview with Cort Durocher, November 23, 2004, author's collection.

[16] Jerry Grey to Tom Crouch, Feb. 11, 2005, author's collection.

[17] "AIAA Fiscal 92 Revenue," *AIAA Annual Report, 1992–1993*, p. 7.

[18] Karen Holloway to Rodger Williams, e-mail, March 22, 2005, author's collection.

[19] A. Thomas Young, "AIAA Headquarters Relocates to Suburbs," *AIAA Bulletin*, March 1996, p. B1.

[20] Young, "AIAA Headquarters Relocates," p. B13.

[21] *AIAA Annual Report for 2003,* AIAA Headquarters Files.

[22] Meeting of the AIAA Board of Directors, May 11, 1983, pp. 16–18, AIAA Headquarters Files.

[23] Meeting of the AIAA Board of Directors, Nov. 1, 1972, p. 10, AIAA Headquarters Files.

[24] Meeting of the AIAA Board of Directors, Jan. 10, 1973, p. 10, AIAA Headquarters Files.

[25] Robert Fuhrman, "1992–93 AIAA President's Annual Report," *AIAA Annual Report, 1992–1993.*

[26] Greater Houston Partnership Press Release, "World Space Congress Comes to Houston," March 1, 2002.

[27] "Education," *AIAA Annual Report, 1994–1995,* p. 3.

[28] "AIAA Member-Built Wright Flyer Given to Museum to Mark 60th Anniversary of History-Making Flight," *Aeronautics & Astronautics,* 1964.

[29] http://www.wrightflyer.org/, cited Aug. 2005.

[30] Celebrating the Evolution of Flight: A Final Report, AIAA, Reston, VA, 2004.

[31] http://dj.inabox.com/MediaMoments/2003Woodland-1.shtml, cited March 29, 2005.

[32] Tim Gaffney, "Forum Examines Space Plight," *Dayton Daily News,* July 16, 2003, p. V2.

[33] Orville Wright to George Spratt, June 7, 1903, The Papers of Wilbur and Orville Wright, edited by Marvin W. McFarland, McGraw-Hill, New York, 1953, p. 313.

Bibliography

A Note on Original Documents Still Held by AIAA

At the time of this writing the American Institute of Aeronautics and Astronautics retains at its headquarters the only complete file of the minutes of the governing bodies of the AIS, ARS, IAS, and AIAA. Filed with those minutes are a great many documents pertaining to issues under discussion at particular meetings. The simplest means of footnoting that material has been to identify the organization and the date of the meeting. If a document other than the minutes is cited in that fashion, it is an item included with the minutes for that date.

The citation of other files retained by the AIAA has been more difficult. Some stray historical files have simply been filed away in a series of cardboard boxes normally stored offsite. Other specific historical files are housed in filing cabinets at AIAA Headquarters but have not been incorporated into any standard office-wide system. Simply put, these are separate files discovered during the course of research. There are, for example, an extensive set of legal records covering the acquisition, handling, and dispersal of the IAS historical collections. Although they cannot be cited by folder, file, or even drawer number, most are included in particular groups of files. Where possible, folder names have been included. AIAA staff members familiar with the records of the organization can assist subsequent researchers in locating any item in question without too much difficulty.

Other Manuscript Collections

National Air and Space Museum Archive:

Jerome Hunsaker Papers, xxxx-0001
The Papers of Andrew G. Haley Papers, xxxx-0200
The Frederick Clark Durant Collection, xxxx-0084
The Charles W. Chillson Papers, xxxx-0008
The Richard Porter Papers, 1997-0037
S. Fred Singer Papers, 1989-0130
Lovell Lawrence, Jr., Collection, xxxx-0010
American Astronautical Society Records, xxxx-0163

In addition, the author has consulted numerous biographical and other folders in the hanging file system maintained by the NASM Archivists.

Johns Hopkins University Library, the Papers of Hugh Dryden
Princeton University Libraries, the Papers of G. Edward Pendray
Library of Congress, Manuscript Division, American Institute of Aeronautics and Astronautics History Collection
Smithsonian Institution Archive, The Papers of S. Paul Johnston

Oral History, Telephone and E-Mail

Transcripts of oral history interviews with Richard Porter and Milton Rosen, National Air and Space Museum, Division of Space History

Author interview with James Harford
Author interview with Cort Durocher
Author interview with Frederick C. Durant, III
Dr. Jerry Grey, e-mail response to author questions
Extended discussions with Dr. John Anderson and Mr. Hal Andrews

Manuscript Items

Howard Eisenberg, "Paul Reveres in Space Suits: The American Rocket Society," manuscript in the G. Edward Pendray Papers, Princeton Univ. Library.

J. Laurence Pritchard, "Lester Durand Gardner," copy in the Lester Durand biographical file, National Air and Space Museum Archive.

James J. Harford, "The American Institute of Aeronautics and Astronautics, 1963–1993," a paper presented at the Space History Symposium of the History of Science and Technology, Moscow, Russia, May 12, 1993.

James Harford, "The American Rocket Society, 1953–1963: A Memoir," IAF-88-605, 39th Congress of the International Astronautical Federation, copy in the author's collection, 1988.

J. C. Hunsaker and Lester D. Gardner, "Background and Incorporation of the Institute of Aeronautical Societies," in Box 2.53, Folder 1952, the Papers of Hugh L. Dryden, Special Collections and Archives Division, Johns Hopkins Univ. Library.

Anthony Springer, "The Development of an Aerospace Society, The AIAA at 70," AIAA Paper 2001-0177, 2001.

Anthony Springer, "Early Experimental Programs of the American Rocket Society, 1930–1941," AIAA Paper 2000-3279, 2000.

Published Sources

This is a select bibliography, listing only the most important publications referenced during the course of research on this book. Readers in search of specific articles appearing in the publications of the AIAA and predecessor organizations, newspapers, and magazines or journals are directed to the chapter notes. The bibliography is limited to those items of broader interest to readers in search of further reading on the subject.

Anonymous, "The American Society of Aeronautic Engineers Established," *Aerial Age*, Vol. 1, No. 20, Aug. 2, 1915.

Anonymous, *Celebrating the Evolution of Flight: A Final Report*, AIAA, Reston, VA, 2004.

Anonymous, *A Descriptive and Historical Catalogue of a Collection of Engravings, Drawings, Portraits, Autographs, and Other Materials Illustrative of the History of Ballooning and Flying*, Messrs. Hodgson & Co., London, 1917.

Anonymous, *Fuhrer durch die historische abteilung der internationalen luftschiffahrt-ausstellung, Frankfurt A.M., 1909*, Druck Der Kunstanstalt Wustrn & Co., Frankfurt A.M., 1909.

Anonymous, *History of the Institute of the Aeronautical Sciences, Tenth Anniversary, 1932–1942*, IAS, New York, 1942.

Anonymous, *The Institute of the Aeronautical Sciences and Aeronautical Library and Museum*, IAS, New York, 1946.

Anonymous, "Rocket Rendezvous," *Newsweek*, Oct. 23, 1961.

Anonymous, "The Skyport," *Journal of the Institute of the Aeronautical Sciences*, Jan. 1934.

Anonymous, *SAE in Aerospace: Reliability and Progress in the Air*, promotional brochure, SAE, 2004.

Anonymous, *This is IAS, IAS Facts and Figures*, IAS, New York, Jan. 1962.

Anderson John, *The History of Aerodynamics and its Impact on Flying Machines*, Cambridge Univ. Press, Cambridge, England, U.K., 1997.

Arthur, George R., "A Few Reflections, 1959–1960," *Twenty-Five Years of the American Astronautical Society, 1954–1974*, edited by Eugene M. Emme, AAS Publications Office, San Diego, 1980.

Bangs, Allan P., "In Appreciation," *Astronautics & Aeronautics*, Sept. 1996, p. B11.

Biddle, Wayne, *Barons of the Sky*, Simon and Schuster, New York, 1991.

Billstein, Roger, *Orders of Magnitude: A History of the NACA and NASA, 1915–1990*, NASA, Washington, DC, 1989.

Blèriot Louis, and Raymond, Edward, *La Gloire des Ailes*, Les Editions de France Paris, 1927.

Burden, W. A. M., *Peggy and I: A Life Too Busy For a Dull Moment*, New York, 1982.

Caidin, Martin, "More Ways Than One," *Twenty-Five Years of the American Astronautical Society, 1954–1974*, edited by Eugene M. Emme, AAS Publications Office, San Diego, 1980.

Chang, Iris, *Thread of the Silkworm*, Basic Books, New York, 1995.

Chanute, Octave, *Progress in Flying Machines*, Chicago, IL, 1894.

Crouch, Tom D., *A Dream of Wings: Americans and the Airplane, 1875–1905*, W.W. Norton, Inc., New York, 1981.

Crouch, Tom D., "IAS: The Origin and Early Years of the Institute of the Aeronautical Sciences," *Astronautics & Aeronautics*, May 1981.

Crouch, Tom D., *Wings: A History of Aviation From Kites to the Space Age*, W.W. Norton, Inc., New York, 2003.

Doherty, E. Jay, "Of Planes They Sang!" *Flying and Popular Aviation*, Feb. 1942.

Durant, III, Frederick C., "Perspectives on the American Astronautical Society," *Twenty-Five Years of the American Astronautical Society, 1954–1974*, edited by Eugene M. Emme, AAS Publications Office, San Diego, 1980.

Ehricke, Krafft, "A National Space Flight Program," *Astronautics*, Jan. 1958.

Eiffel, Gustave, *La résistance de l'air et l'aviation: experiences effectuées au Laboratoire du Champ-de-Mars*, H. Dunod et Pinat, Paris, 1910.

Findley, Earl, "The Institute of Aeronautical Sciences," *U.S. Air Services*, Jan. 1933.

Gernsback, Hugo, "Science Wonder Stories," *Science Wonder Stories*, Vol. 1, No. 1, June 1929.

Glines, C. V., "The Guggenheims: Aviation Visionaries," *Aviation History*, Nov. 1996.

Goddard, Robert H., *A Method of Reaching Extreme Altitudes*, Smithsonian Inst. Press, Washington, DC, 1919.

Goddard, Robert H., *Liquid Propellant Rocket Development*, Smithsonian Inst. Press, Washington, DC, 1936.

Goddard, Robert, *Rockets*, American Rocket Society, New York, 1946.

Goddard, Robert H., to G. Edward Pendray, *Bulletin of the American Interplanetary Society*, No. 10, June–July 1931.

Green Constance, and Lomask, Milton, *Vanguard: A History*, Smithsonian Inst., Washington, DC, 1971.

Grey, Jerry, *Enterprise*, William Morrow, New York, 1979.

R. Cargill, Hall, "Origins of U.S. Space Policy: Eisenhower, Open Skies and the Freedom of Space," *Exploring the Unknown: Selected Documents in the History of the U.S. Space Program*, edited by John Logsdon, NASA, Washington, DC, 1995.

Hallion, Richard, "The Rise of Air and Space," *Astronautics & Aeronautics*, May 1981.

Heslin, James J., "Bella C. Landauer," *Keepers of the Past*, edited by Clifford Lord, Univ. North Carolina Press, Chapel Hill, NC, 1965, pp. 180–189.

Hodgson, J. E., *A History of Aeronautics in Great Britain*, Oxford Univ. Press, Oxford, England, U.K., 1924.

Hunsaker, Jerome C., "Europe's Facilities for Aeronautical Research, I." *Flying*, April 1914.

Hunsaker, Jerome C., "Europe's Facilities for Aeronautical Research, II" *Flying*, May 1914.

Hunsaker Jerome C., and Gardner, Lester D., "Background and Incorporation of the Institute of Aeronautical Sciences," Hugh L. Dryden Papers, Box 253, Folder 1952, Manuscript Division, Johns Hopkins Univ. Libraries.

Killian, James R., *Sputnik, Scientists, and Eisenhower: A Memoir of the First Special Assistant to the President for Science*, MIT Press, Cambridge, MA, 1979.

Landauer, Bella C., *Bookplates: From the Aeronautica Collection of Bela Landauer*, Privately Printed at the Harbor Press, New York, 1930.

Landauer, Bella C., "Collecting and Recollecting," *New-York Historical Society Quarterly*, July 1959.

Landauer, Bella C., "Literary Allusions in American Advertising as Sources of Social History," *New-York Historical Society Quarterly*, July 1947.

Leonard, Jonathan, "Journey Into Space," *Time*, Dec. 8, 1952.

Ley Willy, and von Braun, Wernher, *The Exploration of Mars*, Viking, New York, 1956.

Loening, Grover, *Our Wings Grow Faster*, Doubleday, Doran & Co., New York, 1935.

Logsdon, John M., (ed.), *Exploring the Unknown: Selected Documents in the History of the U.S. Civil Space Program, Volume 1: Organizing for Exploration*, NASA, Washington, DC, 1995.

Logsdon, John, (ed.), *Legislative Origins of the National Aeronautics and Space Act of 1958*, NASA, Washington, DC, 1998.

Malina, Frank J., "America's Long-Range Missile and Space Exploration Program: The ORDCIT Project of the Jet Propulsion Laboratory, 1943–1946: A Memoir," *Essays on the History of Rocketry and Astronautics: Proceedings of the Third Through the Sixth History Symposia of the International Academy of Astronautics, Volume II*, edited by R. Cargill Hall, Conference Publication 2014, NASA, Washington, DC, 1977.

Marsh, C. L., "In Search of Treasure: Thrills and Humours of Twenty Years Spent in Collecting an Aeronautical Library," *Airways*, June 1928.

McAleer, Neil, *Odyssey: The Authorized Biography of Arthur C. Clarke*, Victor Gollancz, Ltd., London, 1993.

McFarland, Marvin W., "Lester Durand Gardner: Elder Statesman of Aviation," *U.S. Air Services*, Oct. 1956.

McFarland, Marvin W., *The Papers of Wilbur and Orville Wright*, McGraw-Hill, New York, 1953.

Millikan, Clark, "Reflections on our First Quarter Century," *Aeronautical Engineering Review*, March 1957, p. 32.

Neufeld, Michael J., "Orbiter, Overflight, and the First Satellite: New Light on the Vanguard Decision," *Reconsidering Sputnik: Forty Years Since the Soviet Satellite*, edited by Roger Launius, John Logsdon, and Robert Smith, Harwood Academic, 2000, pp. 231–251.

Okrent, Daniel, *Great Fortune: The Epic of Rockefeller Center*, Viking, New York, 2003.

Pearson, Henry Greenleaf, *Richard Cockburn Maclaurin: President of the Massachusetts Institute of Technology*, Macmillan, New York, 1937.

Ordway, Frederick I., and Liebermann, Randy, (eds.), *Blueprint for Space: Science Fiction to Science Fact*, Smithsonian Inst. Press, Washington, DC, 1992, pp. 135–146.

Pendray, G. Edward, *The Coming Age of Rocket Power*, Harper, New York, 1947.

Pendray, G. Edward, "Early Rocket Developments of the American Rocket Society," *First Steps Toward Space: Proceedings of the First and Second History Symposia of the International Academy of Astronautics at Belgrade, Yugoslavia, 26 September 1967, and New York, U.S.A., 16 October 1968*, edited by Frederick C. Durant, III and George S. James, Smithsonian Inst. Press, Washington, DC, 1974.

Pendray, G. Edward, "The First Quarter Century of the American Rocket Society," *Jet Propulsion*, Nov. 1955.

Pendray, G. Edward, "The German Rockets," *Bulletin of the American Interplanetary Society*, No. 9, May 1931.

Pendray, G. Edward, "Why Shoot Rockets?" *Journal of the British Interplanetary Society*, Vol. 2, Oct. 1935.

Prescott, Samuel C., *When MIT Was "Boston Tech," 1861–1916*, MIT Press, Cambridge, MA, 1954.

Preston, E., (ed.), *FAA Historical Chronology: Civil Aviation and the Federal Government, 1926–1996*, Federal Aviation Administration, Office of Public Affairs, Washington, DC, 1998.

Pritchard, J. L., "The Society's Exhibit at the International Aero Exhibition, Olympia, July 16th–27th, 1929: A Summary by the Editor," *The Journal of the Royal Aeronautical Society*, Oct. 1929.

Reeve, Harrison F., "ASME Role in Powered Flight," *Mechanical Engineering*, Dec. 1953, pp. 987–999.

Rich, Daniel, AIAA at 50: Yesterday, Today, Tomorrow, AIAA, New York, 1982.

Rich, Daniel, "AIAA Yesterday and Today: Part 1: Profile of the American Rocket Society," *Astronautics & Aeronautics*, Jan. 1971.

Rich, Daniel, "Profile of the Institute of the Aerospace Sciences," *Astronautics & Aeronautics*, Feb. 1971.

Roland, Alex, *Model Research*, NASA, Washington, DC, 1985.

Rosen, Milton, "On the Utility of an Artificial Unmanned Earth Satellite," *Jet Propulsion*, Feb. 1955.

Rosenquist, James, "Founding of the American Astronautical Society, 1953–1954," *Twenty-Five Years of the American Astronautical Society, 1954–1974*, edited by Eugene Emme, AAS Publications Office, San Diego, 1980.

Rae, John, *Climb to Greatness: The American Aircraft Industry, 1920–1960*, MIT Press, Cambridge, MA, 1968.

Ryan, Cornelius, (ed.), *Across the Space Frontier*, Viking, New York, 1952.

Ryan, Cornelius, (ed.), *The Conquest of the Moon*, Viking, New York, 1953.

Simonson, Gene R., (ed.), *The History of the American Aircraft Industry: An Anthology*, MIT Press, Cambridge, MA, 1968.

Singer, S. Fred, "Enduring Challenges of Astronautics," *Twenty-Five Years of the American Astronautical Society, 1954–1974*, edited by Eugene M. Emme, AAS Publications Office, San Diego, 1980, p. 149.

Smith Bernard, as told to Frederick I. Ordway, "Some Vignettes From An Early Rocketeer's Diary: A Memoir," *History of Rocketry and Astronautics: Proceedings of the Seventeenth History Symposium of the International Academy of Astronautics, Budapest, Hungary, 1983*, edited by John L. Sloop.

Smith, David R., "They're Following Our Script: Walt Disney's Trip to Tomorrowland," *Future*, May 1978.

Stehling, Kurt, "Blazing a Trail to the First U.S. Satellites," *Astronautics & Aeronautics*, May 1982.

Thornton, Earl A., "MIT, Jerome C. Hunsaker, and the Origins of Aeronautical Engineering," *Journal of the American Aviation Historical Society*, Winter 1998, p. 309.

Tissandier, Gaston, *Bibliographie aéronautique: catalogue de livres d'histoire, de science, de voyage et de fantasie, traitant de la navigation aérienne ou des aerostats*, Paris: H. Launette et cie, 1887.

Trimble, William, *Jerome Hunsaker and the Rise of American Aeronautics*, Smithsonian Inst., Washington, DC, 2002.

von Kármán Theodore, with Edson, Lee, *The Wind and Beyond*, Little, Brown, Boston, 1967, p. 259.

Wells, Herbert George, "Preface," Kathleen Burn Moore, The Countess of Drogheda, *Catalog of Paintings and Prints of the Earliest and Latest Types of Aircraft . . .*, The Grosvenor Gallery, London, 1917.

Winter, Frank H., "ARS-Founders—Where Are They Now?" *Astronautics & Aeronautics*, May 1981.

Winter, Frank H., *Prelude to the Space Age*, Smithsonian Inst., Washington, DC, 1983.

Winter, Frank, "Bringing Up Betsy," *Air & Space*, Dec. 1988/Jan. 1989.

Zahm, A. F., *Report on European Aeronautical Laboratories*, Smithsonian Miscellaneous Collections, Vol. 2, No. 3, Washington, DC, 1914.

Appendix A— Presidents of the ARS, IAS, and AIAA

1930–31	DavidLasser	(ARS)
1932	G. Edward Pendray	(ARS)
1933	Laurence Manning	(ARS)
	Jerome Hunsaker	(IAS)
1934	G. Edward Pendray	(ARS)
	Charles Lawrance	(IAS)
1935	G. Edward Pendray	(ARS)
	Donald Douglas Jr.	(IAS)
1936	John Shesta	(ARS)
	Glenn Martin	(IAS)
1937	Alfred Africano	(ARS)
	Clark Millikan	(IAS)
1938	Alfred Africano	(ARS)
	Theodore Wright	(IAS)
1939	Alfred Africano	(ARS)
	George Lewis	(IAS)
1940	H. Franklin Pierce	(ARS)
	James Doolittle	(IAS)
1941	H. Franklin Pierce	(ARS)
	Frank Caldwell	(IAS)
1942	Roy Healy	(ARS)
	Hall Hibbard	(IAS)
1943	Cedric Giles	(ARS)
	Hugh Dryden	(IAS)
1944	James Hart Wyld	(ARS)
	Reuben Fleet	(IAS)
1945	James Hart Wyld	(ARS)
	Charles Colvin	(IAS)
1946	Lovell Lawrence Jr.	(ARS)
	Arthur Raymond	(IAS)
1947	Roy Healy	(ARS)
	Preston Bassett	(IAS)
1948	Charles Villiers	(ARS)
	John Northrop	(IAS)
1949	William Gore	(ARS)
	William Burden	(IAS)
1950	William Gore	(ARS)
	J. H. Kindelberger	(IAS)

1951	H. R. J. Grosch	(ARS)
	Lawrence Richardson	(IAS)
1952	C. W. Chillson	(ARS)
	Wellwood Beall	(IAS)
1953	Frederick Durant III	(ARS)
	Charles Mccarthy	(IAS)
1954	Andrew Haley	(ARS)
	John Leland Atwood	(IAS)
1955	Richard Porter	(ARS)
	Robert Gross	(IAS)
1956	Noah Davis Jr.	(ARS)
	Edward Sharp	(IAS)
1957	Robert Truax	(ARS)
	Mundy Peale	(IAS)
1958	George Sutton	(ARS)
	Edward Wells	(IAS)
1959	John Stapp	(ARS)
	William Littlewood	(IAS)
1960	Howard Seifert	(ARS)
	Donald Putt	(IAS)
1961	Harold Ritchey	(ARS)
	H. Guyford Stever	(IAS)
1962	William Pickering	(ARS)
	L. Eugene Root	(IAS)
1963	Martin Summerfield	(ARS)
1963	William Pickering	
1964	Courtland Perkins	
1965	Richard Horner	
1966	Raymond Bisplinghoff	
1967	Harold Luskin	
1968	Floyd Thompson	
1969	Robert Seamans Jr.	
1969–70	Ronald Smelt	
1971	Martin Goland	
1972	Allen Puckett	
1973	Holt Ashley	
1974	Daniel Fink	
1975	Grant Hansen	
1976	Edgar Cortright	
1977	F. Allen Cleveland	
1978	Rene Miller	
1979	George Mueller	
1980	Artur Mager	
1981	Joseph Gavin Jr.	
1982	Michael Yarymovych	
1983	Norman Augustine	
1984	John Mclucas	

1985	Alan Lovelace
1986	Allen Fuhs
1987	Laurence Adams
1988	William Ballhaus Jr.
1989	Bastian Hello
1990	John Swihart
1991	Joseph Shea
1992	Robert Fuhrman
1993	Brian Rowe
1994	Malcolm Currie
1995	A. Thomas Young
1996	L. S. "Skip" Fletcher
1997	Edward "Pete" Aldridge Jr.
1998	Sam Iacobellis
1999	Robert Crippen
2000	Sheila Widnall
2001	Dennis Picard
2002	Kent Kresa
2003	Alan Mullally
2004	Donald Richardson
2005	Roger Simpson

Appendix B—
Award Recipients of the ARS, IAS, and AIAA

Aeroacoustics Award

The Aeroacoustics Award was established in 1973 and is presented for an outstanding technical or scientific achievement resulting from an individual's contribution to the field of aircraft community noise reduction.

1975	Michael J. Lighthill
1976	Herbert S. Ribner
1977	John E. Fowcs Williams
1979	Harvey H. Hubbard
1980	Alan Powell
1981	Thomas G. Sofrin
1983	Marvin E. Goldstein
1984	Geoffrey M. Lilley
1986	David G. Crighton
1987	Christopher K. Tam
1988	Krishnamurty Karamacheti
1990	Philip E. Doak
1992	Alfons Michalke
1993	Krishan K. Ahuja
1995	Donald B. Hanson
1996	Feri Farassat
1997	Albert R. George
1998	John M. Seiner
1999	Philip J. Morris
2000	Hafiz M. Atassi
2001	Michael S. Howe
2002	Christopher L. Morfey
2003	Thomas F. Brooks
2004	Stewart A. Glegg
2005	Michael J. Fisher

Aerodynamic Measurement Technology Award

The Aerodynamic Measurement Technology Award was approved by Board of Directors in 1995 and is presented for continued contributions and achievements toward the advancement of advanced aerodynamics flowfield and surface measurement techniques for research in flight and ground test applications.

1996	Ronald K. Hanson
1998	Robert L. McKenzie
2000	Richard B. Miles
2002	Ronald J. Adrian
2004	James D. Trolinger

Aerodynamics Award

The Aerodynamics Award was approved by the Board of Directors in 1983 and is presented for meritorious achievement in the field of applied aerodynamics, recognizing notable contributions in the development, application, and evaluation of aerodynamic concepts and methods.

1983	John E. Lamar
1984	Lars E. Ericsson
1985	Forrester T. Johnson
1986	Bruce J. Holmes
1987	Robert H. Liebeck
1988	William P. Henderson
1989	Edward N. Tinoco
1990	Hans W. Grellmann
1991	Norman D. Malmuth
1992	Joe F. Thompson
1993	Tuncer Cebeci
1994	Heinz A. Gerhardt
1995	Lawrence E. Putnam
1996	Jan R. Tulinius

1997	Werner K. Dahm	2002	Richard L. Campbell
1998	M. L. Spearman	2003	Thomas J. Mueller
1999	Frank T. Lynch	2004	Jean M. Delery
2000	C. Edward Lan	2005	Robert M. Hall
2001	Gerald L. Winchenbach		

Aerospace Communications Award

The Aerospace Communications Award is presented for an outstanding contribution in the field of aerospace communications.

1968	Harold A. Rosen	1990	John E. Keigler
1968	Donald D. Williams	1992	Tadahiro Sekimoto
1969	Eberhardt Rechtin	1994	Rene C. Collette
1970	Edmund J. Habib	1996	Robert E. Berry
1971	Siegfried Reiger	1998	Ray Leopold
1972	Wilbur L. Pritchard	1998	Bary Bertiger
1974	Arthur C. Clark	1998	Ken Peterson
1976	Robert F. Garbarini	1998	Mikhail F. Reshetnev
1978	Leonard Jaffe	2000	Joseph N. Sivo
1980	Irwin M. Jacobs	2000	Richard T. Gedney
1980	Andrew J. Viterbi	2002	Neil R. Helm
1982	W. Ray Morgan	2002	Takashi Iida
1984	Sidney Metzger	2003	Krishna P. Pande
1986	C. Louis Cuccia	2004	Yasuo Hirata
1988	Burton I. Edelson	2005	Chandra Kudsia

Aerospace Contribution to Society Award

In 1977 the Institute established this award to recognize a notable contribution to society through the application of aerospace technology to societal needs.

1978	Elmer P. Wheaton	1987	E. P. Muntz
1979	Richard S. Johnston	1988	Ronald H. Stivers
1980	Seymour N. Stein	1989	Lee H. Collegeman
1981	Louis Mogavero	1990	Arthur Kantrowitz
1982	William L. Smith	1993	James T. Rose
1983	Richard H. Davies	1995	Geoffrey K. Bentley
1984	Ralph R. Nash	1998	Alan T. Pope
1985	Kenneth D. Taylor	1999	Joseph A. Schetz
1986	Robert D. Giffen	2001	Bradford W. Parkinson

Aerospace Design Engineering Award

Approved by the Board of Directors in 1992, this award was established to recognize design engineers who have made outstanding technical, educational or creative achievements that exemplifies the quality and elements of design engineering.

1993	Ernest L. Thomas	1995	Mason Chew
1994	Paul W. Hoekstra	1996	Joseph C. Burge

1997	Siamak Ghofranian	2002	Norval F. Burgy
2000	George N. Bullen	2004	Maynard L. Stangeland

Aerospace Guidance, Navigation, and Control Award

Approved by the Board of Directors in 1998, this award was established to recognize important contributions in the field of guidance, navigation, and control.

2000	Arthur E. Bryson	2004	Duane T. McRuer
2002	Richard H. Battin		

Aerospace Maintenance Award

This award was first presented in 1987 to an individual who has made a major contribution to the aerospace maintenance discipline (aviation, missile and space) resulting in a significant improvement in operational and cost effectiveness.

1987	James D. A. Van Hoften	1994	James E. Van Laak
1988	Frank J. Cepollina	1995	Ronald Sheffield
1989	Henry D. Hall	1996	David L. Wensits
1990	Larry Anderson	2001	Edward M. Henderson
1991	Frank M. Krantz	2002	Stephen Pauly
1992	Lewis M. Israelitt	2002	Frank Robinson
1993	Martin K. Smith		

Aerospace Power Systems Award

Established in 1981, this award is presented for a significant contribution in the broad field of aerospace power systems, specifically as related to the application of engineering sciences and systems engineering to the production, storage, distribution, and processing of aerospace power.

1982	Theodore J. Ebersole	1995	Richard J. Hemler
1983	Hans S. Rauschenbach	1995	James Lombardo
1984	Stephen J. Gaston	1996	Vincent Truscello
1985	William G. Dunbar	1997	Jack N. Brill
1986	Gary F. Turner	1998	Richard E. Quigley, Jr.
1987	Gerrit Van Ommering	1999	Clyde P. Bankston
1988	Charles Badcock	1999	Perry Bankston
1989	William Billerbeck	2000	Bruce Anspaugh
1991	Frederick E. Betz	2001	Chuck Lurie
1992	Sanjiv Kamath	2002	Ronald J. Sovie
1993	G. Ernest Rodriquez	2004	Dean C. Marvin
1994	Lowell D. Massie	2005	David M. Landis
1995	Gary L. Bennett		

Aerospace Software Engineering Award

The Aerospace Software Engineering Award is presented for outstanding technical and/or management contributions to aeronautical or astronautical software engineering.

1989	Daniel G. McNicholl	1995	Francis E. McGarry
1991	Christine M. Anderson	1999	Merlin Dorfman
1993	Watts S. Humphrey	2001	Donald J. Reifer

AIAA Foundation Award for Excellence

The AIAA Foundation Award for Excellence was established in 1998 to recognize unique contributions and extraordinary accomplishments by organizations or individuals.

2004	John Shalikashvili	2005	Alan Mulally

AIAA Foundation Educator Achievement Award

The AIAA Foundation recognizes up to six outstanding educators for their contributions to the continued study of mathematics, science, and related technical studies among America's youth.

1997	Cyndee Collier	2000	Penny Valentini
1997	Angela DiNapoli	2000	Sara Ewing
1997	Stephen Rocketto	2000	David Weidow
1997	Carol Hodanbosi	2000	Linda Jernigan
1997	Juliet Sisk	2003	Ellen Holmes
1997	Juanita Ryan	2003	Christian Laster
1999	Paul Stengel	2003	Angela Mitchell
1999	Alan Horowitz	2003	Jennifer Linrud Sinsel
1999	Sue McInerney	2003	Joy Reeves
1999	Dan Caron	2003	Tanya Wright
1999	Carolyn Harden	2003	Brett Williams
1999	John Henrici	2005	Paula Leavitt
1999	Anthony Matarazzo	2005	Terry Sue Fanning
1999	Virginia Davis	2005	Bradley Staats
2000	Karen Hall	2005	Dolores Garay
2000	Stephanie Wright	2005	Brendan Casey
2000	Kathleen Foy	2005	Brian Jackson

Air Breathing Propulsion Award

Established in 1975, this award is presented for meritorious accomplishment in the arts, sciences, and technology of air breathing propulsion systems.

1976	Frederick T. Rall	1981	Fred McFee
1977	Edward Woll	1982	William H. Sens
1978	William J. Blatz	1983	Donald W. Bahr
1979	Arthur J. Wennerstrom	1984	Bernard L. Koff
1980	Melvin J. Hartman	1985	Robert D. Hawkins

1986	Frank E. Pickering
1987	Gordon C. Oates
1988	William H. Heiser
1989	Raymond M. Stanahar
1990	John J. Curry
1991	Frank B. Wallace
1992	Donald M. Dix
1993	Thomas L. Dubell
1994	Budugur Lakshminarayana
1995	Thomas A. Auxier
1996	Louis A. Povinelli
1997	H. Lee Beach
1998	Joseph A. Schetz
1999	Allen S. Novick
2000	Charles R. McClinton
2001	John J. Adamczyk
2002	Richard B. Rivir
2003	Ben T. Zinn
2004	Edward T. Curran
2004	Tom Curran
2005	Vigor Yang

Aircraft Design Award

This award was established in 1968 and is given to a design engineer or team for the conception, definition, or development of an original concept leading to a significant advancement in aircraft design or design technology.

1969	Harold W. Adams
1970	Harrison A. Storms
1971	Joseph F. Sutter
1972	Ben R. Rich
1973	Herman D. Barkey
1975	Walter E. Fellers
1976	Kendall Perkins
1977	Howard A. Evans
1978	John K. Wimpress
1979	Harold Raiklen
1980	Elbert L. Rutan
1981	Frank W. Davis
1981	Adolph Burstein
1982	Paul B. MacCready
1984	Kenneth F. Holtby
1984	Philip M. Condit
1984	Everette L. Webb
1985	Harry J. Hillaker
1986	James N. Allburn
1986	Norris Krone
1986	Glenn L. Spacht
1987	Lawrence A. Smith
1988	Michael Pelehach
1989	James E. Kinnu
1989	Irving T. Waaland
1989	John Patierno
1989	John Cashen
1990	Norman Nelson
1990	Alan Brown
1990	C. Richard Cantrell
1990	Ralph H. Shick
1991	Antonio L. Elias
1991	Richard Hardy
1992	David R. Kent
1992	Sherman N. Mullin
1993	Benson Hamlin
1994	Michael R. Robinson
1994	Hannes G. Ross
1994	Helmuth Heumann
1994	John B. Nix
1995	Charles N. Coppi
1996	William A. Norman
1996	Brian L. Hunt
1996	Michael T. Anderson
1996	John V. Chenevey
1996	James B. Godwin
1998	John C. Brizendine
1999	Hermann M. Altmann
2001	Team Raptor
2002	Paul M. Bevilaqua
2002	Paul H. Park
2002	Brian M. Quayle
2002	Rick Rezabek
2004	Charles H. Boccadoro
2004	David H. Graham
2004	Joseph W. Pawlowski

Aircraft Operations Award

Passed by the Board of Directors in 1993, this award is presented to recognize significant accomplishments of one or more individuals that result in increased safety, improved economics, and efficiencies of aircraft operations where the human interface exists.

1994	Herbert D. Kelleher	2000	Frank M. Alexander
1996	Hartwell G. Stoll	2000	James H. Enias
1998	Stephen D. Fulton		

J. Leland Atwood Award

The award is bestowed annually upon an aerospace engineering educator in recognition of outstanding contributions to the profession. The ASEE Aerospace Division and AIAA established the J. Leland "Lee" Atwood Award in 1985. Atwood entered aviation when it was little more than experimentation in a daring sport. Nevertheless, he believed that this new field would be a cornerstone of our national security and serve as a principal medium of world commerce. As an outstanding engineer and a leader of a great corporation, Atwood played a major role in the development of aviation and aerospace technologies for more than 50 years.

The award is bestowed annually upon an aerospace engineering educator in recognition of outstanding contributions to the profession. The award is endowed by Rockwell Collins, Inc. and consists of a $2,000 honorarium and a certificate. In addition, AIAA presents an engraved medal and certificate.

1974	Ben Pollard	1989	John Anderson
1975	Jack Fairchild	1990	Walter Haisler
1976	Barnes McCormick	1991	Robert Nelson
1977	Stanley Lowy	1992	Jack Kerrebrock
1978	Robert Brodsky	1993	Richard Desautel
1979	Donnell Dutton	1995	Robert Culp
1980	Thomas Mueller	1996	Roy Craig
1981	Ira Jacobson	1997	Leland Carlson
1982	L. S. Fletcher	1999	Robert Bishop
1983	Robert Young	2000	Stephen Batill
1984	Wallace Fowler	2002	Peter Torvik
1985	Gordon Oates	2003	John LaGraff
1986	Conrad Newberry	2004	Joseph Schetz
1987	Jan Roskam	2006	Allen Plotkin
1988	John Junkins		

Chanute Flight Award

Originally the Octave Chanute Award, the award honors Octave Chanute, pioneer aeronautical investigator, and is presented for an outstanding contribution made by a pilot or test personnel to the advancement of the art, science, and technology of aeronautics.

1939	Edmund T. Allen	1943	William H. McAvoy
1940	Howard Hughes	1944	Benjamin S. Kelsey
1941	Melvin N. Gough	1945	Elliot Merril
1942	A. L. MacClain	1945	Robert T. Lamson

1946	Ernest A. Cutrell
1947	Lawrence A. Clousing
1948	Herbert H. Hoover
1949	Frederick M. Trapnell
1950	Donald B. MacDiarmid
1951	Marion E. Carl
1952	John C. Seal
1953	W. T. Bridgeman
1954	George E. Cooper
1955	Albert Boyd
1956	A. M. Johnston
1957	Frank K. Everest
1958	A. S. Crossfield
1959	John P. Reeder
1960	Joseph P. Tymczyszyn
1961	Joseph A. Walker
1962	Neil A. Armstrong
1963	Edward J. Bechtold
1964	Robert C. Inis
1964	Fred J. Drinkwater
1965	Alvin S. White
1966	Donald F. McKusker
1967	Milton O. Thompson
1968	William K. Knight
1969	William C. Park
1970	Jerauld R. Gentry
1971	William M. Magruder
1972	Donald R. Segner
1973	Cecil W. Powell
1974	Charles S. Sewell
1975	Jack R. Lousma
1975	Alan L. Bean
1975	Owen K. Garriott
1976	Thomas P. Stafford
1978	T. D. Benefield
1979	Austin J. Bailey
1981	Raymond L. McPherson
1983	S. L. Wallick
1986	George R. Jansen
1988	Russell C. Larson
1990	William G. Schweikhard
1992	Robert A. Hoover
1994	Richard Abrams
1996	Edward T. Schneider
1998	Harold C. Farley
2002	Mike Carriker

Robert J. Collier Trophy Award

In 1944, the association renamed the award the Robert J. Collier Trophy. It is presented annually for the greatest achievement in aeronautics or astronautics in America, with respect to improving the performance, efficiency, or safety of air or space vehicles, the value of which has been thoroughly demonstrated by actual use during the preceding year.

The Robert J. Collier Trophy Award was established in 1912 by Robert J. Collier, publisher and pioneer aviation enthusiast, as the Aero Club of America Trophy. In 1922 the Aero Club of America was incorporated as the National Aeronautic Association. AIAA and other professional societies participate in the selection process.

1911	Glenn Curtiss
1912	Glenn Curtiss
1913	Orville Wright
1914	Elmer Sperry
1915	W. Sterling Burgess
1921	Grover Loening
1925	S. Albert Reed
1926	E. L. Hoffman
1927	Charles Lawrence
1930	Harold Pitcairn
1932	Glenn Martin
1933	Frank Caldwell
1934	Albert Hagenberger
1935	Donald Douglas
1938	Howard Hughes
1940	Sanford Moss
1942	H. H. Arnold
1943	Luis de Florez
1944	Carl Spaatz
1945	Louis Alvarez
1946	Lewis Rodert
1947	Charles Yeager
1947	Lawrence Bell
1947	John Stack
1949	William Lear
1951	John Stack

1952	Leonard Hobbs	1965	Hugh Dryden
1953	Edward Heinemann	1966	James McDonnell
1953	James Kindelberger	1967	Lawrence Hyland
1954	Richard Whitcomb	1968	James Lovell
1955	William Allen	1968	William Anders
1955	Nathan Twining	1968	Frank Borman
1956	Charles McCarthy	1969	Michael Collins
1956	James Russell	1969	Neil Armstrong
1958	Gerhard Neumann	1969	Edwin Aldrin
1958	Neil Burgess	1971	David Scott
1958	Clarence Johnson	1971	Alfred Worden
1960	William Raborn	1971	R. Gilruth
1961	Joseph Walker	1972	Thomas Moorer
1961	Robert White	1973	William Schneider
1961	Forrest Petersen	1974	John Clark
1961	A. Scott Crossfield	1974	Daniel Fink
1962	John Glenn	1975	David Lewis
1962	L. Gordon Cooper	1977	Robert Dixon
1962	Alan Shepard	1978	Sam Williams
1962	Walter Schirra	1979	Paul MacCready
1962	M. Scott Carpenter	1982	Thomas Wilson
1962	Virgil Grissom	1985	Russell Meyer
1962	Donald Slayton	1986	Elbert Rutan
1963	Clarence Johnson	1986	Jeana Yeager
1964	Curtis LeMay	1988	Richard Truly
1965	James Webb	1989	Ben Rich

Walter J. and Angeline Crichlow Trust Prize

This prize was established in 1994 and is presented for excellence in aerospace materials, structural design, structural analysis, or structural dynamics. It is presented every four years.

1995	Robert S. Ryan	2003	Paul C. Paris
1995	Paul C. Paris	2003	Robert S. Ryan
1999	Lucien A. Schmit		

Command, Control, Communication, and Intelligence Award

The Command, Control, Communication, and Intelligence Award is presented for significant contribution to the overall effectiveness of C3I Systems through the development of improved C3I Systems and Systems Technology.

1991	Jerry Tuttle	1997	Robert J. Hermann
1993	John G. Grimes	2000	Arthur K. Cebrowski
1995	Harry H. Heimple	2003	Richard Mayo

Computer-Aided Engineering and Manufacturing Award

This award was first presented in 1988 to an individual who has conceived, defined, or developed an original concept leading to a significant advancement in the use of interactive computer graphics for conceptual design, computer imagery, or computer-aided design and computer-manufacturing.

1988	Cecelia Jankowski	1993	Arthur Kamm
1990	Alan R. Mitchell		

de Florez Award for Flight Simulation

This award is named in honor for the late Admiral Luis de Florez and is presented for an outstanding individual achievement in the application of flight simulation to aerospace training, research, and development.

1965	Lloyd L. Kelly	1984	Walter S. Chambers
1966	Warren J. North	1985	John B. Sinacori
1967	Edwin H. Link	1986	Waldemar O. Breuhaus
1969	Gifford Bull	1987	Edward A. Stark
1970	Harold G. Miller	1988	Bertram W. McFadden
1971	Walter P. Moran	1989	Don R. Gum
1972	James W. Campbell	1990	Jacob A. Houck
1973	Carroll H. Woodling	1991	Frank M. Cardullo
1974	Hugh H. Hurt	1992	Lloyd D. Reid
1975	John C. Dusterberry	1993	Edward M. Boothe
1976	John E. Duberg	1994	Robert T. Galloway
1978	William Hagin	1995	A. L. Ueltschi
1979	James F. Burke	1996	Jack A. Thorpe
1980	Carl B. Shelley	1997	Edward A. Martin
1982	Richard J. Heintzman	1998	Patricia A. Sanders
1983	Robert M. Howe	2001	Richard E. McFarland

Digital Avionics Award

This award, established in 1984, is presented to recognize outstanding achievement in technical management and/or implementation of digital avionics in space or aeronautical systems, including system analysis, design, development or application.

1984	Peter L. Sutcliffe	1992	John M. Borky
1984	Richard A. Peal	1994	Cary R. Spitzer
1984	Robert E. McDonald	1996	Henry J. Dhuyvetter
1984	Richard E. Spardin	1997	A. Tom Smith
1986	Kenneth J. Cox	1999	John C. Ruth
1988	David M. Smith	2003	Reginald D. Varga

Distinguished Service Award

The Board of Directors established this award in 1968 to give unique recognition to an individual member of AIAA who has distinguished himself or herself over a period of years of service to the Institute. The recipient shall be a member in good standing that has shown particular dedication to the interests of the Institute by making significant and continuing contributions over a period of years. This is interpreted as a minimum of five years of identifiable participation, not necessarily consecutive. A short concentrated spurt of activity, regardless of its value, is insufficient to establish eligibility.

Current national officers and directors are exempt from the award. Also, technical contributions or contributions purely to the profession as a whole, or the aeronautical/astronautical/hydronautical disciplines, are not a consideration in this award.

1968	Harvey Cook
1969	Peter Johnson
1970	H. Moran
1971	Frederick Roever
1972	William Chana
1973	H. Abramson
1974	Charles Appleman
1975	Warren Curry
1976	Kenneth Randle
1978	Ann Dickson
1979	Charles Eyres
1980	Tony Armstrong
1981	Gil Moore
1981	George Frankel
1981	Edith Woodward
1981	Herbert Bair
1981	William Simmons
1982	George Mills
1983	Herbert Rosen
1984	Norman Chaffee
1985	William Williams
1986	Thomas Gagnier
1987	Norman Baullinger
1988	James Harford
1988	George Sutton
1989	Andrew Morgan
1990	Donald Stone
1991	Lawrence Stephens
1992	William Macdonald
1993	Velice BetSayad
1994	Robert Justice
1995	Roland Schoenhoff
1996	Patricia Fresh
1997	George Seibert
1998	Doris Lampe
1999	Philip Hattis
2000	Don Fuqua
2002	L. S. Fletcher
2003	John Swihart
2004	Abe Zarem
2005	Michael Francis

Dryden Lectureship in Research Award

This Dryden Lectureship in Research was named in honor of Dr. Hugh L. Dryden in 1967, succeeding the Research Award established in 1960. The lecture emphasizes the great importance of basic research to the advancement in aeronautics and astronautics and is a salute to research scientists and engineers.

1961	James Van Allen
1962	A. Theodore Forrester
1964	Henry Shuey
1965	Wallace Hayes
1966	ShaoChi Lin
1967	Edward Price
1968	Hans Liepmann
1969	Gerard Kuiper
1970	Bernard Budiansky
1971	Coleman Donaldson
1972	John Houbolt
1973	Herbert Friedman
1974	Herbert Hardrath
1975	Antonio Ferri
1976	Anatol Roshko
1977	Abraham Hertzberg
1978	Gerald Soffen
1979	Dean Chapman
1980	Jack Kerrebrock
1981	Herbert Ribner
1982	Laurence Young
1983	Edward Stone
1984	Arthur Bryson
1985	Donald Coles
1986	Irvine Glass
1987	Kenneth Iliff

1988	William Phillips	1997	Heinz Erzberger
1989	George Carrier	1998	Dennis Bushnell
1990	Seymour Bogdonoff	1999	Robert Loewy
1991	Vijaya Shankar	2000	Charles Elachi
1992	Frederick Billig	2001	David Morrison
1993	Joseph Marvin	2003	Wesley Huntress
1994	Eli Reshotko	2004	Ilan Kroo
1995	Jason Speyer	2005	R. John Hansman
1996	John Lumley		

Durand Lectureship Award

The Durand Lectureship, named in honor of William F. Durand, was approved by the Board of Directors in 1983. It is presented for notable achievements by a scientific or technical leader whose contributions have led directly to the understanding and application of the science and technology of aeronautics and astronautics for the betterment of mankind.

1984	Simon Ramo	1994	A. Richard Seebass
1986	Robert Seamans	1996	Sheila Widnall
1988	Donald Hearth	1998	Norman Augustine
1990	Konrad Dannenberg	2004	Robert Bigelow
1992	Eugene Covert		

Energy Systems Award

Established in 1981, the Energy Systems Award is presented for a significant contribution in the broad field of energy systems, specifically as related to the application of engineering sciences and systems engineering to the production, storage, distribution, and conservation of energy.

1981	Kenell J. Touryan	1997	K. N. Bray
1983	Stanford S. Penner	1998	Janos M. Beer
1984	L. S. Fletcher	1999	Chung K. Law
1990	Ashwani K. Gupta	2000	Krishan K. Ahuja
1991	C. Bryon Winn	2000	Adel F. Sarofim
1992	David G. Lilley	2001	Subramanyam R. Gollahalli
1993	Dilip R. Ballal	2002	Merrill K. King
1994	Geoffrey J. Sturgess	2003	Gabriel D. Roy
1995	William D. Jackson	2004	William A. Sirignano
1996	Alexander I. Kalina	2005	James H. Whitelaw
1997	Kenneth Bray		

Engineer of the Year Award

The award is presented "To an individual member of AIAA who has made a recent significant contribution that is worthy of national recognition." The idea of this award is to "capture the moment." The award consists of a medal and certificate of citation.

1989	Bernard Koff	1991	Frank Gillette
1990	Clarence Wesselski	1992	Antonio Elias

1993	Domenic Maglieri	1999	Leonard Weinstein
1994	Roland Bowles	2000	Krishan Ahuja
1995	George Springer	2001	Steven Bauer
1996	Preston Henne	2003	William Parks
1997	David Urie	2004	Paul Munafo
1998	Tommaso Rivellini	2005	Prasun Desai

Excellence in Aerospace Standards Award

This award is presented to recognize substantive contributions by individuals that advance the health of the aerospace community by enabling cooperation, competition, and growth through the standardization process.

2003 William W. Vaughan

Fluid Dynamics Award

This award was approved by the Board of Directors in 1992 and is presented for outstanding contributions to the understanding of the behavior of liquids and gases in motion as related to need in aeronautics and astronautics.

1976	Mark V. Morkovin	1991	Dennis M. Bushnell
1977	Harvard Lomax	1992	Julian D. Cole
1978	Charles E. Treanor	1993	Antony Jameson
1979	Charles H. Kruger	1994	Peter Bradshaw
1980	Eli Reshotko	1995	Philip G. Saffman
1981	Arthur Kantrowitz	1996	Robert W. MacCormack
1982	John L. Lumley	1997	Milton D. Van Dyke
1983	Seymour M. Bogdonoff	1998	Anatol Roshko
1984	Tuncer Cebeci	1999	William C. Reynolds
1985	Edward T. Gerry	2000	Roddam Narasimha
1986	Steven A. Orszag	2001	Israel J. Wygnanski
1987	Gino Moretti	2002	Fazle Hussain
1988	Harold Mirels	2003	William S. Saric
1989	R. Thomas Davis	2004	Alexander J. Smits
1990	Hans W. Liepmann	2005	Jay P. Boris

Gardner-Lasser Aerospace History Literature Award

This award is presented for the best original contribution to the field of aeronautical or astronautical non-fiction literature published in the last five years dealing with the science, technology, and/or impact of aeronautics or astronautics on society.

2002	John Anderson	2004	Michael Gorn
2003	William Trimble	2005	Thomas Crouch
2003	Anne Millbrooke		

Goddard Astronautics Award

The Goddard Astronautics Award is the highest honor AIAA bestows for notable achievement in the field of astronautics. It was endowed by Mrs. Goddard to commemorate her husband, Robert H. Goddard-rocket visionary, pioneer, bold experimentalist, and superb engineer, whose early liquid rocket engine launches set the stage for the development of astronautics. The award received its current form in 1975, when the Institute changed the name and widened the selection criteria of its former Goddard Award (which had been bestowed for contributions in the engineering science of propulsion and energy conversion).

Year	Recipient
1941	Theodore von Kármán
1948	John Shesta
1949	Calvin Bolster
1950	Lovell Lawrence
1951	Robert Truax
1952	Richard Porter
1953	David Young
1954	AMO Smith
1955	Edward Hall
1956	Chandler Ross
1958	Richard Canright
1958	Robert Goddard
1959	James Van Allen
1959	S. Hoffman
1960	Thomas Dixon
1960	Theodore von Kármán
1961	Wernher von Braun
1962	C. Stark Draper
1962	R. Gilruth
1963	Jack James
1964	Hugh Dryden
1965	Wernher von Braun
1965	Frank Whittle
1966	W. Randolph Lovelace
1966	A. Blackman
1966	George Lewis
1966	Hans von Ohain
1967	Irving Johnson
1967	Seymour Lieblein
1967	Robert Bullock
1967	Abe Silverstein
1968	William Pickering
1968	Donald Berkey
1968	Ernest Simpson
1968	James Worsham
1969	George Low
1969	Perry Pratt
1969	Stanley Hooker
1970	Gerhard Neumann
1970	Chris Kraft
1971	Hubertus Strughold
1972	Howard Schumacher
1972	David Hoag
1972	Gary Plourde
1972	Brian Brimelow
1972	Richard Battin
1973	Kurt Debus
1973	Edward Taylor
1974	John Sloop
1974	Rocco Petrone
1974	Paul Castenholz
1974	Richard Mulready
1975	Edward Price
1975	Glynn Lunney
1975	George Rosen
1975	Gordon Holbrook
1977	James Martin
1978	James Stewart
1978	Joseph Charyk
1979	Maxime Faget
1980	Robert Parks
1981	Peter Burr
1981	Kenneth Frost
1982	John Yardley
1983	George Mueller
1984	Krafft Ehricke
1985	Frederic Oder
1986	George Solomon
1987	John McLucas
1988	Norman Augustine
1989	Alan Lovelace
1990	Richard Truly
1991	Eberhardt Rechtin
1992	James Plummer
1993	Malcolm Currie
1994	Joseph Rothenberg
1995	Lew Allen
1996	Aaron Cohen
1997	Albert Wheelon
1998	Jimmie Hill
1999	Peter Wilhelm
2000	John Young

2001	Ivan Getting	2004	Pete Aldridge
2002	A. Thomas Young	2005	John Casani
2003	George Jeffs		

Ground Testing Award

Established in 1975, this award is presented for outstanding achievement in the development or effective utilization of technology, procedures, facilities, or modeling techniques for flight simulation, space simulation, propulsion testing, aerodynamic testing, or other ground testing associated with aeronautics and astronautics.

1976	Bernhard H. Goethert	1993	Raymond J. Stalker
1978	Arthur B. Doty	1994	John W. Davis
1979	Jack D. Whitfield	1995	A. George Havener
1981	Glenn Norfleet	1996	James P. Crowder
1982	Gerald L. Winchenbach	1997	Gary T. Chapman
1983	James G. Mitchell	1998	Ivan E. Beckwith
1984	James C. Young	1999	Lawrence E. Putnam
1985	Billy J. Griffith	2000	Travis W. Binion
1986	Kazimierz J. Orlik-Ruckemann	2001	Donald R. Wilson
1987	Robert E. Smith	2002	John I. Erdos
1988	Charles E. Wittliff	2003	Henry T. Nagamatsu
1989	James M. Cooksey	2004	Hugh W. Coleman
1990	Eugene E. Covert	2004	W. Glenn Steele
1991	Robert L. Trimpi	2005	Blair B. Gloss
1992	Charles G. Miller		

Daniel Guggenheim Medal

The Daniel Guggenheim Medal was established in 1929 for the purpose of honoring persons who make notable achievements in the advancement of aeronautics. This award is jointly sponsored by AIAA, ASME, SAE, and AHS.

1929	Orville Wright	1946	Frank Whittle
1930	Ludwig Prandtl	1947	Lester Gardner
1931	Frederick Lanchester	1948	Leroy Grumman
1932	Juan de la Cierva	1949	Edward Warner
1933	Jerome Hunsaker	1950	Hugh Dryden
1934	William Boeing	1951	Igor Sikorsky
1935	William Durand	1952	Geoffrey De Havilland
1936	George Lewis	1953	Charles Lindbergh
1937	Hugh Eckener	1954	Clarence Howe
1938	Alfred Fedden	1955	Theodore von Kármán
1939	Donald Douglas	1956	Frederick Rentschler
1940	Glenn Martin	1957	Arthur Raymond
1941	Juan Trippe	1958	William Littlewood
1942	James Doolittle	1959	George Edwards
1943	Edmund Allen	1960	Grover Loening
1944	Lawrence Bell	1961	Jerome Lederer
1945	Theodore Wright	1962	James Kindelberger

1963	James McDonnell	1985	Thornton Wilson
1964	Robert Goddard	1986	Hans Liepmann
1965	Sydney Camm	1987	Paul MacCready
1966	Charles Draper	1988	J. R. D. Tata
1967	George Schairer	1989	Fred Weick
1968	H. M. Horner	1990	J. Sutter
1969	H. Julian Allen	1991	Hans von Ohain
1970	Jakob Ackeret	1992	Bernard Koff
1971	Archibald Russell	1993	Ludwig Boelkow
1972	William Mentzer	1994	Helmut Korst
1973	William Allen	1995	Robert Seamans
1974	Floyd Thompson	1996	William Sears
1975	Dwane Wallace	1997	Abe Silverstein
1976	Marcel Dassault	1998	Richard Coar
1977	Cyrus Smith	1999	Frank Marble
1978	Edward Heinemann	2000	William Pickering
1979	Gerhard Neumann	2001	Richard Whitcomb
1980	Edward Wells	2002	John Borger
1981	Clarence Johnson	2003	Holt Ashley
1982	David Lewis	2004	Courtland Perkins
1983	Nicholas Hoff	2006	Eugene Covert
1984	Thomas Davis		

Haley Space Flight Award

This award was established in 1954 as the Astronautics Award and was renamed in 1966 as the Haley Astronautics Award. Today the Haley Space Flight Award, honoring Andrew G. Haley, is presented for outstanding contributions by an astronaut or flight test personnel to the advancement of the art, science or technology of astronautics.

1954	Theodore von Kármán	1969	Walter Cunningham
1955	Wernher von Braun	1969	Walter M. Schirra
1956	Joseph A. Kaplan	1970	James A. Lovell
1957	Krafft Ehricke	1970	Frank Borman
1958	Ivan C. Kincheloe	1970	William A. Anders
1959	Walter R. Dornberger	1971	Fred W. Haise
1960	A. Scott Crossfield	1971	James A. Lowell
1961	Alan C. Shepard	1971	John L. Swigert
1962	John H. Glenn	1972	David R. Scott
1963	Walter M. Schirra	1972	James C. Irwin
1964	Gordon Cooper	1972	Alfred M. Worden
1964	Walter C. Williams	1973	John W. Young
1965	Joseph S. Bleymaier	1973	Charles M. Duke
1966	David R. Scott	1973	Thomas K. Mattingly
1966	Neil A. Armstrong	1974	Paul J. Weitz
1967	Edward G. White	1974	Joseph P. Kerwin
1968	Virgil I. Grissom	1974	Charles Conrad
1969	Donn F. Eisele	1975	Edward G. Gibson

1975	William R. Pogue	1985	Roger Hogan
1975	Gerald P. Carr	1985	Michael Cummings
1976	William H. Dana	1985	Kenneth C. Ward
1978	Donald K. Slayton	1985	David K. Hogan
1978	Thomas P. Stafford	1985	Barbara L. Scott
1978	Vance D. Brand	1985	Gay E. Hilton
1980	Richard H. Truly	1985	Thomas W. Karras
1980	C. Gordon Fullerton	1987	Bruce McCandless
1980	Fred W. Haise	1989	John M. Lounge
1980	Joe H. Engle	1989	Frederick H. Hauck
1982	Robert L. Crippen	1989	George T. Nelson
1982	John W. Young	1989	David C. Hilmers
1984	Owen K. Garriott	1989	Richard O. Covey
1984	John W. Young	1991	Kathryn D. Sullivan
1984	Byron Lichtenberg	1991	Bruce McCandless
1984	Brewster H. Shaw	1991	Loren J. Shriver
1984	Ulf Merbold	1991	Charles F. Bolden
1984	Robert A. Parker	1991	Steven A. Hawley
1985	David Coolidge	1993	Daniel C. Brandenstein

Hap Arnold Award for Excellence in Aeronautical Program Management

The Hap Arnold Award for Excellence in Aeronautical Program Management was approved by the AIAA Board of Directors in 1997 and is presented to an individual for outstanding contributions in the management of a significant aeronautical or aeronautical related program or project.

1992	Robert B. Young	2002	Paul D. Nielsen
1997	Darleen A. Druyun	2003	David J. Bernstorf
1998	Michael M. Sears	2004	Patrick C. Waddick
2001	Preston A. Henne		

History Manuscript Award

The History Manuscript Award is presented for the best historical manuscript dealing with the science, technology, and/or impact of aeronautics and astronautics on society. The purpose of the award is to provide professional recognition to an author who makes a major and original contribution to the history of aeronautics or astronautics.

1969	Milton Lomask	1978	Edward Ezell
1969	Constance McLaughlin Green	1979	Roger Bilstein
1971	Richard Lukas	1980	Richard Hirsh
1972	Richard Smith	1981	E. J. W. Gregory
1973	William Leary	1984	Scott Pace
1975	Richard Hallion	1987	James Hansen
1977	Thomas Crouch	1987	Fred Weick

1993	Donald Elder	2001	Andrew Dunar
1994	E. Ralph Rundell	2001	Stephen Waring
1995	Michael Neufeld	2002	Andrew Butrica
1997	Joseph Tartarewicz	2003	Roger Launius
1998	Asif Siddiqi	2004	Mark Bowles
1999	Deborah Douglas	2004	Erik Conway
2000	William Barry	2004	Virginia Dawson
2001	Dik Daso	2005	Mark Bowles

Information Systems Award

This award is presented for technical and/or management contributions in space and aeronautics computer and sensing aspects of information technology and science.

1977	Albert L. Hopkins	1989	Steven Teitelbaum
1979	Algiradas Avizienis	1991	Vincent L. Pisacane
1979	Barry W. Boehm	1993	Ravi K. Iyer
1981	William A. Whitaker	1995	Nancy G. Leveson
1983	Lynwood C. Dunseith	1999	William Gianopulos
1985	Winston W. Royce	2003	John K. Lytle
1987	Lee B. Holcomb		

International Cooperation Award

Approved by the Board of Directors in 1988, this award is presented to recognize individuals who have made significant contributions to the initiation, organization, implementation and/or management of activities with significant United States involvement that includes extensive international cooperative activities in space, aeronautics, or both.

1989	Gareth Chang	1997	John McElroy
1990	Robert Freitag	1998	Lynn F. Cline
1991	Herbert Friedman	1999	Ian Pryke
1992	Irving Statler	2000	Mireille Gerard
1992	Richard Barnes	2000	Richard Petersen
1993	Brian Rowe	2001	George Schmidt
1993	Arnauld Nicogossian	2002	Richard Kline
1994	Burton Edelson	2003	D. Brent Smith
1995	James Harford	2004	Margaret Finarelli
1996	Arnold Aldrich		

Jeffries Aerospace Medicine and Life Sciences Research Award

Recognizing the importance to aeronautics of scientific endeavors in the field of medicine, the John Jeffries Award was established in 1940 to honor the memory of the American physician who made the earliest recorded scientific observations from the air.

1940	Louis Bauer	1942	Edward C. Schneider
1941	Harry G. Armstrong	1943	Eugene G. Reinartz

1944	Harold E. Wittingham	1975	Lawrence F. Dietlein
1945	John C. Adams	1977	Harald von Beckh
1946	Malcom C. Grow	1978	Heinz S. Fuchs
1947	J. Winifred Tice	1979	William L. Smith
1948	W. Randolph Lovelace	1980	Stephen L. Kimzey
1949	A. D. Tuttle	1981	Sam L. Pool
1950	Otis O. Bensen	1982	Arnauld E. Nicogossian
1951	John R. Poppen	1983	Paul Buchanan
1952	John P. Stapp	1985	William E. Thornton
1953	Charles F. Gell	1986	Charles E. Billings
1954	James P. Henry	1987	John P. Meehan
1955	Wilbur E. Kellum	1988	Carolyn L. Huntoon
1956	Ross A. McFarland	1989	Kent Gillingham
1957	David C. Simons	1989	James O. Houghton
1958	Hubertus Strughold	1990	Charles O. Hopkins
1959	Don D. Flickinger	1991	Frank H. Austin
1960	Joseph W. Kittinger	1993	Victor S. Schneider
1961	Ashton Graybiel	1994	Joan Vernikos
1962	James L. Goddard	1994	Dolores O'Hara
1964	Eugene Konecci	1995	Fredrick E. Guedry
1965	William K. Douglas	1996	Norman E. Thagard
1966	Charles A. Berry	1998	Frank Sulzman
1967	Charles I. Barron	1999	Roscoe G. Bartlett
1968	Loren D. Carlson	2000	Margaret R. Seddon
1969	Frank B. Voris	2001	Maurice Averner
1970	Walton L. Jones	2002	Gautam Badhwar
1971	Richard S. Johnston	2003	Bobby Alford
1972	Roger G. Ireland	2004	William Knott
1973	Karl H. Houghton	2005	Bonnie P. Dalton
1974	Malcolm C. Lancaster		

Theodore W. Knacke Aerodynamic Decelerator Systems Award

This award is presented to recognize significant contributions to the advancement of aeronautical or aerospace systems through research, development and application of the art and science of aerodynamic decelerator technology.

1979	Helmut G. Heinrich	1993	Karl F. Doherr
1981	Theodore W. Knacke	1994	Robert W. Rodier
1984	Herman Engel	1995	Carl W. Peterson
1986	William B. Pepper	1996	Matts J. Lindgren
1986	Domina Jalbert	1997	J. Stephen Lingard
1988	John W. Kiker	1999	Dean F. Wolf
1989	David J. Cockrell	2001	Phillip R. Delurgio
1990	Maurice P. Gionfriddo	2003	Vance L. Behr
1991	Donald W. Johnson	2005	Roy L. Fox
1992	James D. Reuter		

William Littlewood Memorial Lecture Award

The William Littlewood Memorial Lecture provides for an annual lecture dealing with a broad phase of civil air transportation considered of current interest and major importance. The objective is to advance air transport engineering and to recognize those who make personal contributions to the field. The award, jointly sponsored by AIAA and SAE, perpetuates the memory of William Littlewood, the only person ever to be president of both AIAA and SAE. He was renowned for his contributions to the design of, and operational requirements for, civil transport aircraft.

To perpetuate his memory, many of his friends requested SAE and AIAA to sponsor a series of lectures to be known as the "William Littlewood Memorial Lecture". The award consists of an appropriate honorarium decided by the committee and a certificate.

1971	Peter Masefield	1987	Scott Flower
1972	John Borger	1988	Donald Hettermann
1973	Richard Jackson	1989	Ernst Simon
1974	Edward Wells	1990	Joseph Sutter
1975	Gerhard Neumann	1991	Bernard Koff
1976	Franklin Kolk	1992	C. Julian May
1976	Raymond Kelly	1993	James Worsham
1979	Willis Hawkins	1994	R. Dixon Speas
1980	Norman Parmet	1995	Robert Davis
1982	Paul Besson	2000	John Cashman
1983	John Brizendine	2005	Robert Doll
1984	Robert Rummel	2005	John McMasters
1985	Paul Johnstone	2006	Klaus Nittinger
1986	Jon McDonald		

Losey Atmospheric Sciences Award

In 1940, the Robert M. Losey Award was established in memory of Captain Robert M. Losey, a meteorological officer who was killed while serving as an observer for the U.S. Army, the first officer in the service of the United States to die in World War II.

1940	Henry G. Houghton	1955	Robert C. Bundgaard
1941	Horace R. Byers	1956	Ross Gunn
1942	F. W. Reichelderfer	1957	Jule G. Charney
1943	Joseph J. George	1958	P. D. McTaggard Cowan
1944	John C. Bellamy	1959	Herbert Riehl
1945	Harry Wexler	1960	Thomas F. Malone
1946	Carl G. Rossby	1961	Arthur F. Merewether
1947	Benjamin G. Holzman	1962	Jacob A. Bjerknes
1948	Paul A. Humphrey	1964	Robert C. Miller
1949	William Lewis	1965	George P. Cressman
1950	Roscoe R. Braham	1966	David Atlas
1951	Ivan R. Tannehill	1967	Elmar R. Reiter
1952	Vincent J. Schaefer	1969	Robert D. Fletcher
1953	Henry T. Harrison	1970	Newton A. Lieurance
1954	Hermann B. Wobus	1971	Verner Suomi

1972	David Q. Wark	1990	Robert E. Turner
1973	George H. Fichtl	1991	Charles H. Sprinkle
1974	Norman Sissenwine	1992	C. Gordon Little
1975	Paul W. Kadlec	1993	Moustafa T. Chahine
1977	Robert Knollenberg	1994	R. John Hansman
1978	Robert A. McClatchey	1995	James A. Weinman
1979	Allan B. Bailey	1996	Edwin F. Harrison
1980	William W. Vaughan	1997	Orvel E. Smith
1981	Jean T. Lee	1998	Mike B. Bragg
1982	T. Theodore Fujita	1999	Carolyn K. Purvis
1983	Walter Frost	2000	Donald J. Kessler
1984	John H. Enders	2001	John T. Madura
1985	Dennis W. Camp	2002	Thomas P. Ratvasky
1986	John S. Theon	2003	Daniel E. Hastings
1987	John McCarthy	2004	Dale L. Johnson
1988	Shelby Tilford	2005	Dennis W. Newton
1989	James D. Lawrence		

George M. Low Space Transportation Award

Established in 1988, this award honors the achievements in space transportation by George M. Low, who played a leading role in planning and executing all of the Apollo missions, and originated the plans for the first manned lunar orbital flight, Apollo 8.

1988	Eberhard F. Rees	1998	Laurence A. Price
1990	Pete Aldridge	2000	James G. Maser
1992	Alan M. Lovelace	2002	William J. Escher
1994	David W. Thompson	2004	John C. Karas
1996	Thomas D. Burson		

Mechanics and Control Award

This award is presented for an outstanding recent technical or scientific contribution by an individual in the mechanics, guidance, or control of flight in space or the atmosphere.

1967	Derek F. Lawden	1977	Joseph R. Chambers
1968	Robert V. Knox	1978	Richard H. Battin
1969	John P. Mayer	1979	Morris A. Ostgaard
1970	Irving L. Ashkenas	1980	Arthur E. Bryson
1970	Duane T. Mc Ruer	1981	Robert W. Farquhar
1971	Kenneth J. Cox	1982	Angelo Miele
1971	William S. Widnall	1983	John L. Junkins
1971	Gregory W. Cherry	1984	George Leitmann
1972	John V. Breakwell	1985	Jason L. Speyer
1973	Henry J. Kelley	1986	Heinz Erzberger
1974	Harold R. Vaughn	1987	Leonard Meirovitch
1975	Bernard Etkin	1988	William H. Phillips
1976	Charles H. Murphy	1989	Byron D. Tapley
1977	William P. Gilbert	1990	Roger E. Diehl

1991 F. William Nesline
1992 Anthony J. Calise
1993 JerNan Juang
1994 Nguyen X. Vinh
1995 Robert D. Culp
1996 Vinod J. Modi
1997 David K. Schmidt
1998 F. Landis Markley
1998 Kyle T. Alfriend
1999 George H. Born
2000 Ronald A. Hess
2000 Robert F. Stengel
2001 James K. Miller
2002 John E. Prussing
2004 Kevin A. Wise
2005 Michael J. Heller

Missile Systems Award

This award is presented in two categories, Technical and Management. The Technical Award is presented for a significant accomplishment in developing or using technology that is required for missile systems. The candidate must have demonstrated expertise in aerodynamics, guidance, thermophysics, navigation, control, propulsion, or other fundamental technical disciplines that has led to substantial improvement in missile systems.

1982 John B. Buescher
1983 Wayne E. Meyer
1984 Vahey S. Kupelian
1986 Daniel M. Tellep
1987 J. Michael Gorman
1988 Glenwood Clark
1989 Frederick C. Corey
1990 L. David Montague
1992 Mick L. Blackledge
1992 Jack H. Kalish
1996 Peter Betterman
1996 Darell B. Harmon
1998 George R. Schneiter
2000 William C. McCorkle
2004 Tony C. Lin

Multidisciplinary Design Optimization Award

Established in 1993, this award is presented to an individual for outstanding contributions to the development and/or application of techniques of multidisciplinary design optimization in the context of aerospace engineering.

1994 Lucien A. Schmit
1996 Jaroslaw Sobieski
1998 Raphael T. Haftka
2000 Vipperla B. Venkayya
2002 Garret N. Vanderplaats
2004 Prabhat Hajela

National Faculty Advisor Award

The National Faculty Advisor Award is presented to the faculty advisor of a chartered AIAA Student Branch who, in the opinion of student branch members and the AIAA Student Activities Committee, has made outstanding contributions as a student branch faculty advisor, as evidenced by the record of his or her student branch in local, regional, and national activities.

1985 James Marchman
1986 Stanley Lowy
1987 Gerald Gregorek
1988 John Whitesides
1989 John LaGraff
1990 H. Fred Nelson
1991 Paul Hermann
1992 George Matthews
1993 L. Michael Freeman
1994 Russell Cummings
1995 James Wade
1996 Donald Ward
1997 Willy Sadeh
1998 Frederick Lutze

2000	Sheryl Grace	2003	D. Brian Landrum
2001	L. Scott Miller	2004	John Valasek
2002	James McBrayer		

F. E. Newbold V/STOL Award

This award is presented to recognize outstanding creative contributions to the advancement and realization of powered lift flight in one or more of the following areas: initiation, definition and/or management of key V/STOL programs; development of enabling technologies including critical methodology; program engineering and design; and/or other relevant related activities or combinations thereof which have advanced the science of powered lift flight.

1990	Harold W. Blot	2000	Troy M. Gaffy
1993	William J. Scheuren	2000	Richard E. Kuhn
1996	Paul M. Bevilaqua	2002	Barnes W. McCormick
1998	Gordon Lewis	2002	Samuel B. Wilson

Pendray Aerospace Literature Award

The Pendray Aerospace Literature Award is named in honor of Dr. G. Edward Pendray, a founder and past president of the American Rocket Society. The award is presented for an outstanding contribution or contributions to aeronautical and astronautical literature in the relatively recent past. The emphasis should be upon the high quality or major influence of the piece rather than, for example, the importance of the underlying technological contribution. The award is an incentive for aerospace professionals to write eloquently and persuasively about their field and should encompass editorials as well as papers or books.

1951	George P. Sutton	1974	Frederick Ordway
1952	Maurice Zuckrow	1975	William Sears
1953	T. S. Tsien	1976	Stanford Penner
1954	Martin Summerfield	1977	George Leitmann
1955	Walter Dornberger	1978	A. Kuethe
1956	Hermann Oberth	1979	Henry Kelley
1957	Grayson Merrill	1980	Fred Culick
1958	Homer Newell	1981	Robert Hotz
1959	Ali Cambel	1982	Angelo Miele
1960	Luigi Crocco	1983	Marvin Goldstein
1961	Krafft Ehricke	1984	Leonard Meirovitch
1962	Howard Seifert	1985	Warren Strahle
1964	Andrew Haley	1986	Wayne Johnson
1965	Dinsmore Alter	1987	Richard Battin
1966	Antoni Oppenheim	1988	Gordon Oates
1967	Robert Gross	1989	Budugur Lakshimarayana
1968	Arthur Bryson	1990	John Junkins
1970	Wilmont Hess	1991	William Sirignano
1971	Nicholas Hoff	1992	John Przemieniecki
1972	Edward Price	1993	Forman Williams
1973	Marcus Heidmann	1995	Ali Nayfeh
1973	Richard Priem	1996	John Anderson

1997	Joseph Schetz	2002	George P. Sutton
1997	Gordon Oates	2003	Wei Shyy
1998	Satya Atluri	2004	Chung Law
1999	Vinod Modi	2005	Sebastien Candel
2000	Ben Zinn	2006	Christopher Tam

Piper General Aviation Award

Formerly the General Aviation Award, this award is presented for outstanding contributions leading to the advancement of general aviation; honoring William T. Piper, Sr., who made the name Piper synonymous with general aviation.

1979	Roger Winblade	1992	Edward J. King
1980	Donald J. Grommesh	1993	Edwin H. Hooper
1981	William H. Wentz	1993	Richard V. Abbott
1982	Stanley J. Green	1994	Charles R. Trimble
1983	Gerald Gregorek	1995	Daniel S. Goldin
1984	Joseph W. Stickle	1996	Hubert C. Smith
1985	David R. Ellis	1997	Sam B. Williams
1986	Jan Roskam	1999	James Coyne
1987	Donald G. Bigler	2001	William F. Chana
1988	Joseph L. Johnson	2002	Karl H. Bergey
1989	Fred E. Weick	2004	Vern L. Raburn
1990	Harry Zeisloft	2005	Neal J. Pfeiffer
1991	Bruce J. Holmes		

Plasmadynamics and Lasers Award

Approved by the Board of Directors in 1991, this award is presented for outstanding contributions to the understanding of the physical properties and dynamical behavior of matter in the plasma state and lasers as related to need in aeronautics and astronautics.

1997	Theodore A. Jacobs	2002	Yuri P. Raizer
1999	Lee H. Sentman	2004	Joseph J. S. Shang
2000	Wheeler K. McGregor	2005	Wayne C. Solomon
2001	George Emanuel		

Propellants and Combustion Award

First presented in 1990, this award is presented for outstanding technical contributions to aeronautical or astronautical combustion engineering.

1990	Arthur H. Lefebvre	1998	Irvin Glassman
1991	Frank E. Marble	1999	Ashwani K. Gupta
1992	William A. Sirignano	2000	Dilip R. Ballal
1993	Gerard M. Faeth	2001	James E. Peters
1994	Chung K. Law	2002	John D. Buckmaster
1995	Kenneth K. Kuo	2004	Forman A. Williams
1996	Ben T. Zinn	2005	Ronald K. Hanson
1997	Hukam C. Mongia		

Public Service Award

Through this award, AIAA honors a person outside the aerospace community who has shown consistent and visible support for national aviation and space goals.

1986	Gene Roddenberry	1995	George Brown
1987	T. Wendell Butler	1996	Norman Mineta
1988	Barry Goldwater	1997	Michael DeBakey
1989	V. June Scobee	1998	Barbara Mikulski
1990	Walter Cronkite	1999	John Holliman
1991	Douglas Morrow	2000	William Perry
1992	Jake Gam	2001	James Sensenbrenner
1993	Hugh Downs	2004	Robert Walker
1994	Gerald Baliles	2005	Harold Gehman

Reed Aeronautics Award

The Reed Aeronautics Award is the highest award an individual can receive for achievements in the field of aeronautical science and engineering. The award is named after Dr. Sylvanus A. Reed, the aeronautical engineer, designer, and founding member of the Institute of Aeronautical Sciences in 1932. Reed was the first to develop a propeller system composed of metal rather than wood. His aluminum alloy propeller gave Jimmy Doolittle's plane the speed it needed to win the 1925 Schneider Cup race and brought the inventor much credit and many rewards.

1934	G. Rossby	1959	Karel Bossart
1935	Frank Caldwell	1960	John Becker
1936	Edward Taylor	1961	Alfred Eggers
1937	Eastman Jacobs	1962	Walter Williams
1938	Alfred de Forest	1964	Abe Silverstein
1939	George Mead	1965	Arthur Raymond
1940	Hugh Dryden	1966	Clarence Johnson
1942	Igor Sikorsky	1967	Adolph Busemann
1943	Sanford Moss	1968	William Cook
1944	Fred Weick	1969	Rene Miller
1945	Charles Draper	1970	Richard Whitcomb
1946	Robert Jones	1971	Grant Hedrick
1947	Galen Schubauer	1972	Max Munk
1947	Harold Skramstad	1973	I. Edward Garrick
1948	George Brady	1974	Willis Hawkins
1949	George Schairer	1975	Antonio Ferri
1950	R. Gilruth	1976	George Spangenberg
1951	Edward Heinemann	1977	William Dietz
1952	John Stack	1979	Paul MacCready
1953	Ernest Stout	1980	Donald Malvern
1954	Clark Millikan	1981	William Sears
1956	Clarence Johnson	1982	John McLucas
1957	Ross Bisplinghoff	1983	R. Widmer
1958	Victor Carbonera	1984	Frederick Rall

1985	Thomas Jones
1986	Robert Patton
1987	R. Heppe
1988	Brian Rowe
1989	John Patierno
1990	Bernard Koff
1991	Richard Petersen
1992	James Krebs
1993	Charles Zraket
1994	Ben Rich
1995	Wolfgang Herbst
1996	Alan Mulally
1997	George Field
1998	Roy Harris
1999	James "Micky" Blackwell
2000	Sheila Widnall
2001	Elbert Rutan
2002	Robert Mitchell
2003	Dain Hancock
2004	Heinz Erzberger
2005	Ralph Heath

Space Operations and Support Award

Established in 1991, this award is presented for outstanding efforts in overcoming space operations problems and assuring success, and recognizes those teams or individuals whose exceptional contributions were critical to an anomaly recovery, crew rescue, or space failure.

1992	August F. Witt
1993	Neil E. Goodzeit
1993	Bruce Burlton
1993	Edward Haddard
1993	Richard Ledig
1993	Carl H. Hubert
1993	D. Jung
1993	Nicholas A. Martens
1993	George M. Reis
1993	Phillip Olikara
1993	Jack Raisch
1993	Barry Tuner
1995	Story Musgrave
1995	Kathryn Thornton
1995	Richard O. Covey
1995	Claude Nicollier
1995	Tom Akers
1995	Kenneth Bowerssox
1995	Jeffrey Hoffman
1997	William O'Neil
1997	Eugene L. Saragnese
1997	Robert A. Mitchell
1997	James C. Marr
1997	Robert Gershman
1997	John B. McKinley
1997	Gary R. Kunstmann
1997	John Zipse
1997	Erik N. Nilsen
1997	KarMing Cheung
1997	Leslie J. Deutsch
1997	Robert Barry
1997	J. Edmund Riedel
1997	Greg Levanas
1997	Joseph Gleason
1997	N. Talbot Brady
1997	Dean Hardi
1997	Joseph Stanton
1997	David Breda
1997	Neal Ausman
1997	San Erickson
1999	Paul Carr
1999	Carolyn D. Chura
1999	Carol L. Collinsa
1999	Jonathan L. Criss
1999	Owen E. Dudley
1999	Gabriella A. Griffith
1999	Charles C. Hall
1999	Patricia A. Hamilton
1999	Gene A. Heyler
1999	Charles T. Kowal
1999	J. Courtney Ray
1999	Mary K. Reynolds
1999	Rolland A. Riehm
1999	Lisa J. Segal
1999	Dina Tady
1999	T. J. Mulich
1999	Karl F. Whittenburg
1999	Robert S. Bokulic

1999	Jason E. Jenkins	1999	Gary A. Moore
1999	Robert L. Nelson	2001	NEAR Shoemaker Space Craft Mission Ops Team
1999	Mark E. Holdridge		

Space Processing Award

The Space Processing Award is presented for significant contributions in space processing or in furthering the use of microgravity for space processing.

1992	August T. Witt	2000	Robert C. Rhome
1994	Simon Ostrach	2002	William J. Masica
1996	Lawrence DeLucas	2004	Gerard M. Faeth
1998	Martin E. Glicksman		

Space Science Award

The award, originally established in 1961, was presented to an investigator who distinguished himself through the achievement in studies of the physics of atmospheres of celestial bodies, or of the matter, fields, and dynamic and energy transfer processes occurring in space or experienced by space vehicles.

1962	John R. Winkler	1982	James A. Van Allen
1964	Herbert Friedman	1983	Charles A. Barth
1965	Eugene N. Parker	1984	Edward C. Stone
1966	Francis S. Johnson	1985	Gerry Neugebauer
1967	Robert B. Leighton	1986	Richard Johnson
1968	Kinsey A. Anderson	1987	Frederick L. Scarf
1969	Charles P. Sonnett	1988	Thomas M. Donahue
1970	Carl McIlwain	1988	John Dassoulas
1971	William Ian Axford	1989	Frank B. McDonald
1972	Norman F. Ness	1990	Alan Title
1973	Paul W. Gast	1991	Michael B. Duke
1974	John H. Wolfe	1992	Peter A. Sturrock
1975	Murray Dryer	1993	John C. Mather
1976	Riccardo Giacconi	1994	Lennard Fisk
1977	Bruce C. Murray	1995	James A. Westphal
1978	Laurence E. Peterson	1996	Eugene Shoemaker
1979	James B. Pollack	2000	Alan Binder
1980	Donald M. Hunten	2004	Steve Squyres
1981	Peter M. Banks		

Space Systems Award

Formerly the Spacecraft Design Award, the Space Systems Award is presented to recognize outstanding achievements in the architecture, analysis, design, and implementation of space systems.

1969	Otto Bartoe	1972	Thomas Kelly
1970	Maxime Faget	1973	Harold Rosen
1971	Anthony Iorillo	1974	Harold Lassen

1975	Caldwell Johnson	1990	W. Hook
1977	Walter Lowrie	1992	Victor White
1978	Wernher von Braun	1992	Robert Walquist
1979	John Casani	1992	Edward Noneman
1980	Charles Hall	1992	John Jamieson
1981	Krafft Ehricke	1993	Donald Krueger
1982	Angelo Gustaferro	1995	Pierre Madon
1983	Seymour Rubenstein	1996	Robert Minor
1986	Robert Thompson	1997	Ronald Symmes
1988	Michael Griffin	1997	Ron Swanson
1988	Mike Rendine	1998	Frank Culbertson
1989	Arnold Aldrich	2001	William Boeing
1989	Robert Crippen	2004	James Wertz
1989	Richard Kohrs	2005	Craig Dean
1989	Richard Truly		

Speas Airport Award

This award was established in 1983 and is co-sponsored by AIAA and the American Association of Airport Executives, and Airport Consultants Council. It is presented to the person or persons judged to have contributed most outstandingly during the recent past towards achieving compatible relationships between airports and/or heliports and adjacent environments.

Examples of airport-environmental enhancement include noise abatement for aircraft engines; aircraft flight-path modification to reduce time and space requirements for take-off and landing; airport design to improve compatibility with the surrounding area; and airport area ground-transportation systems to increase ease of airport use while minimizing interface with normal lifestyles and the ecosystem. The award consists of a certificate and a $10,000 honorarium to the recipient.

1984	Aubert McPike	1995	Matthew Kundrot
1985	Norman Mineta	1996	Dirk Semenza
1986	Robert Doyle	1997	David Hilton
1987	Paul Barkley	1998	Gordon Lewis
1987	Byron Miller	1999	John-Paul Clarke
1988	Timothy Ward	2000	Christine Klein
1989	John Wesler	2002	Daniel Waggoner
1990	Herbert Godfrey	2003	Daniel McGregor
1991	Andrea Riniker	2004	J. Christopher Nielson
1992	John Shelly	2004	David Brown
1993	James Muldoon	2005	Arlene Mulder
1994	Philip Engle		

Elmer A. Sperry Award

Dr. Sperry's son and daughter established the Elmer A. Sperry Award in 1955. The award commemorates the life and achievements of Dr. Sperry by seeking to encourage progress in the engineering of transportation. The award is given in recognition of a distinguished engineering contribution, which, through application proved in actual service, has advanced the art of transportation whether by land, sea or air. AIAA, IEEE, SAE, ASME, SNAME, and ASCE sponsor the award. The award consists of a bronze medal and a certificate.

1955	William Gibbs	1978	Roberts Puiseup
1956	Donald Douglas	1979	Leslie Clark
1957	Richard Dilworth	1980	Malcolm Stamper
1957	Eugene Ketting	1980	Joseph Sutter
1957	Harold Hamilton	1980	William Allen
1958	Ferdidand Porsche	1980	Everette Webb
1958	Heinz Nordhoff	1981	Edward Wasp
1959	Geoffrey De Havilland	1982	Edmund Muller
1959	Charles Walker	1982	Jorg Brenneisen
1959	Frank Halford	1982	Ehrhard Futterlieb
1960	Frederick Braddo	1982	G. Reiner Nill
1961	Robert Letourneau	1982	Joachim Korber
1962	Lloyd Hibbard	1983	M. Andre Turcatzi
1963	Earl Thompson	1983	Henri Ziegler
1964	Michael Gluhareef	1983	Archibald Russell
1964	Igor Sikorsky	1983	Stanley Hooker
1965	William Cook	1983	George Edwards
1965	Richard Rouzie	1984	Theodore Podgorski
1965	Maynard Pennell	1984	Joseph Killpatrick
1965	Richard Loesch	1984	Warren Macek
1965	John Steiner	1985	Carlton Tripp
1966	Hideo Shima	1985	George Plude
1966	Shigenari Oishi	1985	Richard Quinn
1966	Matsuataro Fujii	1986	George Jeffs
1967	Robert Wolfe	1986	George Page
1967	Herbert Stemmler	1986	George Mueller
1967	Marfred Schultz	1986	John Yardley
1967	Hugh DeHaven	1986	Robert Thompson
1967	Edward Dye	1986	William Lucas
1968	Werner Teich	1987	Harry Wetenkamp
1968	Christopher Cockerell	1988	J. A. Piers
1968	Richard Stanton-Jones	1989	Harold Froehlich
1969	M. Nielsen	1990	James Mollenaurer
1969	Edward Teale	1990	John Keeley
1969	Douglas MacMillan	1990	Richard Hanrahan
1970	Charles Draper	1990	Claude Davis
1971	Sedwig Wight	1991	Malcolm McLean
1971	George Baughman	1992	Daniel Ludwig
1972	Leonard Hobbs	1993	Heinz Lieber
1972	Perry Pratt	1993	Jurgen Gerstenmeier
1975	Frank Nemec	1993	Wolf-Dieter Jonner
1975	James Henry	1994	Russell Alther
1975	Jerome Goldman	1996	Thomas Butler
1977	Hanley Unbach	1996	Richard MacNeal
1977	Clifford Eastberg	1997	Richard MacNeal

Lawrence Sperry Award

This award honors Lawrence B. Sperry, pioneer aviator and inventor, who died in 1923 in a forced landing while attempting a flight across the English Channel. The award is presented for a notable contribution made by a young person to the advancement of aeronautics or astronautics.

1936	William C. Rockefeller	1972	Sheila E. Widnall
1937	Clarence L. Johnson	1973	Dino A. Lorenzini
1938	Russell C. Newhouse	1974	Jan R. Tulinius
1939	Charles M. Kearns	1975	Dennis M. Bushnell
1940	William B. Oswald	1977	Joseph L. Weingarten
1941	Ernest G. Stout	1978	Paul Kutler
1942	Edward C. Wells	1979	David A. Caughey
1943	William B. Bergen	1980	William F. Ballhaus, Jr.
1944	William H. Phillips	1981	Charles W. Boppe
1945	Richard Hutton	1982	Jeffrey N. Cuzzi
1946	Peter R. Murray	1983	Luat T. Nguyen
1947	N. A. M. Gaylor	1984	Chiba Kogyo University
1948	Allen E. Puckett	1984	Fredrick Aronowitz
1949	Alexander H. Flax	1985	Vijaya Shankar
1950	Frank N. Piasecki	1986	Parviz Moin
1951	Robert C. Seamans	1987	James L. Thomas
1952	Dean R. Chapman	1988	David W. Thompson
1953	Donald E. Coles	1989	Case (CP) P. Van Dam
1954	A. Scott Crossfield	1990	Ilan M. Kroo
1955	Giels J. Strickroth	1991	Mark Drela
1956	George F. Jude	1992	John T. Batina
1957	Clarence A. Syvertson	1993	Timothy J. Barth
1958	Robert G. Loewy	1994	W. Kyle Anderson
1959	James E. McCune	1995	William P. Schonberg
1960	Robert B. Howell	1996	Penina Axelrad
1961	Douglas G. Harvey	1997	John Kallinderis
1962	Robert O. Pillard	1998	Iain D. Boyd
1964	Daniel M. Tellep	1999	Robert D. Braun
1965	Rodney C. Wingrove	2000	AnnaMaria R. McGowan
1966	Joe H. Engle	2001	Keith A. Comeaux
1967	Eugene F. Kranz	2002	Ed C. Smith
1968	Roy V. Harris	2003	Myles L. Baker
1969	Edgar C. Lineberry	2004	Jeffrey D. Jordan
1970	Glenn Lunney	2005	Tim C. Lieuwen
1971	Ronald Berry		

Structures, Structural Dynamics, and Materials Award

Established in 1967, this award is presented to an individual who has been responsible for an outstanding recent technical or scientific contribution in aerospace structures, structural dynamics, or materials.

1968	John C. Houbolt	1970	Joseph D. Van Dyke
1969	Holt Ashley	1971	Nicholas J. Hoff

1972	M. Jonathan Turner	1988	Satya N. Atluri
1973	Robert T. Schwartz	1989	Mike J. Dubberly
1974	William D. Cowie	1990	Richard H. Gallagher
1975	Theodore H. Pian	1991	H. N. Abramson
1976	Charles F. Tiffany	1992	Elbert L. Rutan
1978	Warren A. Stauffer	1993	Thomas D. Arthurs
1979	Lucien A. Schmit	1994	Robert S. Ryan
1980	Earl H. Dowell	1996	Peretz P. Friedmann
1981	Richard R. Heldenfels	1997	Chin-Teh Sun
1982	John H. Wykes	1998	Christos C. Chamis
1983	Leonard Meirovitch	2000	George S. Springer
1984	Eric Reissner	2001	Edward F. Crawley
1985	Chintsun Hwang	2002	Inderjit Chopra
1986	Bryan R. Noton	2004	Kumar G. Bhatia
1987	James W. Mar	2005	Terrence A. Weisshaar

Summerfield Book Award

The AIAA Summerfield Book Award is named in honor of Dr. Martin Summerfield, founder and initial editor of the Progress in Astronautics and Aeronautics series of books published by the AIAA. The award is presented to the author of the best book recently published by the AIAA. Criteria for selection include quality and professional acceptance as evidenced by impact on the field, citations, classroom adoptions, and sales.

1997	Gordon Oates	2003	Steven Isakowitz
1999	David Pratt	2003	Joshua Hopkins
1999	William Heiser	2003	Joseph Hopkins
2000	Daniel Raymer	2005	Jack Mattingly
2001	Paul Zarchan	2005	William Heiser
2002	Richard Battin	2005	David Pratt

Support Systems Award

This award was established in 1975 and is presented for significant contributions to the overall effectiveness of aerospace systems through the development of improved support systems technology.

1976	Gene A. Petry	1986	John E. Hart
1977	Thomas A. Ellison	1987	John C. McHaffie
1979	Joseph J. O'Rourke	1988	Lewis M. Israelitt
1980	John W. Kiker	1989	William E. Rogers
1982	George E. Marron	1990	Robert G. Mager
1983	Richard P. Adam	1991	Jeffrey A. Drew
1984	Harry M. Seaman	1992	Robert M. Brown
1985	Oscar W. Sepp	1993	William R. Robertson

Survivability Award

Established in 1993, this award is presented to an individual or a team to recognize outstanding achievement or contribution in design, analysis, implementation and/or education of survivability in an aerospace system.

1994	Dale B. Atkinson
1996	Robert E. Ball
1998	Nikolaos Caravasos
2000	Jerry Wallick
2002	Michael Meyers
2004	Lawrence A. Eusanio

Sustained Service Award

The award is to recognize sustained, significant service and contributions to AIAA by members of the Institute. A maximum of 20 awards are presented each year. The awards are presented at the conference or activity with which the recipient is most associated or active.

2000	Gordon McKinzie
2000	William Chana
2000	John Swihart
2000	Robert Brodsky
2000	Paul Kutler
2000	Robert Sackheim
2000	John Ruth
2000	Kenneth Randle
2000	William Best
2000	Joseph Garrett
2000	Norman Chaffee
2000	Roland Schoenhoff
2000	Woodward Waesche
2000	Neil Blaylock
2000	Joe Garrett
2001	Donald Richardson
2001	Raymond Goskowski
2001	Virgil Smith
2001	Robert Winn
2001	Gisela McClellan
2001	Julie Morrow
2001	Arne, Vernon
2001	Charles Ehresman
2001	Christine Anderson
2001	Anita Gale
2001	Norman Bergrun
2001	Dale Fester
2001	J. Michael Murphy
2001	James Lang
2001	Ashwani Gupta
2001	James Van Kuren
2001	Ben Wada
2001	David Lilley
2001	Bob Noblitt
2002	Ranney Adams
2002	E. Vincent Zoby
2002	Harry Staubs
2002	John LaGraff
2002	Richard Culpepper
2002	Ramesh Agarwal
2002	Ferdinand Grosveld
2002	Joseph J. S. Shang
2002	Teresa Jordan-Culler
2002	John Madden
2002	Roy Harris
2002	David Throckmorton
2002	John Przemieniecki
2002	Gary Park
2002	A. Tom Smith
2002	Roger Simpson
2002	Jerry Hefner
2002	L. S. Fletcher
2002	George Seibert
2003	Charles Chase
2003	Joseph Morano
2003	Oskar Essenwanger
2003	Donald Nash
2003	Dennis Pelaccio
2003	G. P. "Bud" Peterson
2003	Gerry Schneider
2003	Lee Sentman
2003	Paul Fedec
2003	Michael Griffin
2003	George Nield
2003	Allen Plotkin
2003	Daniel Raymer

2003	M. David Rosenberg
2003	Sy Steinberg
2003	Randy Truman
2003	John Whitesides
2003	Bobby Berrier
2003	John Blanton
2003	Thomas Gagnier
2004	Aditi Chattopadhyay
2004	Bryan Palaszewski
2004	Syed Zafar Taqvi
2004	Russell Cummings
2004	William Atwell
2004	Alfred Crosbie
2004	Brice Cassenti
2004	John Lineberry
2004	Anthony Palazotto
2004	Stan Powell
2004	Merri Sanchez
2004	Klaus Dannenberg
2004	Philip Hattis
2004	Anthony Gross
2004	Brett Anderson
2004	P. Cox
2004	Ronald Bengelink
2004	Gerald Pounds
2004	L. Scott Miller
2004	Douglas Allen
2005	Steven Morris
2005	Carol Cash
2005	Lynn Nicole Smith
2005	James Peterson
2005	Ben Thacker
2005	Bilimoria, Karl
2005	Blottner, Frederick
2005	Bruno Cavallo
2005	Thomas Milnes
2005	Peter Bainum
2005	Sivaram Gogineni
2005	Laura Richard
2005	James Atwater
2005	Jesse Keville
2005	Leonard Sugerman
2005	Kenneth Harwell
2005	Fred Culick
2005	Andrew Santangelo
2005	Robert Swaim
2005	Timothy Hanneman

Systems Effectiveness and Safety Award

First presented in 1975, this award is presented for outstanding contributions to the field of system effectiveness and safety or its related disciplines.

1975	Anthony M. Smith
1976	Lawrence Guess
1977	F. Stanley Nowlan
1977	Thomas D. Matteson
1979	Willis Willoughby
1980	I. Pinkel
1981	George Hirschberger
1982	Thomas L. House
1983	John de S. Coutinho
1984	Paul Dick
1985	Gerald T. Katt
1986	Robert L. Peercy
1987	Charles O. Miller
1988	Robert B. Abernethy
1989	Frank S. Goodell
1990	John X. Tsirimokos
1991	Bryan D. O'Connor
1992	Jerrell T. Stracener
1993	George A. Rodney
1994	Walter C. Hoggard
1996	Gary W. Johnson
1996	Boris I. Sotnikov
1997	Pat Clemens

Thermophysics Award

This award was established in 1975, and is presented for an outstanding singular or sustained technical or scientific contribution by an individual in thermophysics, specifically as related to the study and application of the properties and mechanisms involved in thermal energy transfer and the study of environmental effects on such properties and mechanisms.

1976	Donald K. Edwards	1991	Peter A. Gnoffo
1977	Chang L. Tien	1992	L. S. Fletcher
1978	Allie M. Smith	1993	Robert Siegel
1979	Raymond Viskanta	1994	Chul Park
1980	George W. Sutton	1995	Fred R. DeJarnette
1981	Walter B. Olstad	1996	G. P. "Bud" Peterson
1982	Richard P. Bobco	1997	John J. Bertin
1983	Stanford S. Penner	1998	Amir Faghri
1984	Tom J. Love	1999	Hassan A. Hassan
1985	Milan M. Yovanovich	2001	Suren N. Tiwari
1986	John T. Howe	2002	Frederick G. Blottner
1987	Alfred L. Crosbie	2003	Ping Cheng
1988	Graeme A. Bird	2004	Je-Chin Han
1989	James N. Moss	2005	Daniel C. Reda
1990	John R. Howell		

von Braun Award for Excellence in Space Program Management

Approved by the Board of Directors in 1987, this award gives national recognition to an individual(s) for outstanding contributions in the management of a significant space or space-related program or project.

1988	Samuel C. Phillips	1996	James P. Noblitt
1989	Samuel Hoffman	1997	Robert R. Lovell
1990	James R. Thompson	1998	Richard L. Kline
1991	Forrest S. McCartney	1999	Christine M. Anderson
1992	Thomas J. Lee	2000	Robert A. Pattishall
1993	Aaron Cohen	2001	Han Hwangbo
1994	Minoru S. Araki	2003	Brewster H. Shaw
1995	Arthur F. Obenshain		

von Kármán Lectureship in Astronautics Award

In 1975, the Institute changed the name of the von Kármán Lecture to the von Kármán Lectureship in Astronautics. Honoring Theodore von Kármán, world famous authority on aerospace sciences, it recognizes an individual who has performed notably and distinguished himself technically in the field of astronautics.

1964	Arthur Kantrowitz	1972	Eugene Love
1965	Raymond Bisplinghoff	1973	Alan Lovelace
1966	Nicholas Hoff	1974	Harris Schurmeier
1967	Lester Lees	1975	I. Edward Garrick
1968	William Sears	1977	Joseph Charyk
1969	Courtland Perkins	1978	Robert Fuhrman
1970	Erik Mollo-Christensen	1979	Chris Kraft
1971	Irmgard Flugge-Lotz	1980	Daniel Fink

1981	Bruce Murray	1995	Eugene Kranz
1982	Willis Hawkins	1996	Bradford Parkinson
1983	George Jeffs	1997	John Junkins
1984	Aaron Cohen	1998	Thomas Moorman
1985	Eberhardt Rechtin	1999	Edward Stone
1986	Albert Wheelon	1999	John Casani
1987	Lew Allen	2000	John Anderson
1988	John Yardley	2001	Roald Sagdeev
1989	Richard Battin	2002	Earl Dowell
1990	Alvin Seiff	2003	Brewster Shaw
1992	Hans Mark	2004	G. Scott Hubbard
1993	James Abrahamson	2005	William Ballhaus, Jr
1994	Arthur Bryson		

Otto C. Winzen Lifetime Achievement Award

Approved by the Board of Directors in 1993, this award is in memory of Otto C. Winzen, a pioneer of modern day ballooning. The award is presented for outstanding contributions and achievements in the advancement of free flight balloon systems or related technologies.

1991	Jean R. Nelson	1999	James A. Winker
1994	Alfred Shipley	2003	Vincent Lally
1996	Justin H. Smalley	2004	James L. Rand
1997	Charles Moore		

Wright Brothers Lectureship Award

Commemorating the first powered flights made by Orville and Wilbur Wright at Kitty Hawk in 1903, this lectureship emphasizes significant advances in aeronautics by recognizing major leaders and contributors thereto.

1937	B. Melvill Jones	1955	Ross Bisplinghoff
1938	Hugh Dryden	1956	Arnold Hall
1939	Clark Millikan	1957	H. Julian Allen
1940	Sverte Pettersen	1958	Maurice Roy
1941	Richard Southwell	1959	Alexander Flax
1942	Edmund Allen	1960	A. W. Quick
1943	W. S. Farren	1961	Robert Jastrow
1945	H. Roxbee Cox	1962	James Lighthill
1946	Theodore von Kármán	1964	George Schairer
1947	Sydney Goldstein	1965	Gordon Patterson
1948	Abe Silverstein	1966	C. Stark Draper
1949	A. E. Russell	1967	Philippe Poisson-Quinton
1950	William Bollay	1968	Charles Harper
1951	P. B. Walker	1969	Pierre Satre
1952	William Littlewood	1970	F. A. Cleveland
1953	Glenn Martin	1971	Robert Lickley
1954	Bo Lundberg	1972	Franklin Kolk

1973	H. Schlichting	1987	John Swihart
1974	AMO Smith	1988	Ben Rich
1975	Henri Ziegler	1989	Roy Harris
1976	John Atwood	1990	Ivan Yates
1977	Gero Madelung	1991	Irving Waaland
1978	George Litchford	1992	Sherman Mullin
1979	Jack Nielsen	1993	German Zagainov
1980	Bernard Etkin	1994	Robert Buley
1981	Holt Ashley	1994	Paul Rubbert
1982	Jack Steiner	1995	Robert Smith
1983	Edward Polhamus	1996	Philip Condit
1984	George Cooper	1997	Eugene Covert
1984	Robert Harper	1998	Frank Marble
1985	John Utterstrom	1999	Edward Curran
1986	Lewis Brown	2000	Kenneth Szalai
1986	Dwain Deets	2002	Robert Liebeck

Wright Brothers Memorial Trophy Award

The Wright Brothers Memorial Trophy Award was established in 1948 by the Awards Committee of the Aero Club of Washington in order to honor the Wright Brothers annually. The award is presented for significant public service of enduring value to aviation in the United States.

1948	William Durand	1972	John Shaffer
1949	Charles Lindbergh	1973	Barry Goldwater
1950	Grover Loening	1974	Richard Whitcomb
1951	Jerome Hunsaker	1975	Kelly Johnson
1952	James Doolittle	1976	William Patterson
1953	Carl Hinshaw	1977	Ira Eaker
1954	Theodore von Kármán	1978	Jennings Randolph
1955	Hugh Dryden	1979	Thomas Wilson
1956	Edward Warner	1980	Olive Beech
1957	Stuart Symington	1981	Dwane Wallace
1958	John Victory	1982	Willis Hawkins
1959	William MacCracken	1983	John Atwood
1960	Frederick Crawford	1984	David Lewis
1961	A. S. "Mike" Monroney	1985	Harry Combs
1962	John Stack	1986	Joseph Sutter
1963	Donald Douglas	1987	Allen Paulson
1964	Harry Guggenheim	1988	Sam Williams
1965	Jerome Lederer	1989	Thomas Jones
1966	Juan Trippe	1990	Edwin Colodny
1967	Igor Sikorsky	1991	Benjamin Cosgrove
1968	Warren Magnuson	1992	Jake Garn
1969	William Allen	1993	Gerhard Neumann
1970	C. R. Smith	1994	A. L. Ueltschi
1971	Howard Cannon	1996	Frederick Smith

1997	Charles Karman	2001	Neil Armstrong
1998	Edward Stimpson	2002	Paul Poberezny
1999	Delford Smith	2003	John Glenn
2000	Herbert Kelleher	2004	Robert Crandall

Wyld Propulsion Award

The Propulsion Award and the James H. Wyld Memorial Award honoring the developer of the regeneratively-cooled rocket engine were combined in 1964 to become the James H. Wyld Propulsion Award. In 1975 the name was again modified to the Wyld Propulsion Award, which is now presented for outstanding achievement in the development or application of rocket propulsion systems.

1948	Frank Malina	1973	G. W. Elverum
1949	James A. Van Allen	1973	Norman C. Reuel
1950	Leslie Skinner	1974	Clarence W. Schnare
1951	William Avery	1975	James Lazar
1952	A. L. Antonio	1975	Roderick Spence
1953	Charles E. Bartley	1976	Howard Seifert
1954	Milton W. Rosen	1977	Martin Summerfield
1954	Harold W. Ritchey	1978	William C. Rice
1955	John P. Stapp	1979	Derald A. Stuart
1955	D. S. Miller	1980	Howard W. Douglass
1956	Bruce H. Sage	1981	John Kincaid
1956	Louis G. Dunn	1982	Lloyd F. Kohrs
1957	Levering Smith	1983	Gerry R. Makepeace
1957	William H. Pickering	1984	Rudi Beichel
1958	Holger N. Toftoy	1985	Rolf S. Bruenner
1958	Barnet R. Adelman	1985	Adolph Oberth
1959	K. J. Bossard	1986	Dominick Sanchini
1959	Ernest Roberts	1987	A. G. Casey
1960	Robert L. Johnson	1988	David C. Byers
1960	Ernst Stuhlinger	1989	James R. Sides
1961	Harrison A. Storms	1990	J. Michael Murphy
1961	Robert B. Young	1991	Thomas F. Davidson
1962	William F. Raborn	1992	Robert L. Sackheim
1962	S. K. Hoffman	1993	Allan J. McDonald
1963	David Altman	1994	Richard R. Weiss
1963	Joseph S. Bleymaier	1995	John R. Osborn
1965	Werner R. Kirchner	1996	Richard H. Sforzini
1966	Maurice J. Zuckrow	1997	Leonard H. Caveny
1967	Adelbert O. Tischler	1998	Herman Krier
1968	Harold B. Finger	2001	Franklin R. Chang-Diaz
1969	Harold R. Kaufman	2002	Ms. Yvonne C. Brill
1970	Hans G. Paul	2003	Stanley V. Gunn
1970	Joseph G. Thibodaux	2004	Thomas L. Boggs
1971	Luigi M. Crocco	2005	Ghanshyam P. Purohit
1972	Karl Klager		

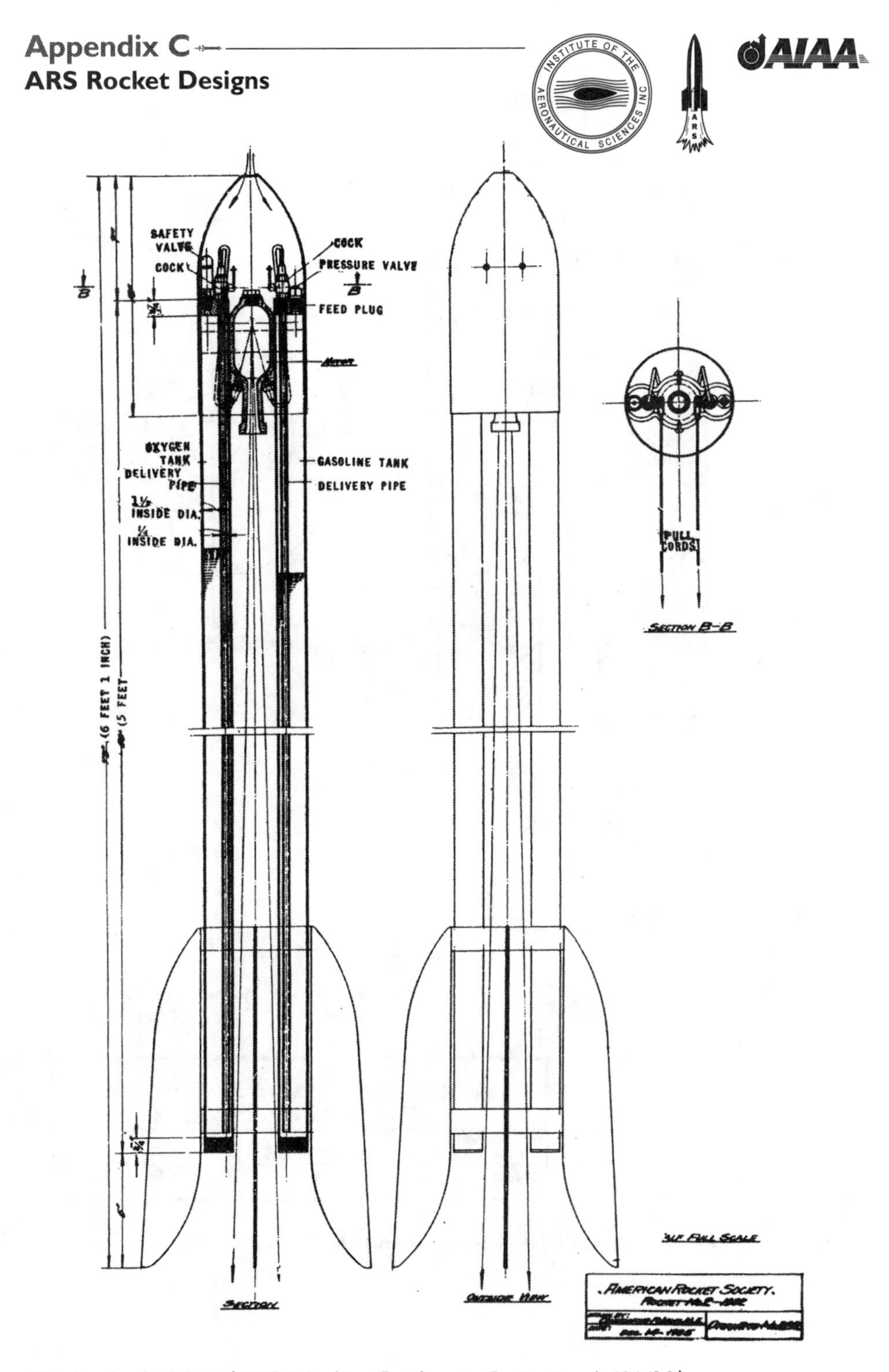

The details of ARS #2. (NASM Archive, Smithsonian Institution, A 4314M.)

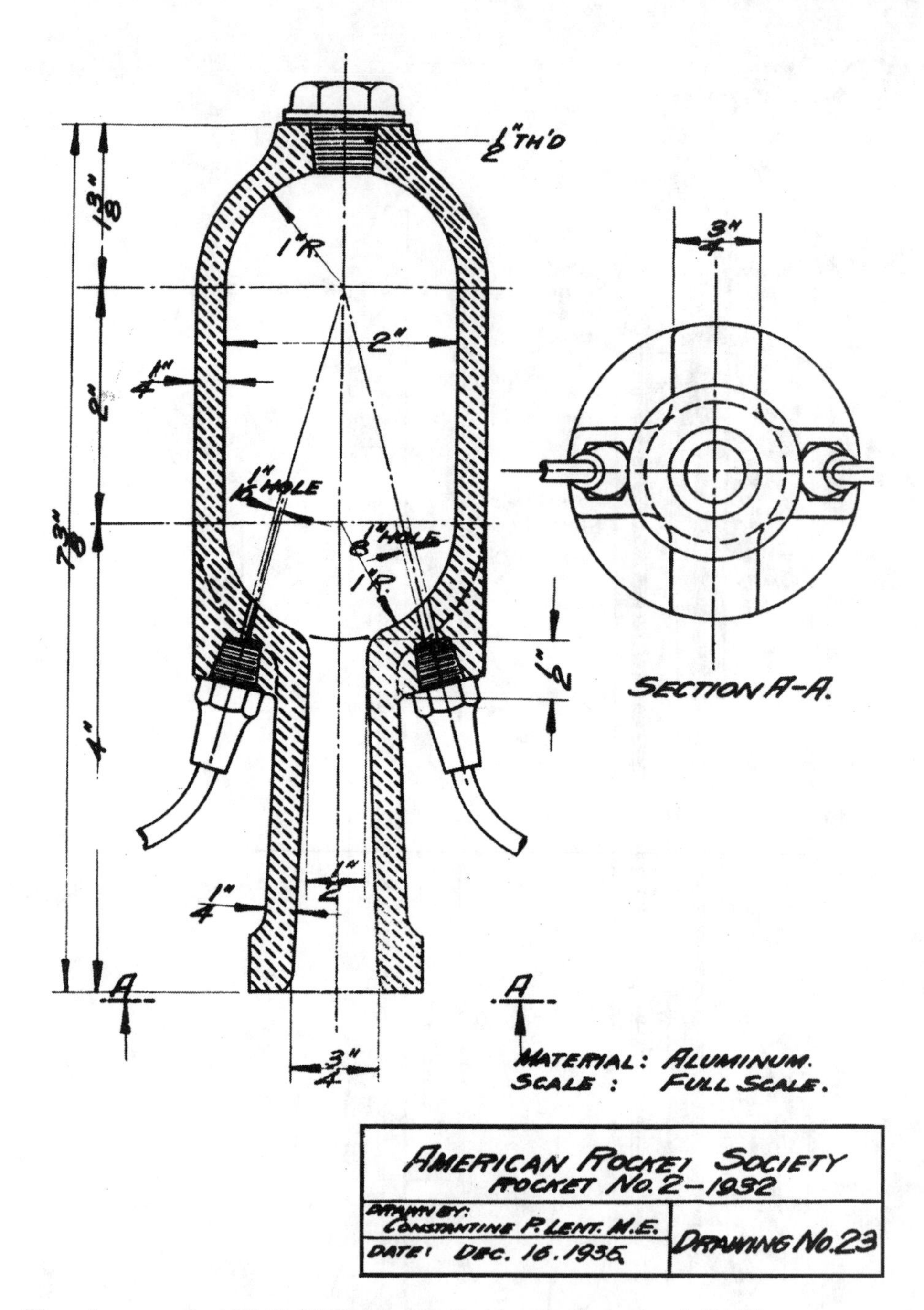

The rocket motor for ARS #2. (NASM Archive, Smithsonian Institution, A 4314J.)

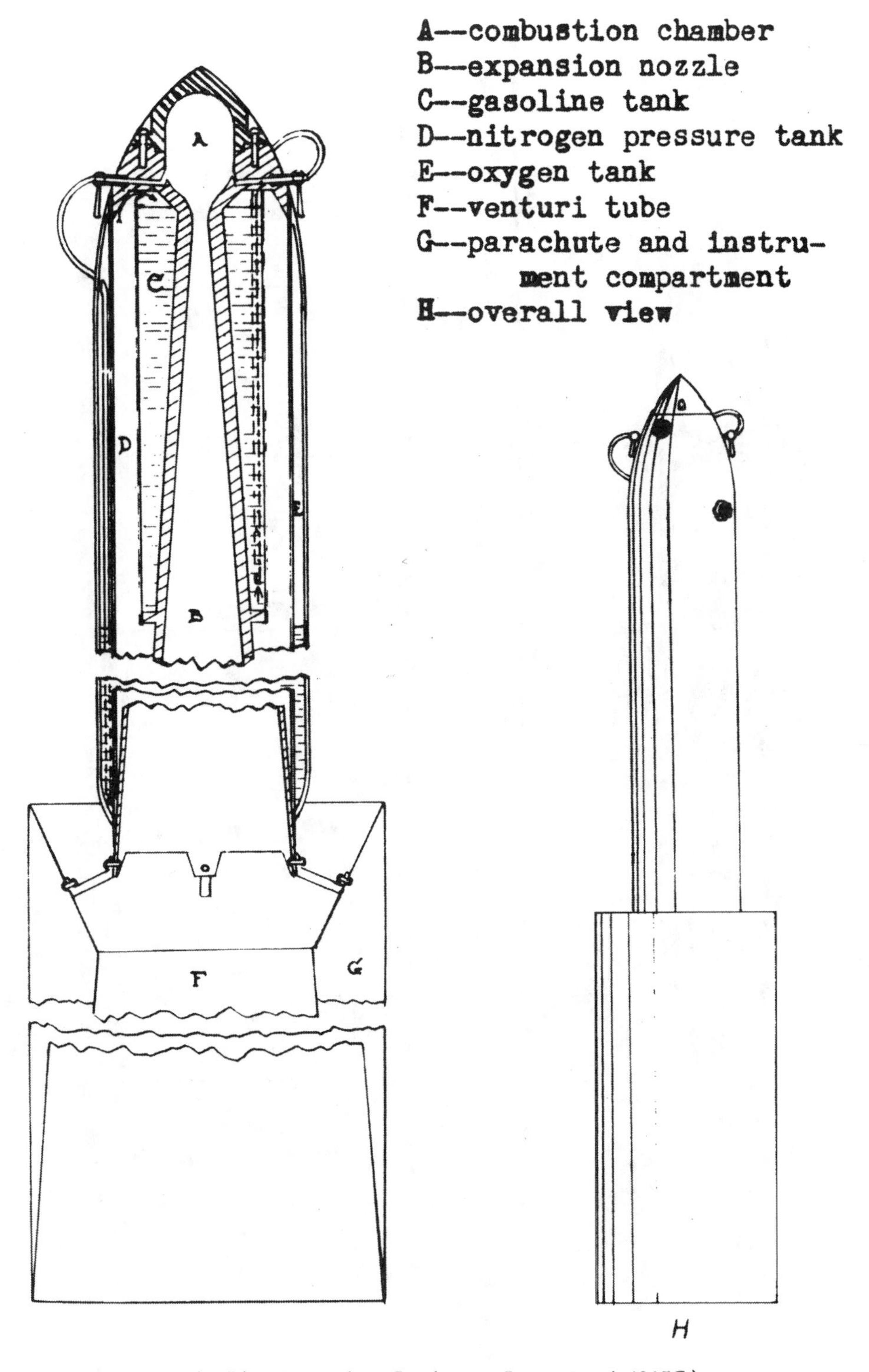

ARS #3, "Ronald Rocket." (NASM Archive, Smithsonian Institution, A 4315C.)

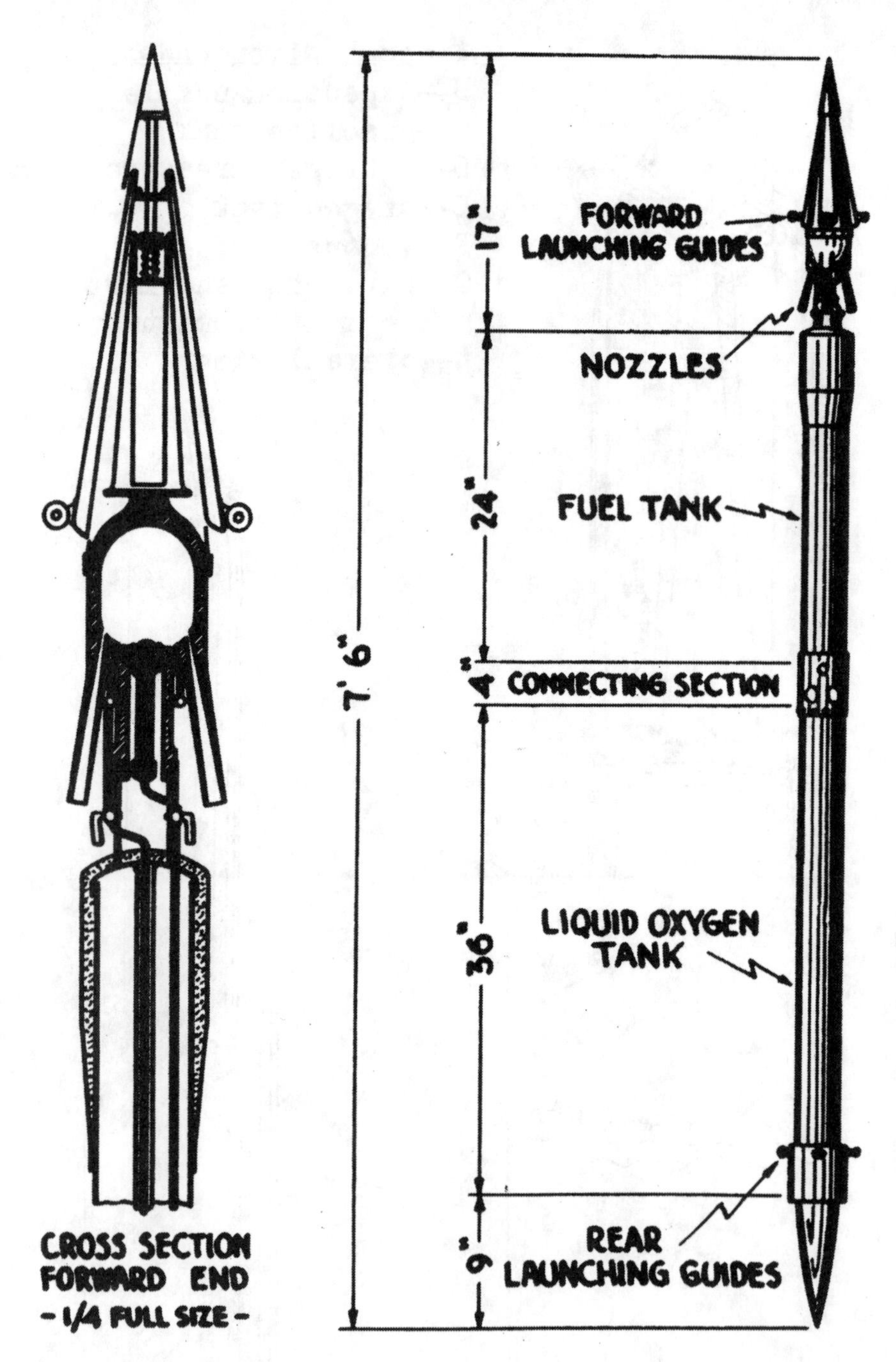

ARS #4. (NASM Archive, Smithsonian Institution, A 4558.)

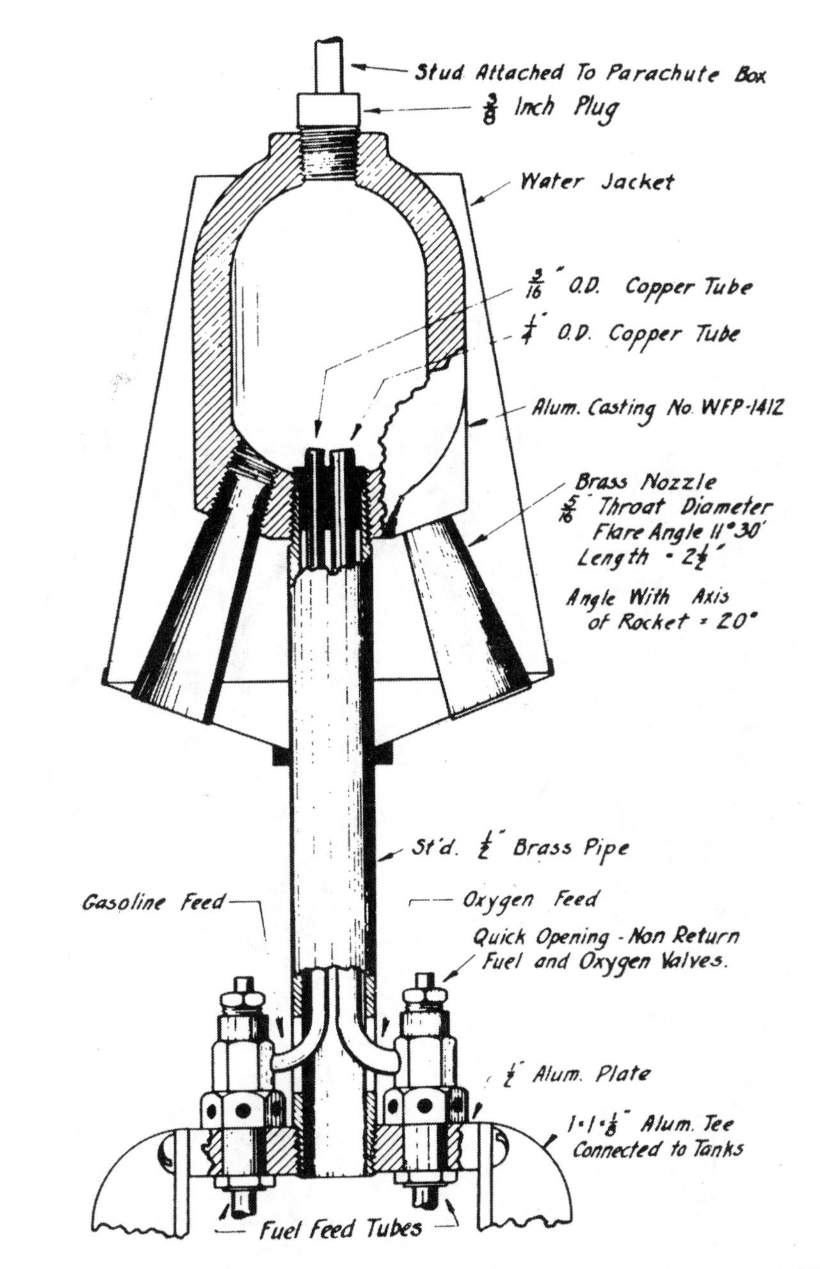

The four-nozzle rocket motor for ARS #4. (NASM Archive, Smithsonian Institution, A 4558A.)

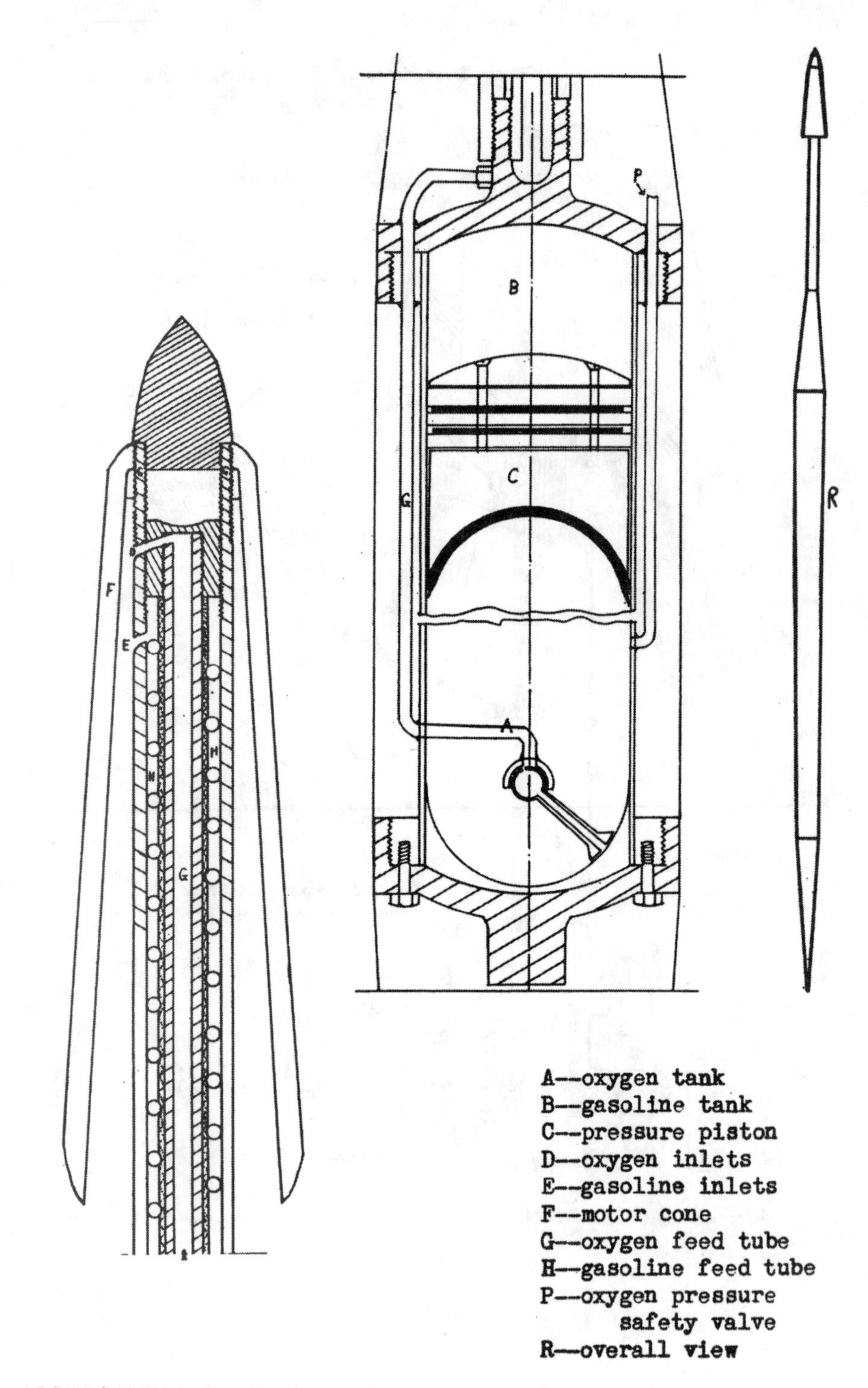

ARS #5. (NASM Archive, Smithsonian Institution, A 4318.)

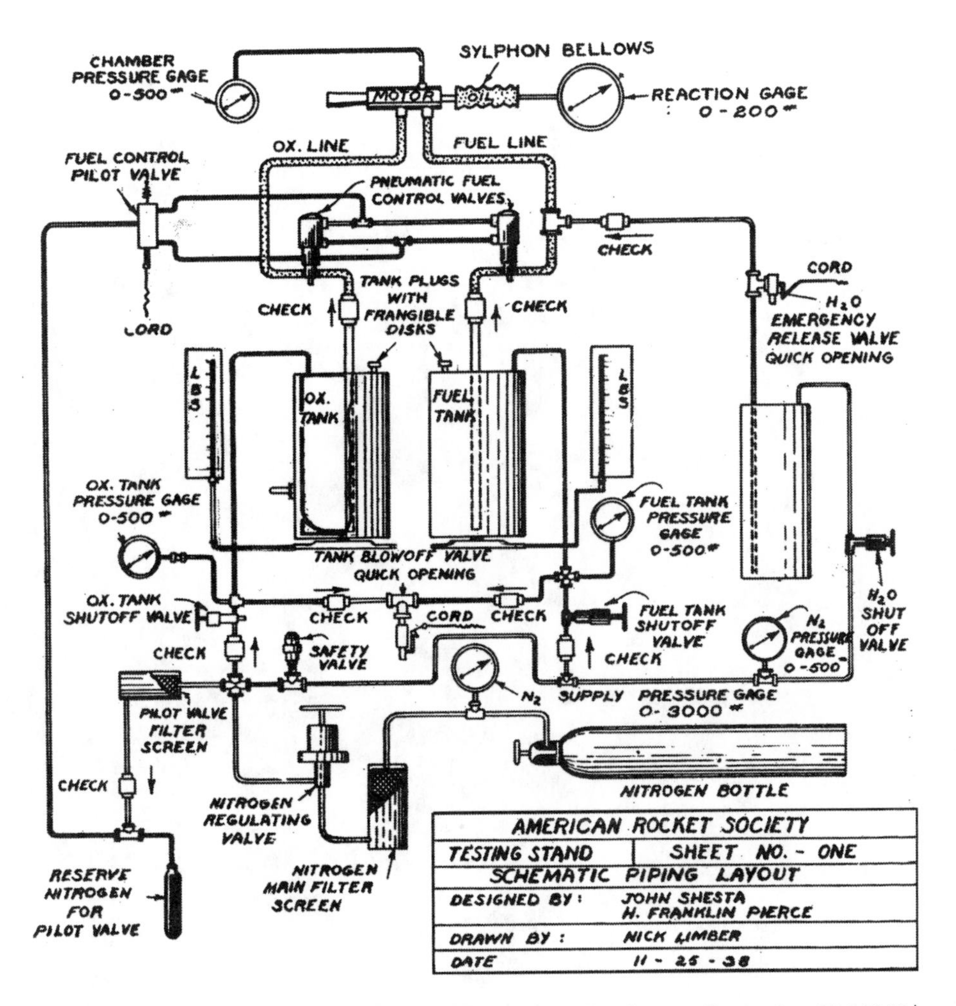

A schematic diagram of ARS test stand #2. (NASM Archive, Smithsonian Institution, 75-10245.)

Index

J

K

L

M

N

O

P